The Ecology of Fungi

Author

William Bridge Cooke

Mycologist (Retired)
Fungus Studies
Robert A. Taft Sanitary Engineering Center
Senior Research Associate
Department of Biological Sciences
University of Cincinnati
Cincinnati, Ohio

CRC Press
Taylor & Francis Group
Boca Raton London New York

CRC Press is an imprint of the
Taylor & Francis Group, an informa business

First published 1979 by CRC Press
Taylor & Francis Group
6000 Broken Sound Parkway NW, Suite 300
Boca Raton, FL 33487-2742

Reissued 2018 by CRC Press

Library of Congress Cataloging in Publication Data

Cooke, William Bridge.
The ecology of the fungi.

Bibliography.
Includes index.
1. Fungi—Ecology. I. Title.
QK604.C637 589'.2'045 78-27812
ISBN 0-8493-5343-2

A Library of Congress record exists under LC control number: 78027812

Publisher's Note
The publisher has gone to great lengths to ensure the quality of this reprint but points out that some imperfections in the original copies may be apparent.

ISBN 13: 978-1-315-89251-1 (hbk)
ISBN 13: 978-1-351-07161-1 (ebk)

Visit the Taylor & Francis Web site at http://www.taylorandfrancis.com and the
CRC Press Web site at http://www.crcpress.com

PREFACE

"Yet it should be remembered that if man does not control the chaos that he has contrived and fails to direct the course of his destiny, if man, through unheeding employment to destructive ends of the tremendous, superhuman powers he has discovered and developed, should finally destroy himself, then the fungi unhindered and unheeding, will continue their many activities undisturbed and will remove the fragments of man's failure, the debris of his disaster and destruction, the remains and the wreckage of his recklessness until they obliterate all traces of man himself."[1222] Dr. W. H. Weston was speaking from personal experience with many kinds of fungi performing many acts of deterioration, mostly of materials useful to man, and in use by man. Seventeen years earlier, B. O. Dodge,[322] a well-known mycologist, specialist in fungus genetics, said: "The fungi, on the contrary, are progressive, ever changing and evolving rapidly in their own way, so that they are capable of becoming adapted to every condition of life. We may rest assured that as green plants and animals disappear one by one from the face of the earth, some of the fungi will always be present to dispose of the last remains."

William B. Cooke

THE AUTHOR

William Bridge Cooke, Ph. D., is a Senior Research Associate in the Department of Biological Sciences, University of Cincinnati, Cincinnati, Ohio, and was formerly in charge of the Fungus Studies Laboratory, Robert A. Taft Sanitary Engineering Center, Cincinnati, Ohio, from which he has retired.

Dr. Cooke graduated in 1937 from the University of Cincinnati with a B. A. degree in Botany. He obtained his M. S. in Mycology and Plant Pathology from Oregon State University, and his Ph. D. degree in Botany from Washington State University with emphasis on the ecology of the fungi.

Dr. Cooke is a Fellow of the American Association for the Advancement of Science and the Ohio Academy of Science, and is a member of the Mycological Society of America (Chairman of the Foray Committee), the American Institute of Biological Sciences, the Botanical Society of America, the American Society for Plant Taxonomy, the California Botanical Society, the British Mycological Society, the International Society for Human and Animal Mycology, and the International Lichenological Society. Dr. Cooke has received the Superior Service Award of the United States Public Health Service and the Award for Excellence of the Federal Water Pollution Control Administration, United States Department of the Interior, and was elected to membership in the Sigma Xi Scientific Research Society. He is Chairman of the Joint Task Group for the 15th and 16th editions of *Standard Methods for the Examination of Water and Waste Water.*

Dr. Cooke has published more than 175 papers on the taxonomy and ecology of fungi, and the relation of fungi to water pollution. He has also published catalogues of the flora and the fungi of Mount Shasta, and has listed the fungi of Lassen Volcanic National Park.

ACKNOWLEDGMENTS

The undertaking of a review of a subject as broad as *The Ecology of the Fungi* has a number of pitfalls. A truism known to every researcher who tries to keep up with the literature is that for every paper published or every book written, there is research completed or in progress, known or unknown to the reader, which makes that paper or book, no matter how apparently far out when spoken or written, out of date. In addition, literature sources have proliferated to such an extent that it is not practical to be acquainted with the contents of every issue of every journal, with every book which may be related to one's subject area, or every journal the contents of which should be familiar to him.

Without the help of correspondents who have sent the products of their research, editors who have allowed us to review current literature, and librarians who have allowed the perusal of literature in their care, much of the material in the following pages could not have been studied.

At least two of the cited books may be out in revised editions by the time this manuscript is published. Subject matter presented in series in journals has been used in part, which is not to say that the remaining parts should not have been used. Selection of materials has been the responsibility of the author. Others may have made different selections, used other emphases, or even deleted some of the subject areas considered here. For these inclusions or lapses there may be no excuse, and none is offered.

Acknowledgment is made to the publishers and authors of books and papers from which figures and tables used in the text were obtained. The libraries of the Environmental Protection Agency Cincinnati Laboratories, the University of Cincinnati, the Public Library of Cincinnati and Hamilton County, the Lloyd Library, and the Kananaskis Environmental Sciences Centre, were consulted and materials obtained from each.

DEDICATION

To my wife, Vivian,
without whose help and forebearance
this book could not have been written

TABLE OF CONTENTS

Chapter 1

INTRODUCTION

Ecology has been defined by Daubenmire[293] as the study of the reciprocal relations between organisms and their environment. Fungi are heterotrophic organisms which cannot manufacture their basic food requirements and so are dependent on food materials produced by other organisms either as saprobes or parasites. They can thus be placed in four of the six categories included in Daubenmire's[293] classification of the types of symbiosis, which is defined as embracing all kinds of interrelationships between organisms. Disjunctive and conjunctive symbioses are distinguished by the fact that in the former the associated organisms are not in constant contact. Each of these two types of symbionts may be divided into social and nutritive groupings. Both social disjunctive and conjunctive symbioses include nonnutritive substrate relations of organisms, but since fungi are dependent on the organism or organic matter with which they are associated, no fungi can be included here. Antagonistic disjunctive symbiosis includes those fungi which are predatory on nematodes, amoebae, protozoans, and other microorganisms, while antagonistic conjunctive symbiosis includes those fungi which, during part or all of their life cycles, are parasitic upon plants or animals. Reciprocal disjunctive symbiosis includes those fungi which are cultured and disseminated by insects and other organisms which use them as food; this classification also includes those fungi used by man as food or for the production of specialized byproducts. Reciprocal conjunctive symbiosis includes those fungi which participate in the formation of mycorrhizae or of lichens. From this general analysis it will be seen that most fungi are nutritive disjunctive symbionts or antagonistic conjunctive symbionts.

For Cooke,[209] symbiotic fungi demonstrate antagonism, neutralism, or mutualism in their relations with each other or with other organisms, and the relationship may be facultative or obligate. As far as nutrition relationships are concerned, fungi may be saprotrophic, deriving nutrients from dead organic matter; necrotrophic, deriving nutrients from dead cells which the organism has killed itself; or biotrophic, deriving nutrients from living cells. Six groups of symbionts are recognized: those which are facultatively or obligately antagonistic, those which are facultatively or obligately neutral, and those which are facultatively or obligately mutualistic.

Most of the fungi have a multicellular thallus or mycelium composed of many fine threads called "hyphae". In some groups, the mycelium is not divided into cells but is elongate, branched, and coenocytic. In other groups, the thallus is unicellular, and each cell is a minute single entity. Those portions of the thallus in which the cell wall has not become thickened are in intimate contact with their environment at all times and on all surfaces, except where formed into hyphal strands, rhizomorphs, or sclerotia. Such cells or hyphae may be compared with the root hairs of higher plants in their youngest stages. They are absorptive organs, may develop their own rhizosphere effects, and in addition may develop certain antagonistic effects between other organisms or other fungi in the associations of which they are members.

Spore germination, mycelial development, fruiting body production, spore discharge and dissemination, and the development of rhizomorphs, sclerotia, and resting cells, all fall within the province of fungus ecology. In addition, cytoplasm and nuclei within the cell walls live in their own environments and help to control the environments in which the fungi possessing them live. It is thought that certain phases of the study of the phylogeny and taxonomy of fungi have application in, or may be basic to, ecology. Therefore, certain applications of these fields will be included in the following chapters. Basically, this book is an elaboration of an earlier review[225] of the subject and will include some material from that review.

I. HISTORICAL

To a large extent, fungi comprise a group of ubiquitous and omnivorous organisms. Their existence has been recognized almost since the beginning of man's recorded experiences and impressions of nature.[10] They number 89,000 to 100,000 species.[9] They are found wherever they are searched for, and their spores are found anywhere that spore traps[301,660] are exposed or soil samples are plated.[225]

At first, the study of fungi was devoted to their use as food, medicine, or intoxicants.[10] The adaptability of fleshy fungi for use as food was and still is[465,466,1071] a matter of trial and error, since poisonous species which look edible occur in all parts of the world. In many areas where fleshy fungi of necessity form a food supplement, the population has learned to recognize many of the edible species. Mushroom culture has been an industry for a long time.[10,225]

As the binomial system of nomenclature became adapted to the uses of botanists, zoologists, and paleontologists, those working with fungi began to recognize more and more species of fungi. At first, this recognition was based largely on macroscopic characters.[10,225,401,895] As the microscope became adaptable to the study of ever smaller structures, including spores, the larger fungi became better understood, and the mold fungi more easily studied.[225,279,637,1163] Except for nutrition studies in the yeasts, certain uses of chemistry, scanning electron microscopy, and certain applications of numerical taxonomy, more sophisticated techniques have not yet become adapted to the uses of fungal taxonomy.

The early history of mycology consisted largely of the cataloguing of the various populations of the macrofungi and plant pathogens in which a student or collector was interested. These were mainly along regional lines, sometimes on national, and sometimes on continental lines. Catalogues could be based on limited numbers of specimens, and these would be supplemented by lists of few to many novelties, or unusual species, or range extensions. This phase of mycology continues to the present day[1028,1029,1112] and should not be discouraged because an inventory of natural resources is important for any community. To a limited extent, but with certain improvements as the realization comes to the worker that more information is necessary, such lists and reports include a minimum amount of information concerning the ecology of the species under consideration. Such information is usually autecologic in nature and may refer only to a single instance in the distribution of a species which may ultimately prove to be ubiquitous and omnivorous or highly restricted in range and fastidious in nutrient requirements.[106,319]

Parallel with the development of systematic mycology, culminating in such universal compendia as those of Saccardo[991] and Oudemans,[848] was the realization that certain fungus species are associated with, or cause, plant diseases. The description of this development[677] and that of plant pathology forms an important portion of the history of fungus ecology.

A very recent development in the systematization of fungi was the suggestion[1052] that standard key characters can be initialed, the initials to be built into words which can be used as code names for species. Another suggestion[454] was for the numerical ordering of genera and species in line with already developed systems of numbered families and orders in such a way that entries can be made in a punch card system for locating a species. The system developed by Gould[455] for families was found by Cooke and Hawksworth[262] to be incomplete, at least for the fungi. Kendrick and Weresub[640] made a study to determine the usefulness of numerical taxonomy at the ordinal level for certain Basidiomycetes. They found that insufficient information was available for the particular group chosen. However, de Hoog and Hermanides-Nijhoff[562] have applied numerical taxonomy to a group of imperfect fungi with some apparent success. In forest pathology, rapid identification of species by cultural characters has progressed to a high degree on a numerical basis. Such sets of numbered characters[832-835] can be readily transposed to punch cards. With the increasing use of punch card filing systems, it is neither visionary nor unscientific to consider such measures applicable in other fields of mycology, provided sufficient data become available.

General surveys of the history of mycology have been presented by Ingold[587] and Martin.[756] The most comprehensive survey of the history of the development of most areas in the field of mycology has been written by Ainsworth.[10]

As man has become more knowledgeable concerning the world about him through advances in various areas, his uses of various phenomena and his control of others have improved and have become integrated in major efforts. That this is true in mycology as well as in other fields was emphasized by Smith[1071,1074] who has described a variety of ways in which mycology developed during and after World War II. These included differentiation between poisonous and edible mushrooms as food sources; control of plant diseases; development of processes of manufacture of food and feed yeasts; mold-proofing of cotton fabrics; control of molds on optical equipment; and development of antibiotics. Christensen[193] and Gray[461] have developed volumes describing activities of fungi related to man and his needs and desires.

II. CLASSIFICATION AND TAXONOMY: KINGDOM FUNGI

Fungi are heterotrophic, eukaryotic organisms without the capacity to produce their own food supplies and are thus completely dependent on preformed organic matter. They have neither photosynthetic nor chemosynthetic pigments nor capabilities for these processes. Their nuclei are organized and usually surrounded by a nuclear membrane. They must obtain food supplies by absorption. In order to utilize certain fractions of potentially nutritive materials in their environment, they must secrete enzymes into their surroundings. These exoenzymes degrade those substances to which they are adapted, and the resulting atoms, ions, or compounds must be absorbed through the wall of the cell. They must pass through cytoplasmic membranes before reaching the cytoplasm in which they are carried to locations in which the endoenzymes react with them, reorganizing them into forms in which they can be combined with other elements, ions, or compounds to produce new protoplasm and cell wall material.

There are three principal groups of organisms on the basis of production and utilization of food materials. The first group is the *producers,* of which a few possess chemosynthetic pigments but most possess photosynthetic pigments which, with the aid of sunlight, are capable of combining water and carbon dioxide to form sugars. Macro- and micronutrient elements and ions are absorbed through the root hairs or mycorrhizae to complete the complement of foods required for the production of tissues composing the mature organism. The *consumer* ingests those parts of the producer organism it is capable of digesting, at least in part, and other consumer organisms in turn consume all or part of the primary consumer. Finally, the *decomposer* organism reduces the bodies of the producers and the consumers and the products of their metabolism to the original chemicals from which they were formed, ideally. However, certain combinations of elements do not decompose readily, accounting for the large amounts of humic acids present in the soil which decompose very slowly. The reducer or decomposer organisms include the bacteria and the fungi. In this book, we are concerned only with the fungi and their activities. All saprobic or saprotrophic fungi are reducer organisms throughout their life cycle. Many parasitic fungi have stages in their life cycle in which they act as reducer organisms. Few parasitic or biotrophic fungi whose life cycle is carried out completely in association with a living organism can be considered as reducer organisms, except that they, too, are completely dependent on preformed organic matter.

At least on the basis of the necessity for fungi to absorb their nutrients which are already in the form of usable organic compounds and their lack of photosynthetic pigmentation, the fungi are now thought to belong to a separate kingdom of organisms, The Fungi. It is suggested that the reader consult Whittaker[1237,1238] for a discussion of this matter. It may be mentioned that many people, including Ingold,[590] assume that fungi form a group of organisms distinct from the Plant Kingdom. It should be noted that, in the following paragraphs, the Myxomycetes may be mentioned, but this is because the Myxomycetes have been considered as fungi traditionally. Here, following Olive[846] and others, the Myxomycetes are thought to belong to the Kingdom Protoctista (Protista).

One of the larger hurdles encountered by the student of fungi is the problem of classification. Within certain groups, classification has become stabilized to a degree mostly because of the dominance of one leader in a school which has gained high rank in the study of the group or because of the degree of cooperation between two or more leaders and their students. Such work is best exemplified in studies of the Myxomycetes by Martin and Alexopoulos,[757] of the yeasts by Lodder and her collaborators,[719] of the fresh-water Phycomycetes (Mastigomycotina) by Sparrow,[1089] and of the marine fungi by Johnson and Sparrow.[608] While few compendia above the genus or family levels are available for other groups, summaries have been prepared for all groups of fungi in the form of keys to many, if not most, of the genera and higher categories in two books edited by Ainsworth, Sparrow and Sussman.[11,12] The keys presented are not completely accepted by all students of the groups concerned. Disagreement over interpretation of certain portions of the International Code of Botanical Nomenclature[676] has hampered progress in certain areas of this field, and students in one school or country do not always agree with conclusions of students in another.[332] Problems of nomenclature in the Homobasidiomycetes have been corrected, except for the older literature, so far as the controversy between adherents of the American and International Codes of Botanical Nomenclature is concerned, but classification of the whole class has been in a state of flux because of the relatively slow development and acceptance of revised generic concepts. Today the results of this rate of development are reflected in the older nomenclature used in certain standard texts and in the newer literature more usually used for reference. As current research is absorbed in the newer literature, results of this development will become more significant. In general, in the following work the nomenclature of major groups will follow that used in Ainsworth, Sparrow and Sussman.[11,12]

In the Fungi Imperfecti (Deuteromycotina) certain problems of taxonomy and classification seem to be almost insurmountable. The changing concepts of classification[576,637,991] tend to clarify concepts of apparent relation between form genera and species, but they do not help to solve the puzzles generated by lengthy species lists. When one species within a genus on such a list is based on different cultural appearance, different natural substrata, or different geographical source, and when many species assigned to a genus can be shown under uniform conditions[54,180,719,948,949] also to be one species, the validity of other lengthy species lists can be readily questioned. However, depending on varying points of view, such consolidations may be too extensive.[118,1083] This situation is most confusing to the student unfamiliar with mycological practice and results in discussions dealing with the treatment of one or another of the synonymous species as separate species or, more commonly, with population groups which are assumed to react as one species, rather than in work which could lead to a broader understanding of the genus or group of species in question.

In the development of our knowledge of fungi, both morphologically and ecologically, it is important that the taxonomy of the group be understandable to all. This and other facts concerning the taxonomy of the fungi have been emphasized by Kern[641] in a plea for the rational development of this science.

Many texts have been written on the taxonomy and morphology of the fungi. Depending on the author's background, each text differs in detail or in fundamental principle. From Switzerland has come a well-known morphology[412] which has been enlarged in an American translation[416] and of which a second edition has been published.[415] A work initiated in France, elaborated in Belgium,[674] and translated into English[675] developed a different approach in which all fungi are to be considered unicellular and coenocytic because of the perforated nature of the cross wall. A detailed American work[102] strongly developed the theme that the fungi have arisen from the algae. Recent texts from England[590,1215] and from the U.S.,[26] including a German translation,[373] give straightforward descriptions of the morphology of the several groups of fungi as exemplified by selected examples from each group. Several recent surveys of the fungi are concerned not only with the morphology of these organisms but also with their importance in medicine and certain industrial processes.[638,674,817,1053]

Other workers have developed concepts of relationship between groups of species and genera of fungi as a result of lengthy studies in culture and in nature. Of such studies, two may be mentioned, namely, the development of a picture of the biology and phylogeny of the fungi[1217] which produce their fruiting receptacles in stromata on living or decaying wood and a morphological study of a group of Discomycetes, the Sclerotiniaceae,[1226] whose sclerotia were considered a specialized type of stroma usually associated with the disease of specific host plants.

At the generic level, many monographs have been written. A few of these are pertinent to this discussion. In the development of monographic studies[1219] in the genera *Leptosphaeria, Pleospora,* and *Clathrospora,* familiarity with the many species found in nature, culminating in such studies as one of species found in Mount Rainier National Park,[1218] has been coupled with an understanding of the species as they behave in pure culture in the laboratory.[1038,1039]

After many years of study and work with *Aspergillus* [948] and *Penicillium,* [949] one student of these economically important genera[1141] found it necessary to discuss the development of a species concept within the group. *Endogone* [435] is a unique group in that zygospores are developed within a fleshy receptacle produced beneath the surface of the soil (hypogaeous) like a truffle. Members of the family based on this genus have been shown to form endomycorrhizae with a variety of vascular plants, but they are collected infrequently. Species of *Septobasidium* [282] form mutualistic symbioses with scale insects on tree bark[209] and inhabit the warmer parts of the world. On the bark of trees they simulate a thin patch of lichen thallus. The species of the genus *Typhula* [957] develop small sclerotia whose peridial cells have very characteristic outlines. They occur on living or dead plant tissues and cause one type of "snow-mold" of grains and other grasses, producing diseases of turf grasses in lawns and golf courses. The life histories and speciation of certain groups of the genus *Allomyces* have been thoroughly studied.[355,1274] The monotypic genus *Aqualinderella* [359,360] has been shown to occupy a most unusual ecological niche. Reasons for its restriction to stagnant water have been explored thoroughly in the laboratory. Within these and other generic studies are features of interest to the ecologist and to which he may be able to supply additional information.

Burnett[165] has presented a picture of the fungi without emphasizing the several classes which are usually basic to discussions by other writers. Major areas of consideration included structure, function, recombination, speciation, and evolution. Under structure and growth were considered fine structure, apical growth, development and assimilative structure, development of reproductive structures, and spore liberation, dispersal, and germination. Among functions considered were general aspects of nutrition and metabolism, transport processes, translocation, transpiration, and the metabolism of accumulated and synthesized products. Various aspects of sexual and parasexual processes were considered among phenomena of recombination.

III. SOME PHYLOGENETIC IMPLICATIONS

Professor E. J. H. Corner,[281] well known to mycologists for his contributions to fungus anatomy, taxonomy, and ecology, is equally well known to phanerogamists. The following statement, equally adaptable to the fungi, is taken from his book *The Natural History of Palms*:

> A living organism is a state of protoplasm manifested by the way in which it works. Given a living and undifferentiated cell, no biologist could prophesy what were its powers unless he could recognize the cell as that of a particular organism. The cell must be grown in order to discover into what it may turn. The protoplasm in this cell has existed since the beginning of life and has reproduced itself generation by generation through geological time. During this enormously long time the cell, owing to its very complexity, has varied and developed its powers of adaptation to the environment. Thus we speak of its evolution in which both the biochemical alteration of the protoplasm is implied and its expression as an organism. Some cells have made little progress and, unicellular or in microscopic aggregates, persist in the humble, primitive, and aquatic situations. At the other extreme there has been the mighty progress which in land plants has led to the flowering tree, climer, palm, and herb. They exist, but their ancestors are dead; their offspring may continue to change. In considering the evolution of higher organisms, we must think of their past, their present, and their future. The past should carry us right back to the beginning of life, but it is too long a journey to book from one station. We may have a rough map on a very small scale, but the number of lines, stations, halts, bridges, and intersections on this railway of life, is so vast and they have been laid, relaid, closed, side-tracked, and reopened so many times that we do not know how they are to be reached.

Charles and Knight[189] have edited the papers presented in a symposium on *Organization and Control in Prokaryotic and Eukaryotic Organisms*. For our purposes, the fungi include only eukaryotic organisms, and material in this volume will help in understanding some of their characteristics.

The problem of the monophyletic or polyphyletic origin of the fungi is not a pressing one for most workers studying the synecology and many phases of the autecology of fungi. However, when one considers the various living states in which organisms assigned to the fungi[26,102,415] are found, the problem cannot be ignored and could become of prime importance. Among mycologists it may be given incidental mention,[102] but Copeland[277,278] proposed a system based on the polyphyletic origin of organisms. Here organisms usually treated as fungi are thought to have arisen at at least four different places in the Kingdom Protista. Lines of connection with ancestral forms are thought to have been lost or completely obscured. The Myxomycetes are thought to be related to amoeboid organisms, the Oomycetes (biflagellate water molds) to be more closely related to heterokont algae than to other fungi, while the chytrids (uniflagellate water molds) occupy a distinct phyllum, as do the remaining fungi. Whether this system will hold up under the critical scrutiny of mycology is a question, but in some ways it expresses habitat factors and morphological similarities better than other systems which have been proposed. (Copeland's work formed the basis for Whittaker's[1237,1238] discussions.)

Rogers[975] has suggested that fungi should be maintained as a group biologically distinct from plants and animals on the basis of such factors as method of restriction of water loss in terrestrial organisms, reproduction without nuclear fusion, nutrition, an extensive generation of cells with a double chromosome complement (although the two sets are in separate but conjugate nuclei), and increased certainty of association of gametes. Zuck[1293] would separate three kingdoms on the basis of nutrition alone: the autotrophic Phyta, the phagotrophic Zoa, and the lysotrophic Mycetes. On the other hand, Walton[1202] has suggested that, on the basis of cellular physiology, there is only one kingdom of organisms in the biosphere. The question "Are fungi plants?" has been asked by Martin.[755] A number of arguments are presented for both points of view, and the reader is asked to exercise his own critical judgment in developing an answer. Whittaker,[1237,1238] developing Copeland's thesis, established the Kingdom Fungi, although for Klein and Cronquist[649] the fungi remain plants.

Within the fungi, a parallelism between life cycles of certain groups is presented in a demonstration of the apparent similarities — and, thus, potential relationships — between the rusts and the red algae.[307,308,597] Various arguments have been presented[532] for using phylogeny to develop a natural classification of the fungi. Singer[1047] has shown how he would derive the agarics from gastroid secotiaceous fungi, and Bessey[102] has presented arguments opposing this point of view. Savile[1007,1008] has proposed 13 principles on which to base a theory of evolution in the fungi. He considered parasitism primitive among the fungi and a state from which saprobism has been developed or derived. On the other hand, Cain,[169] although he accepted most of Savile's principles, could not derive saprobism from parasitism. For Raper,[945] the critical events in the history of the fungi are saprobism, incompatibility, heterokaryosis, and the dikaryon.

Using synthetic capacities, biosynthetic pathways, and deoxyribonucleic acid (DNA) as indicators of relationship, several studies have been reported. Cantino[176] suggested that, in those Phycomycetes now considered in the Chytridiomycetes of the Mastigomycotina, the synthetic capacity of the Blastocladiales is at a more highly evolved level than that of the Chytridiales and that, in the Oomycetes, the nutritional characteristics of the Leptomitales and aquatc Peronosporales appear to relate these orders more closely to one another than to the Saprolegniales. Based on the enzymes involved in the tryptophan synthetic pathway, Hütter and DeMoss[572] postulate a closer relationship between the Chytridiales and the Aspergillales, eliminating the Zygomycetes and the Endomycetales as probable intermediates, these groups being considered sidelines. Their data appear to support the idea of a polyphyletic origin of the Phycomycetes and suggest that those anascosporogenous yeasts tested are related to the Heterobasidiomycetes rather than to the Endomycetales. Hall[492] reviewed additional approaches to taxonomy based on molecular approaches, while Tyrell[1165] surveyed biochemical approaches to the taxonomy of the fungi. Storck and Alexopoulos[1114] found that insufficient information was available on DNA content of genomes of species of fungi to derive a final appraisal of phylogeny in the fungi, although they listed systematically the fungi for which information was available.

As a result of the espousal of the phylogenetic system in which the fungi are set aside in a separate kingdom, certain semantic practices will be adhered to in the following chapters. I will use the term "assimilative" rather than the more usually encountered "vegetative" for all mycelium in the nonreproductive portions of the life cycle. I will refer to all stages in the life cycle as those of fungi rather than plants. These usages will apply to paraphrases of other writers but not to direct quotations.

IV. PALEOMYCOLOGY

According to Ponnamperuma,[918] the bacteria could have originated as early as 3.25 billion years ago and the blue-green algae 2.7 billion years ago, although there is an indication that blue-greens can be dated as early as 3.1 billion years[1018] ago. Both these groups are prokaryotes. Eukaryotic organisms may not have developed until about 1 billion years ago, in late Precambrian time, with the fungi and the green algae originating at about the same time.

Because of the rarity of fossil fungi, probably a result of their readily decomposable tissues, many writers have found it difficult to formulate phylogenies based on the fossil record and have resorted to theories based on intrapolation and extrapolation from present populations and cultural developmental patterns. This technique is subject to the criticism offered by Gilbert White in 1769 and quoted by Sporne:[1093] "Ingenious men will readily advance plausible arguments to support whatever theory they shall choose to maintain; but then the misfortune is, everyone's hypothesis is each as good as another's since they are all founded on conjecture."

Opinion is divided as to the significance of fossil fungi, even regarding their existence. Since woody plants have a long history, and their remains are consolidated only in areas providing special conditions of preservation, it may be assumed that, among microorganisms which reduce wood to its component elements, the fungi must have been important in the geological past. The enzymes necessary for the breakdown of cellulose, lignin, and other compounds probably evolved almost at the time the erect land plants evolved and possibly shortly after, if not concomitantly with, the development of these compounds, although Corner[280] has been unable to associate significant wood decay organisms with modern tree ferns in areas like New Guinea. Like the decomposition of plant materials that make up coal beds,[222] fungi have played a role in the decomposition of wood or woodlike material prior to its deposition.[1016] A fossil woodrot occurs in petrified logs of *Araucarioxylon* in Petrified Forest National Park which is similar to a type of rot found in *Taxodium* wood today, indicating possibly that the same or a similar species of *Stereum* has existed since the middle of Jurassic time, 150 to 200 million years ago. Large conks of identifiable polypore fruit bodies[32] have been found in volcanic ash and lake sediments in the Pliocene or later deposits near Mountain Home, Idaho. *Fomes idahoensis* Brown has been found to be much like modern specimens of *Fomitopsis pinicola* (Sw. ex Fr.) Karst. The fragment of the specimen studied was 15 cm in diameter. Fossil materials indicate that the flora of the region at that time was largely hardwood, while at the present time, with the area in Great Basin scrub, conifers predominate in the adjacent forests. A large conk of *Fomes applanatus*, which may have occurred on *Umbellularia californica*, has been found in the Pleistocene flora of the Tomales formation north of San Francisco.[761] This specimen measured 15 × 10 × 3 cm and had four distinctly separate layers of tubes; its modern representatives, occurring on the same host in the region, are known today as *Ganoderma brownii* (Murr.) Gilb. In the Eocene of Wheeler County, Oregon,[1019] a fungus assigned to the Aspergillaceae (Trichocomataceae) was detected in the lumina of vessels of two unrelated woods tentatively assigned to *Magnolia* and *Castanopsis*. In a book on microfossils[612] is the following statement: "The organisms in this phyllum (Eumycophyta of the Kingdom Protista) are the parasitic fungi; they are found as fossils only in association with other woods as studied in this section. No significant microfossils are known. Devonian — Recent."

Pirozynski and Malloch[917] have suggested that symbiosis between a semiaquatic ancestral green alga and an oomycete-like fungus was necessary before terrestrial green plants could be evolved. Such a symbiotic or mutualistic partnership was necessary before the Siluro-Devonian colonization of land could have taken place; such partnerships were equipped to cope with the problems of desiccation and starvation associated with terrestrial existance. Other organisms may have been necessary to such evolution. Vescicular-arbuscular partnerships may have been necessary to the root system of these plants, as they are to the present time, and indeed have been found in the roots of Rhynie fossils as well as in many kinds of modern plants.

For Corner,[280] the fact that few resupinate members of the Aphyllophorales are associated with living or dead material of tree ferns in their natural habitat indicates that such fungi are derived from pileate species, rather than being primitive as is usually thought.

Chapter 2

MYCOGEOGRAPHY

I. INTRODUCTION

The first statements leading to the development of concepts of mycogeography were made in the U.S. by Diehl[319] and in England by Bisby.[106] Certain groups of species, such as the Phallales and the Xylariaceae, have been shown to be more abundant in tropical than in temperate or arctic regions, while other groups are better known in temperate regions. Apparently more micromycetes[700] than macromycetes[1046] thrive in the arctic regions. A definite series of fungi appears to be associated with melting snowbanks in subalpine western North America[217] and in alpine Switzerland.[375]

Within a particular group of fungi, I know of only one study in North America in which a specialist has considered state-by-state and region-by-region distribution of the fungi of his specialty. This is the description by Overholts[850] of the areas in which members of the Polyporaceae have been found in the U.S.

In general, much more must be known about local fungus populations and about the distribution of individual species in relation to habitat variations before a general fungus geography can be developed beyond the stage of *a priori* generalizations.

One of the earlier attempts to tie fungus distribution into the concept of plate tectonics and the resulting continental drift is the explanation by Demoulin[306] of the occurrence of three pairs of species of *Lycoperdon,* one of the members of each pair occurring in Europe, the other in eastern North America.

On the basis of edaphic, climatic, and other factors, and geographic barriers, Pirozynski[916] has discussed the geographic distribution of marine and fresh-water aquatic fungi, soil fungi, root-inhabiting fungi, saprobic and parasitic land fungi, and fungi in the atmosphere. Under the heading of endemism and the activity of man, six species of parasitic fungi introduced throughout the world with man's crops and two species of ringworm fungi carried by Europeans to areas they have colonized and settled are considered.

Aside from monographic studies of genera and higher categories at various regional levels and general lists prepared for one or another purpose, studies have been made relating distribution of various groups of fungi to local conditions. Among the fresh-water aquatic fungi, a study by Rooney and McKnight[979] of populations in a subalpine lake in Utah during two May to November ice-free periods yielded a total of 34 species of aquatic fungi of relatively unrestricted range. An intensive study of aquatic Hyphomycetes in California by Ranzoni[936] has yielded a list of 22 species from a variety of habitats including vernal ponds. Starting with 39 species isolated from waters in Sweden, Nilsson[831] brought together all the species known to 1964 of aquatic Hyphomycetes. Of these, three major categories were recognized. The first included species which are very common, have a world-wide distribution, of which 6 species are mainly tropical and 18 species are mainly temperate and cold water; the second category includes species which are common and occur in several countries; and the third has a limited distribution, mostly known only from the type locality.

A monographic study of marine fungi in which biological as well as morphological factors are considered has been published by Johnson and Sparrow.[608] An annotated bibliography of halophilous and halolimnic fungi has been prepared by Johnson and Meyers.[607]

The geographic distribution of soil fungi is best summarized by Gilman[445] whose manual lists areas from which soil fungi have been reported. Apinis[41] did not list species but considered conditions which permitted the almost universal appearance of soil fungi in the world, even into the sediments in the bottoms of the oceans. In a more restricted area, Domsch and Gams with an English translation by Hudson[328] listed the fungi of soils of northern Europe, and Meyer[784] listed those fungi he found in soils in the Belgian Congo (Zaire). Hunt and Durrell[581] and Durrell and Shields[347] found populations of species dominated by those whose spore walls contained melanin-like pigments in desert soils in southern Nevada and in Death Valley National Monument, Calif.

New or extended records of micromycetes and macromycetes from the Hawaiian Islands, Tonga, the Line Islands, the Society Islands, and the Marshall Islands by Baker and Meeker[65] demonstrate the "dependence of fungi in geographic distribution upon mycologically interested collectors". Habitats sampled included soil, organic substrata, forest litter, aquatic habitats, and the phylloplane. Records were discussed on the basis of known distributions and help to demonstrate wider distribution than previously known for a number of species of fungi.

A checklist of the fungi which have been found in studies based on the occurrence of fungi in sewage, polluted waters, and sewage treatment systems has been prepared.[257] The list includes species isolated from sewage treatment systems in North America and from streams in North America, Europe, and Japan.

In the late 16th century, Clusius produced the first list of fungi prepared for any region of the world. The area studied was in Vas County, western Hungary, and Ubrizsy,[1170] who continues to study the fungi of that region, has called attention to the similarities of the fungal populations 400 years after the studies of Clusius. Since that time, local and regional catalogues of fungi have been produced for most regions of the world. Such catalogues and regional lists in recent times range from lists based on literature references and herbarium specimens, as in the Benjamin and Slot[93] list for Haiti, to the study of Stevenson[1112] on the fungi of Puerto Rico and the American Virgin Islands based on field collections, herbarium specimens, and literature reports in which host ranges and geographic distributions are given for the cited species.

II. DISTRIBUTION PATTERNS

At the level of smaller groups of fungi, Lowe[722] has surveyed the development of the systematization of pore fungi with suggestions for an integrated manual of the species. In a single genus, Thiers[1139] has surveyed the populations of species of *Suillus* in the U.S. where 24 species occur in the northeast, 19 in the southeast, 27 in the midwest, 10 in the southwest, 6 in the Rocky Mountain region, and 37 on the Pacific Coast. Of the 68 species known to occur in the U.S. and Canada in 1975, 3 species are known in all six regions, 1 species was common to five regions, 3 to four regions, 10 to three regions, 7 to two regions, and 43 species occurred in only one of the six regions. While these regions are correlated with forest types, emphases are still based on areas of intensive collection, and the relationship to the development of conifer forest plantations is still incompletely analyzed.

The comparative distribution of fungi in relation to different regions has been considered by Guzmán[487] who compared fungal populations in Mexico and the U.S. Based on over 200 species, four types of relationship were established, the first two being the stronger: species in the coniferous forests of the northwestern U.S. and coniferous forests of Mexico, mainly in northern Mexico; species in deciduous forests of the eastern U.S. and those of eastern Mexico, mainly along the Gulf Coast; species in the eastern deciduous forests of the U.S. and those of the tropical forests of Mexico; and species of the deciduous forests of the U.S. and those of the coniferous forests of central Mexico.

Based on a biogeographical analysis of *Sclerotinia sclerotiorum,* Reichert[956] has suggested that the fungi of plant diseases are subject to the same biographical factors to which higher plants are subject. In relating fungi and plant diseases to geographical factors, the terms mycogeography and phytopathogeography were proposed. Reichert considered the phytogeographical method as capable of determining the ecological character of the pathogen so that conclusions can be drawn concerning the manner of prevention of the disease by being able to predict its appearance and by selecting the place and period least favorable to its development. In determining the paleogenetic centers of the pathogen in question, the existence of resistant varieties of the host plant may be discovered as well as the organism or organisms antagonistic to the pathogen.

The distribution of the fungi is largely habitat controlled. A report may cite a fungus as occurring in a certain location, region, or continent, but without considering the preferred habitat of the fungus, such statements are relatively meaningless. Among saprobic fungi, some species may have specific habitat preferences, others may be omnivorous. These preferences are based on enzyme systems adaptable to use by the fungus of specific fractions of the substratum such as simple sugars, cellulose, lignin, starch, chitin, and other natural compounds. Among parasitic fungi, those restricted to one species of host appear to have the greatest geographic restriction if that host is geographically restricted. However, where a host is widespread as a result of cultivation by man, either as a crop or an adventive, its parasites may become geographically widespread. Those parasitic fungi which are capable of living saprobically with the dead remains of the host and those parasitic fungi that require two different hosts for the completion of their life cycle may have wider distribution than those species dependent on only one host whose range, in the wild or in cultivation, is restricted.

The apparent distribution of fungi may be associated with the interests of collectors and observers. A person based in a community such as a university campus may collect intensively within a specific range of that campus based on his own personal schedule and capabilities, or a person living in an area and not being able to leave may collect the area intensively leaving those reading his reports to assume restricted distribution for the species collected or observed, while actually the fungus may have a much wider distribution.

Within a particular research area, such as plant pathology or forest pathology, the individual may search for fungi causing specific diseases to the exclusion of other fungi, the results of his study appearing to preclude the existence of other fungi which may have contributed directly or indirectly to the problem. The same may be true of food microbiologists and even medical mycologists and pollution mycologists. The mycophagist will be looking only for edible fleshy fungus fruit bodies, finding the toxic ones by accident, and the photographer will be looking only for photogenic camera studies. In each instance, a bias of a type restricted to the individual searcher will color the reported distribution patterns of the concerned fungi.

On the other hand, the mycoecologist or mycosociologist may develop programs designed to study a variety of habitats, and he will probably include most, if not all, of the fungi in a plot area in his study. If he works with only a fraction of these fungi, saprobes, mycorrhiza formers, pathogens, etc., he will tend to develop broader outlooks on the distribution of the studied fungi, not only on his plots but also wherever they are known. In the case of some species, the broader outlook is important since the fungus under consideration may be specifically distinct from a similar appearing but microscopically different fungus.

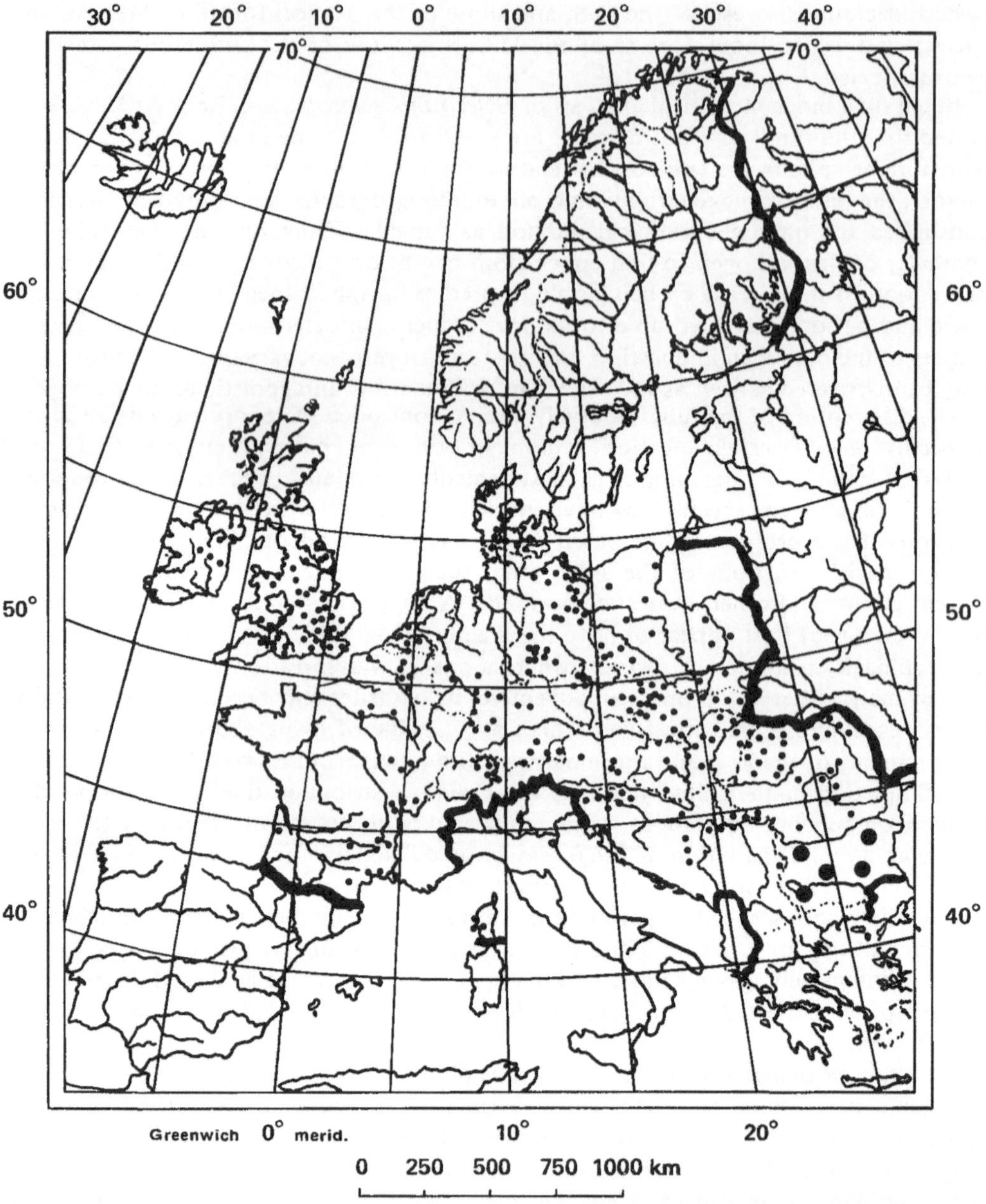

FIGURE 1. Distribution of *Hirneola auricula-judae* (Bull. ex St.-Am.) Berk. in Europe. Large dots indicate "common all over the country." (From Lange, L., *Dan. Bot. Ark.*, 30(1), 55, 1974. With permission.)

III. MAPPING SCHEMES

In at least one area it is thought that sufficient information is available concerning the distribution of fungi of that area, including Europe except Portugal, Spain, Italy, Albania, and Russia, that distribution maps can be prepared delimiting the range of chosen fungi. For 50 of the first 100 chosen species, maps (Figure 1) have been published by Lange.[671] A committee of five members of the European Mycological Congress was set up in Prague. Each of the five members was to be responsible for a region and would report his findings to the secretary of the committee. All continental Euro-

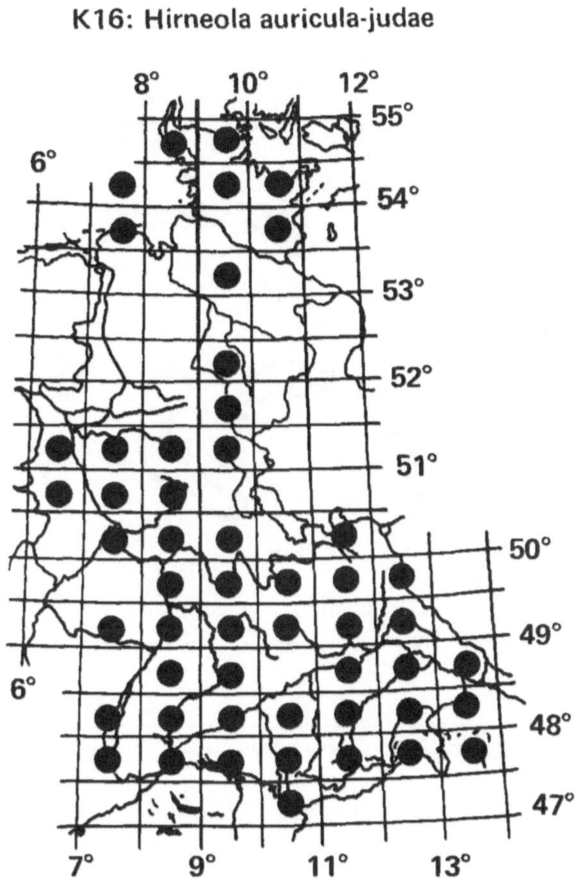

FIGURE 2. Distribution of *Hirneola auricula-judae* in West Germany (BDR). (From Bresinsky, A. and Dichtel, B., *Z. Pilzkd.*, 37, 101, 1971. With permission.)

pean countries, except those cited, plus Great Britain and Iceland were included in the scheme. To supplement lists of collections to be considered, certain fungus biotas were also used. An attempt has been made to eliminate the bias of locating the centers of distribution related to centers of mycological activity. Three groups of species were based on seasonal variation: species where the majority of finds were made in the same few months in all of Europe; species with a long fruiting season in all of Europe; and species fruiting in the spring. Seven distribution types were considered: (1) species which reach or exceed the northern timberline, (2) species with northern limits between 65° N and 68° N, (3) species whose primary distribution is south of Limes Nordlandicus and with only a few finds north of this, (4) species with northern limit at Limes Nordlandicus, (5) species with a northern limit coinciding with that of *Fagus,* (6) species associated with the natural distribution area of *Abies alba* (corresponding to a northern limit at about 52° N), and (7) species with a northern limit approximately coinciding with the northern limit of *Vitis vinifera.* For West Germany, Bresinsky and Dichtel[130] have developed their own grid system and have mapped the distribution (Figure 2) of 73 species of fungi in that country.

Among other mapping schemes is that of the Commonwealth Mycological Institute,[206] Kew, England which has published a series of maps (Figure 3) showing the distribution of important plant diseases and their fungi throughout the world. As con-

COMMONWEALTH MYCOLOGICAL INSTITUTE DISTRIBUTION MAPS OF PLANT DISEASES Map No. 31 Edition 4 Issued 1.x.1977	Pathogen: Fusarium oxysporum Schlecht. ex Fr. f. sp. cubense (E. F. Sm.) Snyd. & Hansen Hosts: Banana and other Musa spp.

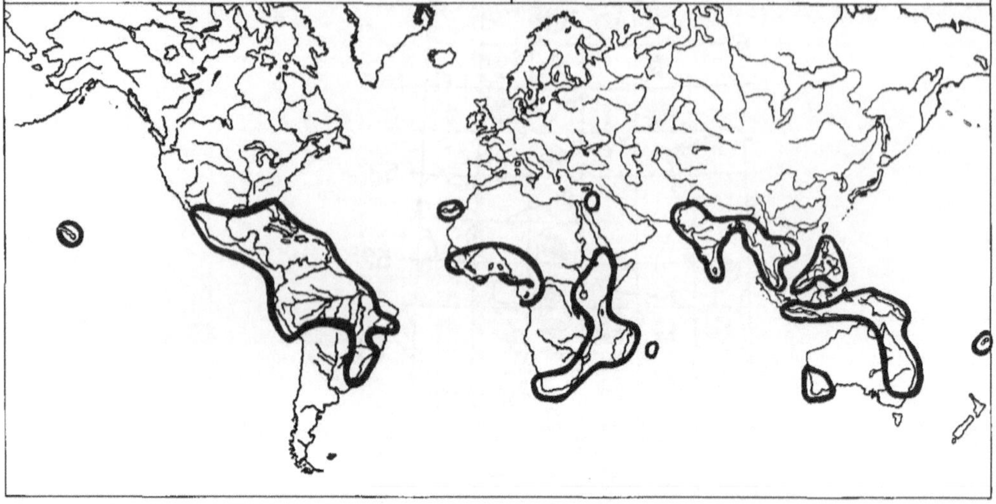

For list of countries in which this disease has been recorded, see overleaf

Map No. 31 Edition 4

AFRICA
- *Cameroon [36:638; 40:148]
- *Canary Islands [42:36]
- Congo [Stover, 41:610]
- Ethiopia [48, 748]
- *Ghana [19:262; Dade, 20:315]
- Guinea [Stover, loc. cit.]
- *Kenya [35:93]
- Malagasy Rep. [41:472]
- Mauritius [Wiehe, 28:472]
- Mozambique [50, 3049]
- *Nigeria
- Sierra Leone [8:291]
- South Africa [43:167]
- *Tanzania [32:324]
- *Uganda [32:423]

ASIA
- Brunei [51, 3059]
- Burma [13:586]
- India (Bengal, Tamil Nadu) [Butler & Bisby, 41:18], (Bihar) [32:324], (Maharashtra, Andhra Pradesh) [35:690]; (Mysore) [36:518]
- Indonesia (Java, Molucca Islands, Banda Island) [13:586; Stover loc. cit.], (West Irian) [41:764]
- *Israel
- Malaysia (W., Sarawak) [32:303; Johnston, 40:395]
- Pakistan [FAO host list, 42:622; 44, 2333]
- Philippines [29:154]
- Sri Lanka [31:132]
- Thailand [6:42; 13:586]
- Vietnam [46, 3031]

AUSTRALASIA & OCEANIA
- Australia (NSW, Qd., W. Aust.) [43, 2003; 27:221; 44, 87]
- Fiji [FAO Rept., 40:582; but doubt has been expressed, 52, 659]
- Hawaii [42:396]
- Papua New Guinea [50, 191]
- Trust Territory Pac. Is. [56, 1212]

NORTH AMERICA
- Mexico [11:463; 12:303]
- USA (Fla.) [Agric. Handb. USDA 165, 1960 (40:511)]

CENTRAL AMERICA & WEST INDIES
- Barbados [53, 4734]
- Belize [29:12]
- Costa Rica [14:387; 43:169]
- Cuba [Phytopathology, 9:345, 1919]
- Dominica [26:146]
- Dominican Rep. [38:450]
- Grand Cayman Is. [17:728]
- Grenada [Wardlaw, 14:322]
- Guadeloupe [17:191]
- Guatemala [43:169]
- Haiti [18:327]; 19:661; 20:125]
- Honduras [33:307]
- *Jamaica [29:351; 37:174]
- Martinique [41:321]
- Nicaragua [Wardlaw, 14:322]
- Panama [43:169]
- Puerto Rico [10:324, 777]
- Salvador [39:605]
- St. Lucia [24:8]
- St. Vincent [16:368]
- Trinidad [25:3]
- Virgin Is. [39:605]

SOUTH AMERICA
- Brazil (Amazonas, Bahia, Para, Rio Grande do Sul, Sao Paulo) [26:33; 33:614; 27:77; 19:105]
- Colombia [33:282]
- Ecuador [14:159]
- French Guiana [49, 2396]
- Guyana [9:796; 10:362]
- Peru [Bazan de Segura, 40:201]
- Surinam [9:232]
- Venezuela [14:397; 21:324]

NOTE: See CMI Descript. 214

*Specimens in Herb. IMI

Numbers in square brackets e.g. [54, 1234] refer to abstracts in the Review of Plant Pathology

FIGURE 3. World distribution of *Fusarium oxysporum*. (From Distribution Maps of Plant Diseases Commonwealth Mycological Institute, Kew, England, October 1, 1977, Map 31, Edition 4. With permission.)

sidered necessary, the maps are brought up to date. Map points are based on reports published in the *Review of Plant Pathology*.

IV. THE NUMBERS OF FUNGI

Among better estimates of the numbers of species of fungi are those of Ainsworth[9] which have developed out of concepts suggested by Bisby and Ainsworth[107] and Martin.[754] Allowing for the existence in the literature of a number of synonyms and assuming that all species in existence have not yet been recognized, it is thought that there are more than 100,000 species of fungi in the world.

Chapter 3

GENETICS

I. GENETICS, SEXUALITY, VARIATION

General summaries of work on genetics and sexual mechanisms in the fungi have been developed in both North America[185,943] and Europe.[166] A general discussion of natural selection in the microbes included discussion of certain fungi[1181] and the elaboration of principles applicable to fungus studies.

In the group of organisms commonly brought together to form the Phycomycetes there is a wide diversity of sexual phenomena, and some of these have been made the bases for studies of genetics in the fungi.

In the Blastocladiales (Chytridiomycetes) the life histories, sexuality, and variability of several genera have been studied intensively.[355,356,513,514] Morphological variability of a species in the Saprolegniaceae (Oomycetes) has been observed in which many sporangial types could be produced under varying conditions of the culture medium. The fact that there are several sexual hormones in the genus *Achlya,* each active in a different phase of sexual development, has been demonstrated by Raper.[940-942] Most genera of the mucoraceous (Zygomycetes) fungi include heterothallic species requiring two opposing strains before sexual processes can occur.

In the Ascomycotina, sexual phenomena are developed in quite different ways. Five principal groups of fungi have been studied in this subdivision with reference to genetic phenomena. Nuclear behavior has been reviewed earlier.[845] The yeasts are fairly simple, and some of the work with them in North America has been published from St. Louis[706] and Davis, California.[904] In England the biology of yeasts[592] has been brought up to 1955, and developments in cytology and genetics have been included. Continuing studies on *Glomerella*[1224] in Louisiana, in which several hormones have been demonstrated,[333] and on *Venturia*[117] in Wisconsin have been used for various types of genetic analyses, including chromosome-mapping techniques; such studies are related to the genetics of the pathogenicity of the strains of those fungi used. *Pyronema* has been used, together with species in related genera, in the study of sexuality and nuclear condition in various phases of the life cycle and in investigations of homothallism and heterothallism. *Neurospora* has been considered in a large number of experiments dealing with genetic variation, with special emphasis on variants deficient in the ability to grow in one type or another of nutrient in the absence of one type or another of required nutrient source. The genetics of *Ascobolus stercorarius*[108] has been described.

Studies in the sexuality of the Basidiomycotina have been of several types. Many of those on rusts have been summarized,[155] including both the kinds of rusts which require more than one host upon which to complete their life cycle — such as the white pine blister rust (*Cronartium ribicola*) and the black stem rust of wheat (*Puccinia graminis*) — and those kinds which complete their life cycle on only one host. Heterothallism and its manifestations in the smuts have been summarized.[1236] The genus *Schizophyllum*[944] has received attention in genetic studies of various kinds. A variety of studies dealing with bipolar and tetrapolar sexuality in the Basidiomycetes[834,1233-1235] has been presented. The types of tetrapolar sexuality in the genus *Coprinus,* in which as many as 16 combinations of monosporous mycelia can be made within one species, have been worked out in Canada. Included in the studies on social organization in the Hymenomycetes[149-152] was work on diploidization in the higher fungi which has been reviewed by Buller.[156] The pairing of monosporous mycelia derived from two apparently different fungi[816,1182] is commonly used to determine whether two strains are members of one species in the Homobasidiomycetes. Work with *Hirschioporus*, *Pycnoporus*, *Coriolellus*, and other fungi has demonstrated the usefulness of the methods involved. In contrast, the component strains of dikaryotic mycelia can be recovered[944] by special techniques.

In the Fungi Imperfecti, one set of studies has resulted in a completely different species alignment in the genus *Fusarium*,[1084] while another set of studies by Booth,[118] based on different cultural techniques, has resulted in a clarification of an earlier concept. In both cases, the concept of the dual phenomenon,[494] associated with heterokaryosis[495] in these fungi has been taken into consideration, as well as perfect state relationships. Work wih heterokaryosis has been carried over into studies in *Trichophyton*,[1247] *Botrytis cinerea*,[496] species of *Penicillium*,[63] and other genera. Work with the fungus which produces penicillin, *Penicillium chrysogenum*, in which methods of producing high yield strains of the fungus are described, has been summarized in a geneaology.[61] Additional manipulation with heterokaryosis, leading to the elaboration by Pontecorvo[919] of the theory of the parasexual cycle in fungi has been described.

The steps in the parasexual cycle in the fungi[919] have been described as: (1) fusion of two unlike haploid nuclei in a heterokaryon, (2) multiplication of the resulting diploid heterozygous nucleus side by side with the parent haploid nuclei in a heterokarvyotic condition, (3) eventual sorting out of a homokaryotic diploid mycelium which may become established as a strain, (4) mitotic crossing over during multiplication of diploid nuclei, and (5) vegetative haploidization of diploid nuclei. The most important events here are considered to be the fusion of unlike nuclei, mitotic crossing over, and haploidization. There are three important results of such a phenomenon: (1) production of haploid strains like the starting ones, (2) production of haploid strains which recombine in all possible ways the chromosomes and chromosome parts of the initial strains, and (3) production of a small proportion of diploid strains homozygous and heterozygous for all possible associations and combination of the members of Items 1 and 2.

A summary of the study of mycogenetics has been written by Burnett.[166] Following a discussion of fungi as organisms for genetic study, three major areas are considered. Under formal genetics are discussed genetic markers; recombination, segregation, and linkage in mitotic and meiotic systems; recombination and segregation of nuclei and of extrachromosomal elements; and quantitative inheritance. Under population genetics are considered the generation of variation, experimental, and natural selection and the mechanism of isolation. Among areas of application of fungal genetics are considered industrial applications, genetic aspects of fungal pathogenicity, and such aspects of the relation of fungi to general genetics as recombination and gene action and regulation.

In a series of studies conducted in England, several problems in the distribution of higher fungi have been worked out by Parker-Rhodes on a statistical-genetic basis. Most of these studies have been conducted on species of fungi in the population of Skokholm Island, a small island off the coast of Wales. In most cases, the species on the island are compared with similar material on the mainland. Two species, *Panaeolus campanulatus* and *Leptonia solstitialis*, were studied.[867] To determine whether spore size could be used to indicate the presence of a distinct insular biota, in favorable cases a comparison of the variance and higher cumulants of distribution of spore measurements from different populations might indicate the degree of inbreeding in an area. In both species evidence was found that the insular populations were more highly inbred than those on the mainland. It was thought that some barrier exists to prevent the free interchange of genes between the insular and mainland populations, and on this basis it is suggested that small animals play more of a role in the distribution of these species than has been commonly suspected. It was suggested that this type of separation may result in the appearance of insular subspecies in certain species.

In a study of *Panaeolus papilonaceus*,[868] an analysis of spore breadth, using tetrasporic clusters for analysis of related species, was developed. The population of Skokholm Island was significantly more homozygous than the mainland population, and the specimens examined were probably heterozygous for at least one pair of allels having a major effect on spore breadth. It was concluded that probably the 2 mi of water separating the island from the mainland produced a satisfactory barrier to wind distribution of spores, at least in this species.

A species of *Psilocybe* occurring on rabbit and horse dung on Skokholm Island was intermediate between two previously well-known species.[869] By statistical analysis of spore tetrads from the hybrid and its suspected parents, it was thought that the new hybrid had become stabilized in recent times and that it is a distinct genetic entity, although capable of back crossing to the parent species.

Intramycelial variation in the agaric *Hygrocybe turunda* var. *lepida* collected in Orchid Bog on Skokholm Island has been investigated.[873] From each of a number of basidiocarps 100 spores were measured in length and breadth. Characters employed in two statistical analyses of the data were spore length and shape; the latter was measured as the ratio of length to breadth. As a result of this analysis it was concluded that the mycelium was genetically heterogenous, containing more than one diplont.

Mycena galopus has been chosen[874] as an example of a Basidiomycete in which to demonstrate deme structure in the fungi. From a number of collections obtained in England, 100 or multiples of 100 spores were measured as to length and shape ratios. The collections studied represented a series of phenodemes which, on a statistical basis, were distinct but which could not be distinguished morphologically, even by the measurement of a few selected spores. Following a discussion of the significance of genetic barriers between colonies of apparently identical individuals and the possibility of the lack of barriers between other individuals, the following statement was made:[874] "It is one thing to discover what is *possible*, by laboratory experiments, but quite another to work out how these possibilities actually work out in nature, which is usually what most closely concerns us."

For Shepherd[1031] the *genotype* is the individual's genetic constitution which describes its genetic formula and breeding behavior. The *phenotype* is a term referring to the external appearance produced by an organism with a given genotype in a given environment. Within the term *genome* are included chromosomal genes and particles in the nucleus or cytoplasm having the properties of hereditary determination and an autonomy in replication. The term *ecosystem* includes the interrelationships between the organism as a whole and its immediate physicochemical and biotic environment, both internal and external, and also includes alterations of these relationships with time. With this in mind, *ecology* may be redefined as a study of genome-controlled phenotypes in relation to their environments.

Four groups of interactions between genome and environment were recognized: (1) the fixed genome in a fixed environment (of no interest in ecological studies since there is no interaction), (2) the fixed genome in a variable environment (for example, the formation of carotenoid pigments in *Fusarium oxysporum* in light but not in darkness), (3) the variable genome in a fixed environment (heterokaryosis was given as an example overlapping into a fourth group), and (4) the variable genome in a variable environment.

Genetic adjustment to the environment was discussed by Person[894] in terms of adaptation mechanisms for either saprobic or parasitic situations. Such adaptation may be reversible when the enzyme systems which were utilized developed in response to certain types of available nutrients, or it may be irreversible in cases where gene or chromosomal changes acted as the principal response of the fungus to a changed environment or nutrient supply.

One type of reaction to the environment is the development in the life cycle of an imperfect state, either in the monokaryotic mycelium or in the dikaryotic mycelium. The imperfect state permits the survival over a single or a large number of generations of a fungus growing in a habitat subject to rapid changes in nutrient supply, moisture, temperature, or some other factor. It is possible that, should heterokaryosis become involved with a phenomenon which occurs sometimes such as mitotic crossing over, permanently incompatible nuclei could be produced resulting in a life cycle which remains in the imperfect state, thus giving rise to a true moniliaceous fungus as has been suggested by Cooke[226,232,268] in the case of at least certain strains of *Aureobasidium pullulans*.

Multiple inoculation of a substratum,[946] whether it is over a period of time in wood-rotting forms such as *Schizophyllum commune* or simultaneously by exposure of preinoculated substrates to conditions favorable for coprophilous forms such as *Coprinus* species, results in physiologically unified dikaryons constituted of many independent associations of many diverse origins. The product is a genetic mosaic, all parts of which may produce fruiting bodies and liberate basidiospores. The entire integrated dikaryotic mycelium may be very extensive and is, in fact, a mosaic population capable of harboring enormous variability. An experimental analysis of a demonstrated natural mosaic population has yet to be made.

In plant pathology, Flor developed the gene-for-gene hypothesis in 1955. This hypothesis has been used to advantage in the study of more than 18 host-pathogen relations[388] and was considered by Wheeler[1225] to be a significant advance in phytopathology. The theory, as stated by Person and quoted by Burnett,[166] is: "A gene-for-gene relationship exists when the presence of a gene in one population is contingent upon the presence of a gene in another population, and where the interaction between the two genes leads to a single phenotypic expression by which the presence or absence of the relevant gene in either organism may be recognized." It was developed from the observation that in flax rust (*Melampsora lini*) the inheritance of the rust reaction within the host is paralleled by the inheritance of virulence in the pathogen, and their interaction is conditioned by the two corresponding pairs of genes. Flor[388] noted that "the gene-for-gene concept has been useful as a tool to identify the roles of hybridization, mutation, heterokaryosis, and somatic hybridization in pathogenic variation of pathogenic fungi. Also it has served to survey and identify the resistance germ plasm in the host, to study induced mutations to resistance, and to develop multiline and multigenic varieties."

II. TERMINOLOGY OF MATING SYSTEMS

Few people have tried to integrate the study of sexual systems in the diverse organs called fungi. In an early work, Link[710] tried to develop a terminology of the sexual systems as they were understood in 1929. In the intervening 48 years, a large amount of information has been developed on the genetics and physiology of the processes involved in mating in fungi, and a new set of terms was proposed by Burnett.[164] Only 28 of the 171 references cited at that time were published in 1929 and earlier. Burnett's terminology follows:

Heteromixis is the condition where sexual reproduction only results from the fusion of genetically different nuclei normally derived from different thalli. It includes: (a) *Dimixis*, the heteromictic condition where there are two and only two types of complementary nuclei which control mating. The nuclear types are determined by two allelomorphs at a single locus. (b) *Dimorphomixis*, the heteromictic condition where several types of complementary nuclei occur which control mating. The nuclear types are determined by multiple allelomorphs at one or two loci, the bipolar and tetrapolar conditions respectively. (c) *Homoheteromixis*, this is essentially a heteromictic condition where sexual reproduction results only from the fusion of genetically different nuclei derived normally from the same thallus. Such forms are derived from dimictic or diaphoromictic forms, hence the derived terms "homodimictic" and "homodiaphoromictic." *Homomixis* is the condition where sexual reproduction can result from the fusion of genetically similar nuclei derived normally from the same thallus. *Amixis* is the condition where the essential events of sexual reproduction are lacking but the pre-conjugation and post-meiotic events normally associated with sexual reproduction may occur.

"Amixis" is chosen here rather than "apomixis" which is a term usually associated with flowering plants.

Numerous preexisting terms are tabulated in comparison with the newly suggested terminology, and examples of fungi demonstrating each term are given. The new terms have the advantage that, in addition to being well documented, they have easily understood bases in stem, suffix, and prefix words from the Greek which have universal application in genetic and botanical literature. As many as seven preexisting terms may be replaced by one new one. The older terms tend to compartmentalize the phenomena they describe unnecessarily, are composed of multiple words, in some cases, and in other cases too loosely define the phenomena they describe by the use of analogy with conditions as they occur in higher plants which may not be truly comparable.

Chapter 4

AUTECOLOGY AND SYNECOLOGY

I. AUTECOLOGY

Autecology is the study of the individual in relation to its environment. Studies may be largely physiological in relation to single fungal species. Synecology is the study of communities of species, is rarely physiological, and includes studies of the structure, development, function, and causes of distribution of fungi, and their ecosystem relationships.

Methods and techniques for studying autecology are considered in later chapters in relation to nutrition and physical and chemical factors of the environment. Work largely in autecology has been summarized in a series of papers presented in a symposium edited by Williams and Spicer.[1252] A description[92] of the relation of taxonomy to technology has been given, and the sources of fungus cultures and their maintenance in culture collections, systems of classification, and the methods of naming organisms were included in the discussion. In addition, Smith[1073] was concerned with fungi both inimical to and beneficial to man and his activities. Among methods of protecting industrial products considered were sterilization, limitation of moisture, and antisepsis. Beneficial activities cited included brewing, baking, cheese manufacture, the preparation of various organic acids, glycerol, fats, proteins, vitamins, antibiotics, and the modification of stereoids for therapeutic use.

The presentation of information about various fungi to mycology classes is considered by a number of teachers. Emerson[357] discussed techniques useful for work with Phycomycetes, Koch[651] has developed a useful laboratory manual, and Stevens and his committee[1111] have produced a guidebook for use in all kinds of mycological instruction and demonstration.

Ecological life history studies bring together a variety of data published in usually scattered reports. A proposed outline for such a life history was published by Cooke.[212] In following such an outline it has been found that certain features need greater emphasis than has been given in published studies, and other areas may be overemphasized. Such a life history has been prepared by Cooke[226] for *Aureobasidium pullulans*, a ubiquitous fungus. A similar study has been published, summarizing available information, for *Phycomyces*[98] in a symposium-like format.

Bergman[98] has edited summarized knowledge concerning the genus *Phycomyces*, based largely on publications dealing with the species *P. blakesleeanus*. The study, or review, was introduced with a survey of the systematic position of *Phycomyces* and its natural history. Under life cycle are considered spore germination; mycelial growth; sporangiophores, including stages of development and their dimensions, effects of temperature and humidity, the growth zone, incorporation of matter from the mycelium, growth in water, and spore liberation; the sexual cycle; experimental studies of growth and regeneration; plucked sporangiophores including techniques and growth of whole sporangiophores; and oxygen consumption including rate and ATP level in sporangiophores, rate of oxygen consumption in the mycelium, the respiratory quotient, and dependence of sporangiophores upon oxygen. Among items of interest in cytology are the kinds of particulates and their distribution, and intracellular centrifugal separation of organelles in state I sporangiophores, including techniques, description of layers, microsurgery, and the search for a photoreceptor. A number of topics are considered under sensory physiology based on light, including growth responses to light; light and dark adaptation based on sunrise experiments; dioptric effect; sporangiophores immersed in fluids of refractive index of *n* or neutral; phototropic neutrality at the appropriate wavelength; a detailed analysis of the growing zone (longitudinal autonomy vs. integration); temporary inversions of tropic responses; and effects of light in the high-intensity range, including indifference to light, loss of growth rate regulation, and tropic hunting. Items discussed under receptor pigment include action spectra under which, following general remarks, are considered action spectrum of growth response, action spectrum of tropic response, unity of receptor pigment, orientation of receptor pigment, comparison with other blue action spectra, and the possible occurrence of receptor pigment in the mycelium; transmission spectrophotometry and the estimation of the upper and lower limits of the concentration of receptor pigment; and the receptor pigment. Other types of sensory physiology include stretch, gravity, and the avoidance response. Under stretch, the mechanical properties of the sporangiophores and response to stretch are discussed; under gravity, geotropism in the absence of blue light and effect of blue light on geotropism are discussed. Genetics is considered in two ways: under asexual genetics are considered mutants, heterokaryons, karyology of the asexual cycle, and quantitative segregation; under sexual genetics are considered sex determination, karyology of the sexual cycle, and crosses. Finally, under biochemistry are considered nutrient requirements; base composition of nucleic acids, proteins, and ferritin; lipids including glycerides and phospholipids, stereols and carotenes; phenolic acids; cell walls of sporangiophores; and indole-acetic acid.

In reference to various limits of growth of soil fungi, Apinis[41] noted that many such limits are edaphic and climatic but sometimes biotic as in the case of the presence or absence of roots. Such limits include soil moisture, especially where well-aerated soils meet water-logged soils in marshes; ocean shore soils where salinity factors enter but do not inhibit all fungi; and soil depth where, except for the vicinity of roots, organic matter at greater depth becomes limiting. Other limiting factors considered included special populations and nutrient and other environmental factors.

For Hawker[516] an important direction of study in fungus ecology is the study of factors influencing the survival of fungi in natural habitats. For survival of the individual, alteration of the assimilative cells and production of special resting bodies are considerations; for survival of the species, conditions of importance are escape from an unsuitable habitat through rapid growth of the mycelium, rapid formation and efficient dispersion of spores, and the advantage of rapid spore germination; the ability to colonize unusual or borderline habitats and the ability to alter the habitat in a direction unfavorable to other organisms; and variability.

Recent texts on microorganisms including fungi have considered a number of ecological factors and the relation of fungi to them. Brock[139] has treated the biology of microbes and their ecology.[141] Burnett[165] has related a number of factors to the development of fungi. Physical factors have been emphasized by Griffin,[482] and Hudson[567] has discussed the relation of fungi to the saprobic systems of nature.

II. SYNECOLOGY

To a large extent, the study of fungus ecology has been the study of fungus autecology. Such work as has been carried out in the field of synecology has been confined to the larger fungi whose fruit bodies are visible and readily obtainable. Most of the work published on synecology up to 1952 was reviewed by Cooke.[210,213] Some of this material and additional work has been reviewed in Holland[570] and in Austria.[554] Work in the correlation of fungus and vascular plant associations has been published in

Czechoslovakia.[658] However, work with fungus sociology has failed to impress most students outside of eastern Europe.

According to Garrett,[427] such observations and studies are superficial. He compares them with the study of the vascular plants of a meadow collected by scanning a haystack. Having carried out work of this type myself,[215] I can admit the partial validity of this criticism, although I can see where one not accustomed to working with the larger fungi, confronted with a mixed population in a woodland in the "fungus season", sees a complex mass of fruit bodies of edible and poisonous fungi arranged without order. If these were collected by sampling the area without regard to speciation, the criticism would be completely valid, but the synecologist separates species in the field, usually without much regard to positioning of mycelium which is invisible, except that in the case of some species mycorrhizal habits are known, in the case of others, preference for different types of litter, and different types of substrates are known or possibly suspected. For comparative surveys of fungus populations of fields and forests, or of forests of different types on various types of soil parent materials, it is thought that this type of study is an adequate tool. This is especially true if the technique is not equated with apparently similar techniques used for flowering plants and if it is supplemented by studies of the whole fungus population and by studies of the type suggested by Warcup.[1204,1205] It is true that fewer and fewer climax habitats are available for synecological studies. Even in such situations conditions are subject to change, although more slowly than in more seral woodlands or plantations which initially are more like uniculture fields. For Hawker,[516] "The problem of the distribution and succession of fungi in the natural habitat, which is seldom constant for more than a brief interval, is also the problem of the survival of different individuals and different species under changing conditions. A knowledge of the methods by which fungi survive under adverse conditions is essential to the interpretation of the distribution of species in time and space."

Because of their strict dependence on the vascular plant flora, fungi never form independent associations, according to Ubrizsy.[1166] The apparent fungus associations depend on the seasons, altering as the aspects: if they are stable, mycoaspect; if ephemeral, phase aspect. The maximum aspect occurs in the autumn. For an area there is an initial aspect and a terminal or climax aspect depending on the developmental history. Ubrizsy[1166] gave a life form classification for fungi patterned after those of Raunkiaer and Braun-Blanquet and included nine major categories. Use of this classification has not been made in recent publications, and other workers have not taken advantage of it.

Zaleski and co-workers[1287] collected the fungi of a state forest in Poland in 1946 and 1947. The more important correlations between fungus occurrence and external factors included mechanical injuries from World War II activities, other wound factors, old solitary trees undergoing decay, time periods of greater precipitation correlated with temperature conditions, soil qualities such as humus content and moisture relations, and phytosociological relationships.

Höfler's[553] use of the term "fungus aspect" described the appearance in the same forest of essentially the same groups of fungi at approximately the same season of the year. Thus, each forest may have its own series of fungus aspects throughout the year, and each forest may have different series of fungus aspects, or each fungus aspect may be characterized by different series of species.

In his review of techniques in fungus sociology, Höfler[554] reemphasized statements made earlier and quoted by Hueck:[570] "On the basis of theoretical considerations it is argued that the community of higher fungi, in the present stage of investigation, must be studied independently and not as part of a known association of higher plants."

Since the fungi and bacteria are heterotrophic and dependent on higher plants, "we arrive at the conclusion that autotrophic and heterotrophic plants must be considered as a bi-entity in which the components are like two poles, fundamentally different, often antagonizing each other, yet indivisibly united. The separate study therefore of communities of fungi and of higher plants each according to their own peculiarities must ncessarily, in a later stage of investigation, lead to a synthesis of both aspects of the plant community."

Lange,[672] working in Greenland, found a fairly large number of larger fungi. These were largely of temperate affinity, and those occurring in larger number were associated with thickets of *Salix glauca*. Arctic-boreal elements were low in number.

Nespiak[829] studied fungus and forest associations in a national park in Poland. He used 100-m^2 plots in all associations. From his data he found it possible to demonstrate analogies between populations of fungi and of higher plants. The development and fruiting of fungi depended on the ecological conditions of the association as a whole. The 100-m^2 square quadrats were too small for adequate use in forest associations but adequate for use in pine-*Sphagnum*, and *Vaccinium-Sphagnum* associations. In many cases, the species found inside the quadrat had to be supplemented with species from outside the quadrat for more accurate delimitation of populations. From tables compiled for every plot, the similarities of the investigated associations were determined using Jaccard's and Steinhau's[829] formulae. The frequency with which species common in one association occurred in other associations was shown using Cooke's[215] diagrams (Figure 1).

Coenological analyses of Hungarian forest associations by Ubrizsy[1168] have shown them to have fluctuations from year to year, but in permanent 100-m^2 plots there was a certain uniformity. Regular examination of permanent quadrats over at least a 5-year period was required using coenological methods to describe the so-called fungus aspect and determine the seasonal rhythm. Optimal weather conditions occurring in the period from July to September helped establish the maximal aspect. The remaining aspects found in less favorable seasons complete the picture of the total production as determined by the study of the reference stations.

Bohus and Babos[113,114] developed a coenological study of terricolous macroscopic fungi in certain deciduous forests in Hungary. For each type of information considered, they analysed the previous usage as presented in the earlier literature, then defined the concept as it would be used in their study. Dominance was expressed as the percentage drawn from the number of specimens (fruit bodies) of a species as related to the total number of specimens (fruit bodies) of all species found in a biocoenose or biocoenose portion. Eudominants were species with large fruit bodies, dominants were species with medium fruit bodies, and subdominants were species with small fruit bodies. The index of dominance is D, determined by the formula $D = a/b \cdot 100$ where a is the number of fruit bodies of a given species and b is the number of fruit bodies of all species. Weight dominance is the percentage based on the weight of the fruit bodies of a given species as related to the total weight of the fruit bodies of all species found in an investigated biocoenose or in a biocoenose portion. Ten weight dominance classes were established, from class 1, representing values from 1 to 10, to class 10, representing values from 91 to 100. For a group of plots, the weight dominance class numbers are averaged, or the average could be taken for a number of surveys for each plot.

The complex dominance and the complex weight dominance for ecological groups or taxonomic groups of species is the average of results from all plots in all associations of more or less comparable nature. Examples were given of results from eight and from four associations, and comparisons could be made from tables prepared from such data.

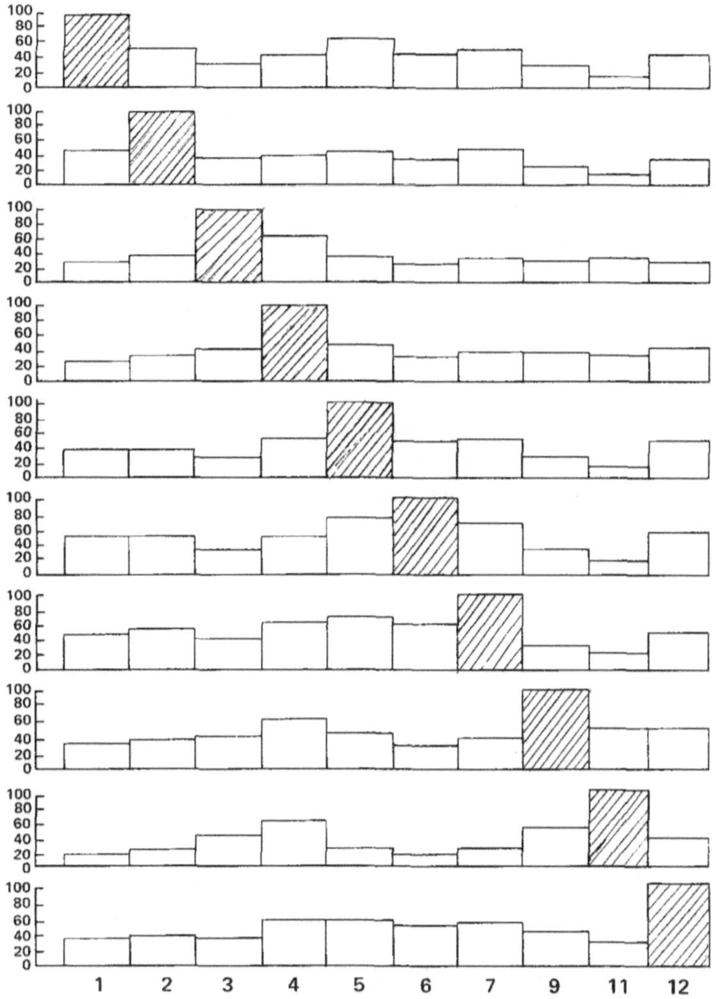

FIGURE 1. Species fidelity chart. "Bridge Cooke species identity graph in the forest types of the model research area in the Bükk Mountains, Hungary, related to a unit area of 500 m² on the basis of 7 surveys in 1955—57." In each line, the shaded block represents 100% of species, the remainder represent the percent species present in that area. (From Bohus, G. and Babos, M., *Bot. Jb.*, 80(1), 68, 1960. With permission.)

Abundance is the term chosen for the average specimen number of a species, or a species group, occurring in a unit area of an investigated biocoenose. Production, and complex production, is the weight of fungus fruit bodies in one, or in all studied associations. This is independent of the quality of the soil; maximal or near maximal production may occur on soils of rather low quality or on those with an acidic character. From the point of view of fungal factors, the mycocoenose index is given by the formula $M = (A + P)/(30 \cdot N)$ where A is the complex dominance, P is the complex production, and N is the number of surveys. For species groups, the value obtained in this way may be termed the Prosperity Value. Species saturation is used for the numerical representation of the species of a given fungus group (certain genera, ecological groups) in an area unit of an investigated biocoenose. The higher the value of species saturation, the more varied is the biocoenosis under study; the more species will find their life conditions in it. "Constancy" is the percentage of occurrence of a given species of fungus per unit as related to the total number of investigated plot units. Species present in 80 to 100% of the examined area units are constant species, those present in 60 to 80% of the units are subconstant species, and species present in 0 to 60% of the units are accessory species. Four degrees of constancy are recognized: (1) plots in a small homogeneous area of a single biocoenose give homogeneous constancy or frequency; (2) plots staked within a larger area so that a single quadrat within the area, when surveyed, shows that every variant within the stand gives analytical constancy; (3) plots in various stands of the biocoenose give regional constancy; and (4) plots within the areas of various subassociations of a given association give subassociation constance. Character species is restricted to a given association. Differential species is a term Bohus and Babos chose not to use until it is better understood.

Jaccard numbers, obtained for two biocoenoses or two areas within a biocoenose by dividing the number of species present in both by the total number found in both, and Cooke species identity graphs (Figure 1), obtained by letting 100 equal the number of species found in *all* plots and comparing it with the number of species found *in* each, are used for the analyses of agreements and deviations within the mycocoenoses. Identity numbers show very strikingly the agreement of mycocoenoses of closely related forest types and dissimilarities of distant forest types. The distances between areas of the biocoenoses have little or no influence on these values. Coenological relations between forest types may, to a certain degree, be analysed by species identity number on the basis of mycocoenoses.

The relation of terricolous macroscopic fungi and the plant associations of deciduous forests may be based on seven categories of fungal species: (1) character species; (2) species restricted to a certain association but occurring, sometimes only sporadically, in other associations; (3) species occurring in more than one association but always absent in certain other associations; (4) association-indifferent species; (5) species more or less restricted to a plant species; (6) species restricted to plant or other detritus; and (7) species occurring rarely in forest associations. Accidental species may also be present.

Singer and Moser[1048] studied the distribution of fungi in the Cordillera Pelada, Chile, with special reference to ectotrophic mycorrhiza formers. Distribution was correlated with climatic conditions and with the distribution of forests in which acceptable host species occurred. There were no ectotroph-formers among the fungi in areas without *Nothofagus.* Ectotroph-free forests above 500 m above sea level had a mycobiota similar to that of austral-andine forests with *Nothofagus* and were considered a product of alteration of the original vegetation by man. Ectotroph-free forests below 500 m are true rain forests whose mycobiota differed basically from that of the ectotrophic forests.

Meyer[784,785] characterized microfungal communities under an equatorial rain forest near Yangambi, Belgian Congo (Zaire) on the basis of the principles and methods of the Zürich-Montpelier school of plant sociology. The 231 species isolated fell into three major groups: (1) common species found everywhere, including species with a high and fairly constant density, and species which are widespread but which have higher densities in certain sites; (2) characteristic species with a high incidence in a given environment; and (3) accidental species, including species found irregularly and at low densities, including true soil fungi or contaminants from the air or from plant residues.

Studying the fungi found in association with pine, oak, and elm woodlands and openings in the Schott plain near Augsberg, Bavaria, Stangl[1098] described the vascular plant associations and, with pictograms, charted the occurrence of the fungi according to the woods in which they were found in the 3 years of the study. Geologic, physiographic, temperature, and precipitation data were also charted.

In comparing macrofungal populations in oak woodlands on slate and on limestone soils, Hering[535] used 12 100-m^2 permanent quadrats. He found similar total numbers of fruit bodies, more abundant mycorrhizal species on slate, and a maximum fresh weight production of 95 kg/ha on slate, while on limestone the maximum fresh weight production was only 37 kg/ha.

Ubrizsy[1169] characterized the fungus population of the Bükk Mountains of Hungary both as to forest associations and the fungi accompanying them. The aspects of the mycocoenoses in the various seasons of the year were correlated with seasonal periods and were determined using observations made in each of a number of successive years. In fruit body formation the annual distribution of the "stenothermic" species, those depending on temperature demands, was closely correlated with seasonal phenomena. Ubrizsy's studies of the R factor, the soil temperature plus the tenfold quantity of the soil moisture, have shown the value of this characteristic for developing the quantitative and the qualitative composition of the fungus population in all biotopes and in every case.

Hering[537] grouped fungi decomposing broad-leaved litter in three associations: those colonizing leaves on the tree, those colonizing new fallen litter, and those colonizing older litter on the soil. An overall estimate of the importance of a given species could be obtained by combining data on abundance, duration of activity, and decomposing ability of sterile litter. The fungal populations in high-altitude oak woodlands were shown to be significantly different from those at low altitudes.

Based on site type, rather than forest species type, Kalamees[616] has made extensive mycocoenological studies in Estonia. Most species occur in site types with optimal or slightly excessive humidity conditions, rather than in extremely wet or extremely dry conditions. The species occurring in more than 25% of analyses were regarded as common in the corresponding site types or stand groups. The species which occurred with the highest frequency in site types or stand groups were regarded as characteristic. The richest in species of the five site types studied in detail was the *Oxalis*-type with 175 species, followed by the *Aegopodium-Mercurialis*-type with 142 species, the *Myrtillium*-type with 131 species, the swampy *Myrtillus*-type with 85 species, and the *Vaccinium*-type with 76 species. As to the fungus cover, the *Aegopodium-Mercurialis* site type, characterized by 21 species, is easiest to be characterized out of the five site types studied, the *Myrtillus*-type being the most difficult one with only four characteristic species.

Using mycocoenological methods, Kalamees[614] has analysed seven east Estonian forest types in three districts. Starting with 900-m^2 plots, these were modified to 30 × 2 m strips separated by 0.5-m passageways. On the whole, different forest types could be compared only partially, since observation periods differed in length and observations were carried out in different years. Emphasis was placed on species considered edible.

For Kalamees[618,619] it is not correct to treat fungal groups in biocoenoses as independent communities or to refer to them as associations. Neither do such groups represent synusiae. Fungal groupings are considered more restricted spatially and temporally include quite different morphological elements subject to a different rhythm than synusiae. The role of fungal groupings in biocoenoses can best be characterized as consortia, that is, functional structural elements of a biocoenosis, the relationship being largely trophic or nutritional.

From a synecological point of view, Maldagne and Hilger[743] conducted studies on microbial, faunistic, and respirometer data in four biotopes of the central Congo forest area, considering the influence of secondary ecological factors such as flood sediments from the Congo River, dung supply, nitrogen enrichment of the soil by legumes, pH, and other soil characteristics which varied while the dominant ecological factors remained constant. From the synecological point of view, no direct relations seem to exist between soil fauna and soil micropopulations; variations appear to be a response to environmental factors; the response of the two groups may be parallel rather than interdependent.

A demonstration was given by Franz[399] of the usefulness of the Sorensen similarity quotient in the biocoenotic analysis of all phases of the population of a study area. While the agreement it gives is not complete between soil biocoenoses and plant communities, the statistical analyses of a large mass of biocoenotic data give important indications for the problems of interrelations between biological and abiotic factors.

III. COMMUNITY ADJUSTMENTS

Two principal types of activities may be considered here. In the normal course of activity, when groups of organisms live together, certain degrees of associative ability are experienced. The fungi may form a complex of hyphae through the soil, in the litter, or in a mass of decaying wood, without interfering with the activities of each other, or some species may secrete substances inhibitory to the activity of other fungi or other organisms. A certain amount of symbiotic activity also occurs among the fungi. Species expressing this type of activity may live exclusively with another organism as in a lichen, or their activity may be less exclusive as in mycorrhizae, where the mycelium may be in active or passive competition with other fungi in the soil. Here active competition is considered as competition for nutrients, but many mycorrhizal fungi do not obtain nutrients from the soil but from their mycorrhizal partners; passive competition is considered as competition for space in which to produce fruit bodies and to move between hosts.

With the development[947] of the antibiotic penicillin, produced by such fungi as *Penicillium chrysogenum* and *P. notatum*, searches have been instituted by workers all over the world for antibiotics which may be produced by other fungi as useful, more useful, or supplemental to the first "wonder drug". The various antibiotic products of fungi have been reviewed.[560] Compendia devoted to a listing of antibiotics[909] and their properties have appeared on both sides of the Atlantic. The problems encountered in the search for antibiotics, using screening procedures for all fungi found in habitat samples,[987] have been outlined by one group of industrial workers. American workers[973] screened large numbers of wood decay fungus cultures for their ability to produce antibiotics in culture. In Portugal[913] it was noted that at different stages in their life cycle certain Hymenomycetes were capable of producing larger amounts or lesser amounts of antibiotic substances. English workers[134,1248] screened large numbers of fungus cultures, listing variations in technique as the screening process was developed. In Holland it was found that certain Discomycetes[480] were able to produce antibiotic substances.

In tests with *Trichoderma viride*,[1280] a gliotoxin-like antibiotic was produced in two types of soil in England. Both soils had been sterilized and inoculated with *T. viride*. In uninoculated sterilized soil no antibiotic was produced. The beneficial effect of autoclaving the soil on production of the antibiotic assumed to be gliotoxin was analyzed and separated into three distinct effects: elimination of the microbiota, increase in availability of nitrogen compounds, and increase in available carbon compounds. The latter was thought to be of greatest significance. A chromatographic bioassay technique showed that the antibiotic produced in the soil was gliotoxin.

The study of the fungi producing antibiotics and of their ability to yield these substances in increasing amounts, as genetic manipulation[61] was used to change strain characteristics, has developed into a major field of effort on the part of mycologists[1101] and chemists.[893] Paper chromatography has a place[29] in the study of various kinds of antifungal antibiotics.

The study of the production of antibiotics under natural conditions in the soil and the use of antibiotics in the control of forest tree disease[110] has been developed. Using very sensitive techniques it has been found that, under special conditions (as in sterile soil without competing organisms), strains of *T. viride* which actively produce gliotoxin in the laboratory could produce this substance in the soil. The amount and the type of gliotoxin produced by *T. viride* have been investigated in England. It appears that it may be possible eventually to purify this antibiotic so that it may be used without the toxic effects now observed. Certain fungi are able to produce antibiotics in English heath soils. Proponents of the idea that fungi, actinomycetes, and bacteria can produce effective antibiotics in soil[427] suggest that the amount of such substances required by the fungi is very small and need be secreted only in the immediate vicinity of the mycelium producing them.

The subjects of fungus antibiotics in relation to the occurrence of the organisms producing them in association and in competition with other soil microorganisms[1220] and in relation to fungus antagonisms[920] have been covered in earlier reviews. In the therapy of fungus diseases of man, a number of papers have been published in symposium form[1106] in which newly developed antibiotics were discussed in relation to the control of several of the more difficultly treated mycotic infections.

Factors affecting natural antibiotic production in the soil[135] include availability of carbon sources, the effect of other soil microbiota on antibiotic production, adsorption, and chemical and biological breakdown of antibiotics in the soil. There are several ways in which antibiotic production may be related to microbial antagonism. The sites where competition between root parasites and soil saprobes is most intense are in fragments of organic matter, in the rhizosphere, and in the spermatosphere. In experiments where root parasites and antibiotic producers have been inoculated into sterile soil together, the antibiotic producers have frequently been shown to reduce the survival or infective vigor of the parasite. Organisms shown to be antagonistic to root pathogens in soil have been found to produce antibiotics in soil, where antibiotic-producing strains of a given species have been compared with given strains producing no antibiotics as antagonistic to root pathogens in soil; the antibiotic-producing strains have been found to be the more effective antagonists. Soil conditions unfavorable to the accumulation of a given antibiotic have been shown to reduce the antagonistic activity of the organisms producing the antibiotic. There is evidence that antibiotics, or antibiotic-like substances, are produced in water as well as by a few plant pathogens.

Brian[135] did not expect all soil saprobes to be antibiotic producers. Antibiotic production is one aspect of the fitness of an organism, as are resistance to antibiotics and rapid growth rate.

Chapter 5

PHYSIOLOGY

I. NUTRIENT REQUIREMENTS

The filamentous fungi are organisms with restricted or elaborate hyphal systems whose cell walls are formed either of cellulose (Oomycetes) or chitin, the latter usually referred to as "fungus chitin".[111,721] They are aerobic, only rarely anaerobic, and then (except for one or two species) only during part of their life cycle, require carbon, hydrogen, nitrogen, and the macro- and microelements to varying degrees. They are set apart from other organisms[975] in that they do not ingest food as animals do, nor do they produce their own food through chemo- or photosynthetic pigments, but they secrete enzymes into their environments which break down bulk materials into substances readily absorbable through the fungus cell walls. Preferred habitats are acidic, but they will grow in substrates with pH levels higher than neutral, and some species have been known to grow at pH 11. Many species are able to produce their own vitamin supplies, but others require an outside source of vitamins for growth. A plea has been made for the integration of research on metabolism in microorganisms.[1277] Too much work has been done on either one or another aspect of physiology without attempts at correlation, which could be useful.

It is thought to be of interest to indicate the distribution of carbon and nitrogen in the biosphere as reported by Delwicke[304] in a symposium on "Microbiology and Soil Fertility" (Table 1). The amounts listed are presumed to be the total amounts of these basic elements available to living organisms.

Studies of general or specific nutrient requirements have been made on specific fungi or groups of species. Some of these studies will be mentioned here, although most of the studies that have been made have been summarized in general reviews and texts. The general subject of the growth of fungi in nutrient solutions has been summarized in two reviews.[1103,1104] It has been pointed out that certain microelements are present as impurities in many chemical compounds of high reported purity, and these compounds must be refined before final evaluations of nutrient requirements can be made. As a result of the presence of chemical impurities, certain compounds[240] need not be added to a culture medium to supply required elements in culture solutions. In studying the growth of wood decay fungi, among others, a technique was developed in which submerged growth in liquid media was used with success.[600] Here, as in most nutrition studies, the correct combination of nutrient materials can spell the difference between success and failure. Four unrelated species of cellulose-decomposing fungi[1045] were studied in determining the optimum environmental and nutrient requirements of such fungi. It has been found that nutrient factors and vitamin supplies may be critical[687,1217] in promoting pycnidium formation in some species of Sphaeropsidales. In other cases, no factors have yet been revealed which will promote the formation of pycnidia or even spores in sterile mycelia isolated from many habitats. Aside from the yeasts, for which it is important to know a wide variety of nutritional needs for classification purposes, studies have been made on carbon sources available to filamentous fungi[223,699,1186] and on other nutrient sources, although in the filamentous fungi little of this information is used in classification.

Many species of fungi in such diverse groups as the Basidiomycetes,[970] the Fungi Imperfecti,[972] and the dermatophytes[971] have been studied in large-scale screening experiments to determine their ability to use vitamins and for their inability to grow

TABLE 1

Distribution of the Carbon and Nitrogen of the Earth

Region	g C/cm^2	g N/cm^2
Atmosphere	0.11	755.0
Biosphere	0.7	0.036
Hydrosphere	10.0	0.022[a]
Crust (excluding sediments)	22,000.0	2,500.0
Sediments		
Sandstones and schists	300.0	
Calcites and dolomites	1,450.0	
Peat, coal, bitumen	800.0	

[a] Excluding dissolved N_2.

From Delwicke, C. C., in *Microbiology and Soil Fertility*, Oregon State University Press, Corvallis, 1965, 31 and 40.

without vitamins in culture media. Other studies on vitamin requirements include additional factors influencing the growth of thiamin-requiring fungi[688] and the relation of vitamin deficiencies to growth rates of fungi.[697] The requirements for growth factors in the nutrition of fungi have been reviewed, and a classification has been proposed[404] based on requirements for vitamins and combinations of vitamins, Following studies with a number of species of Hymenomycetes,[1284] their requirements for essential metabolites were found. A wide variety of fungal species[699] was screened for their ability to use a number of different simple to complex sugars. Whether or not fungi could use three different oils as carbon sources[952] was determined by a series of experiments using filamentous fungi known to use cellulose as a carbon source.

The aquatic fungus *Leptomitus lacteus* has been used[1009,1010] in a series of tests to determine its specialized nutrient requirements. It used amino acids rather than ammonia or inorganic nitrogen sources. This characteristic may be related to the environment in which it is most commonly found — waters polluted by sewages. That this condition reflects a relatively derived state of existence has been shown in a recent survey of the nutrient requirements of Phycomycetes[176] in relation to their possible phylogeny.

The subgenus *Euallomyces* of the genus *Allomyces,* including several species of aquatic fungi, has been studied intensively by Machlis, and work on its physiology has been carried out to discover its growth factor requirements in optimal and minimal nutrient media,[730] the optimal composition of a minimal medium in which it can be grown extensively,[731] and the sources of carbon used most readily by the strains of the fungus under study.[732]

In the study of the nutrient requirements of the fungi, two approaches can be used. These form the basis of two papers, each of which serves a different purpose. For a general survey, the techniques in a study of the nutrient requirements of two species of aquatic Hyphomycetes[935] can be useful as a guide. A more intensive study has been reported in connection with experiments on the species *Memnoniella echinata* and *Stachybotrys atra.*[891]

In the use of complex substances as food, the fungi secrete a series of exoenzymes which break down the more complex, naturally occurring materials into simpler substances which can be absorbed into the fungal cells. Within the cells, endoenzymes continue the metabolic process. In early work on enzyme sytems[1290] at least 22 different enzymes were found in the wood decay fungus *Gloeophyllum (Lenzites) sepiarium.*

This work was carried out before such metabolic processes as the Krebs cycle was discovered, and other enzymes may also be involved.

Spore germination presents numerous interesting problems in physiology. Many of these have been reviewed[453] and discussed by Sussman and Halvorson.[1124] In many cases[402,403] the germination of spores and development of mycelium of those species of soil Hymenomycetes and Gasteromycetes which may be mycorrhizal, but which apparently are not obligate mycorrhiza-formers, are encouraged by the presence of yeasts on the culture plates. Species of red yeasts, members of the genus *Rhodotorula*, were found to be most effective.

Nutrients are required by fungi in varying degrees. Summaries of carbon and nitrogen requirements have been presented in texts,[201,516,698] while elsewhere[1050,1197] additional carbon sources which can be attacked by cellulolytic fungi have been listed. Reese and Downing[953] have reported on the ability of species of *Aspergillus* to use cellulose, cellulose derivatives, and wool. Nitrogen requirements are summarized in the cited texts as are requirements for macro- and microelements. The nitrogen metabolism[405] of microorganisms has been described in a book from England. Recent reviews have summarized the requirements of fungi for microelements,[892] their heavy metal nutrition,[394] and their requirements for vitamins. Books on the chemistry of microorganisms[123] and the chemical activities of the fungi[395] have been published in England and in the U.S.

II. SOME HEAVY METAL RELATIONSHIPS

Ashida[55] has noted that certain fungi can develop resistance to toxic metals through mutation or adaptation. He concluded:

> The nature of metal toxicity has not yet been well elucidated. It might then be supposed that this is too early to study the mechanism of resistance to metals. What can be studied, however, is whether metal ions are detoxified by some substance or whether permeability to metals is lower although determinations free of exceptions are not easy to make. When the metabolic pathways which are specifically developed in resistant cells are studied in parallel with the reaction steps which, in sensitive cells, are most sensitive to the metal in question, as done by Murayama, there will be a way to elucidate the metal toxicity hand in hand with the resistance mechanism.

Zjic and Chiu[1291] worked with two species of *Penicillium* which were resistant to such heavy metals as uranium and strontium. In culture solutions using a medium with 100 ppm uranium, a concentration toxic to many organisms, significant growth occurred, and 70% of the uranium was removed from solution by absorption or adsorption or by a combination of both. It has been indicated that "at least part of the ion taken up is transported to the cell membrane to form some type of complex with the membrane structure."

Analyses based on destructive neutron activation techniques, carried out by Byrne and co-workers,[167] yielded information concerning concentrations of ten trace elements — As, Br, Cd, Cu, Hg, I, Mn, Se, V, and Zn — in up to 27 species of higher fungi from several sites in Slovenia, Yugoslavia. High levels of mercury were found in many species, and cadmium was accumulated by many fungi, reaching an average of 5 ppm. Species of *Amanita* accumulated bromide, and *Boletus edulis* accumulated selenium. Correlation analyses between all pairs of trace elements gave values for r of from 0.75 to 0.43 for seven pairs, including Cu and Hg (0.75) and Se and As (0.69). It was suggested that these values demonstrate a possible use in the study of environmental pollution.

III. NUTRITIONAL GROUPS OF SOIL FUNGI

The fact that numerous goups of soil fungi can be based on nutritional requirements[57,716] indicates that such fungi form a large and heterogenous assemblage. Such groups can go beyond the soil as such into the litter, dead wood, and almost any material in or on which fungi grow. Prior to 1970, it was convenient to combine these groupings under "soil fungi", since this is the substratum in which they have been most intensively studied.[57,427] A symposium edited by Dickinson and Pugh[318] has shown that certain fungi (usually thought of as soil fungi) are more importantly associated with litter decomposition.

The simplest of these groups, as to carbon requirements, includes the most widespread and abundant in numbers of colonies encountered. This group has been referred to as the "saprophytic sugar fungi". The species of this group use simple sugars, especially hexoses, and other simple carbohydrates.

Cellulose-decomposing fungi form a large group[1050] capable of reducing the cellulose molecule to simpler carbohydrates, thus reacting with a substance which has formed the basis of many manufactured textiles in common use throughout the world, as military supplies, clothing, tentage, and rope. A large amount of work has been done in an effort to prevent the activities of such fungi, at least on man-made materials.

A number of substances in nature are composed of pectin or cutin and pectin- or cutin-like materials. Leaf and stem surfaces are covered with a protective layer of such materials which bind cells together and may include cutin or pectin. The nature of such materials and the layers they form was reviewed in a series of symposia edited by Preese and Dickinson[926] and by Dickinson and Preece.[317] Apple and citrus fruits include pectin substances together with cutins in the surface layer of leaves and fruit. In dew-retting of flax,[1243] pectin-decomposing fungi play an important role. Techniques for the study of such fungi[1242] and the distribution of such substances in nature[1241] have been described. Such materials are considered to be prevalent[1064] in litter, and the organisms which are primary invaders of litter are thought to be those capable of using cutin, for instance, *Aureobasidium pullulans*. This fungus can use other materials of this general type in nature.[226] In a study of the microbiological populations of buds,[632] this was one of the more important fungi in such populations. The fungus is also very important in the deterioration of paints and plastic surfaces in Florida[959] and elsewhere.

Lignin, a cementing substance, comprises 15 to 25% of woody tissue. This material does not lend itself to ready analysis so that its chemistry is still doubtful, although a large supplemented text[128,129] on the subject presented a number of hypotheses concerning its origin, development, possible constituents, and degradation. A large number of wood decay fungi[222] are known to use lignin slowly in nature, but few fungi and no bacteria are known to use it in the laboratory in pure culture, except very slowly over periods of months. In Sweden it has been found that many soil and litter Hymenomycetes use the lignin present in litter materials.[701] A basic study method on the use of lignin and other materials in the soil has been developed[138] in West Virginia. Here litter materials were incubated over long periods of time with suspensions of forest floor materials, and rates of decomposition were determined during the exposure period. Certain soil fungi[533] are able to use some of the building blocks suspected of eventually forming lignin, such as para-hydrobenzaldehyde, ferrulic acid, syringaldehyde, and vanillin. One of the more recent summaries of knowledge about microbial, and especially mycological, degradation of lignin is that of Kirk[645] who, following a description of the structure of lignin, considered litter-decomposing Basidiomycetes (white rot fungi, soft rot fungi) and the degradation of lignin by microorganisms in pulp mill wastes.

Keratin is a common substance in many soils, especially in the vicinity of cemetaries, in soils in which animal remains are incorporated, and in other places. A special group of fungi is more or less closely restricted to this substance as a nutrient substrate.[16,628] It is possible that fungi growing on many types of proteinaceous materials should be placed here, and those fungi like *Onygena* which grow on hooves, feathers, owl pellets, and other substrates involving skin and similar habitats also belong here. The fungus *Keratinomyces,* whose spores strongly resemble those of the *Microsporum*-type of ringworm fungus, isolated from soils in Belgium,[1177] the U.S., and elsewhere, is a member of this group.

Predacious fungi[209] are usually considered a group with specialized nutrition. Their habitat, aside from the soil surface and litter in which they are usually found, is the bodies of protozoans which they entrap with special appendages and then proceed to devour. Such fungi are found among the Zygomycotina and the Deuteromycotina. At least one imperfect species has yielded an ascomycetous perfect state, and at least one species has been found with clamp connections, making its assignment to the Basidiomycotina necessary. Entrapment mechanisms may be of two types. A three-celled lasso is formed by one type, or a multicellular net is formed. A nematode moving through the lasso or net is caught by the rapid enlargement of the cells forming the lasso or by a sticky substance secreted by such cells. In the second type, special appendages may trap emoeboid organisms and rotifers by secreting a sticky substance which holds the organism while the fungus mycelium penetrates the host body and consumes its substance.[340]

The rhizosphere is a highly specialized habitat in the soil. Here, exudates from root cells and sloughing of root tissues form a nutrient or toxic complex to which many bacteria and some fungi are constantly exposed. Summaries of the biology of this habitat have been published from Canada[878] and England.[427] Work in England[1212] with plant succession on sand dunes has demonstrated the existence of a succession of microorganisms associated with the rhizosphere of the various successive groups of colonizing plants.

The fungi which form mycorrhizae with higher plants constitute another group in a special nutritional environment. In addition to utilizing carbohydrates and possibly lignin in the soil, these fungi form mycorrhizae with roots in the soil. These structures are depicted in detail in a book published in Germany early in 1941 on the anatomy of the fungi.[721] Recent work in England[500,502] and in Sweden[109] has given more recent views regarding the function and types of mycorrhizal development. Some Hymenomycetes which form mycorrhizae with *Pinus strobus* have been identified and proved as mycorrhiza formers.[512] In a recent volume[427] on root-infecting fungi, mycorrhiza-forming fungi are considered to be parasitic on their host plant. Some of these are weak parasites which can be obtained in culture; others, difficult to culture, are considered obligate parasites. The weak parasite can exist without the plant with which it forms a union, and it can be brought into pure culture. The obligate parasite can be brought into pure culture with great difficulty, if at all, and in nature appears to require the presence of the plant with which it makes its mycorrhizal union. Both types can form either ectotrophic or endotrophic mycorrhizae. Lohwag[721] noted that the host plant produces defensive mechanisms without arresting the natural flow of nutrients between the two organisms. Harley[504] indicated that the carbon nutrition of the mycorrhizal fungus is related more to sources of supply from the host than to its own ability to convert organic matter in the soil to usable carbon-containing compounds. Papers presented in two recent symposia, one on ectomycorrhiza[750] and one on endomycorrhiza,[1005] bring most areas of mycorrhizal research up to the dates of their publication.

A. Based on Carbon Requirements

1. Sugar, Cellulose, Lignin, Humus, Hydrocarbons

It has been found in Sweden[702] that while *Clavaria gracilis* used only the lignin fraction in the litter tested, *Collybia butyracea, Clitocybe alexandri, C. nebularis, Clavaria dendroides, C. ligula, Marasmius peronatus, M. pusillus,* and *M. androsaceus* used a small part of the cellulose as well as the lignin, while species of *Marasmius, Clitocybe, Collybia,* and *Mycena* attacked the lignin and the cellulose to an equal extent without substantial enrichment of the solution in which growth was taking place.

In eastern France, Kiffer and Mangenot[644] have noted that cellulose degradation in the eight forest soils studied was not proportional to the total biologic activity but is in direct relation with the coefficient of carbon mineralization. There seemed to be some competition between certain members of the mesofauna and the fungi for cellulose-containing nutrients.

Rhizoctonia solani[831] can produce saprobic mycelial growth in the soil in the absence of living roots or fresh organic matter. It has been shown that this fungus can competitively colonize and decompose cellulose in the soil, and isolates grow well on filter-paper cellulose in culture.

Using 300 freshly isolated strains of soil fungi, Domsch and Gams[327] checked capacities for enzyme production for degradation of pectin, xylan, and carboxymethylcellulose (CMC). Pectin was most rapidly decomposed by *Penicillium expansum,* xylan by *Phoma eupyrena,* and CMC by *Truncatella truncata.* The most active of the ten species common to all three compounds were *Penicillium* cf. *atrovenetum, Fusarium culmorum, F. solani,* and *Aureobasidium bolleyi.*

A wide range of fungus cultures has been tested by Nyms, Auquière, and Wiaux[839] to determine their ability to assimilate such hydrocarbons as *n*-paraffins, aromatic hydrocarbons, and petroleum fractions. The ability to assimilate hydrocarbons was found to be scattered through the fungi but was largely in the Mucorales and the Moniliales, with *Penicillium* and *Aspergillus* having the larger number of species capable of this. Hydrocarbon assimilation is not necessarily common to related species nor the property of one species, but is is more the property of individual strains. Different strains belonging to the same species differ in metabolic activity when tested against a series of hydrocarbons. The property of assimilating hydrocarbons appears to lack taxonomic value. Species in the same genus show only a tendency to behave in the same way, e. g., *Penicillium* strains usually assimilate *n*-decane and light gas oil, whereas *Aspergillus* strains seldom do. *Aspergillus* strains sporulate better on long chain *n*-paraffins; on some hydrocarbons they develop special pigments. *n*-Paraffins with at least ten carbon atoms support better growth than petroleum fractions. Individual strains are very sensitive to minor changes in hydrocarbon composition or structure. Only sparse delayed growth was observed on aromatic hydrocarbons, *n*-heptane, petroleum ether, naphtha, and kerosene which are often toxic, whereas aromatic hydrocarbons are usually nontoxic.

While the deep sea appears to be poor in fungi, sets of wood panels[652,654] exposed at three Pacific and Atlantic Ocean locations at 1615 and 5315-m depths yielded Ascomycete and Deuteromycete fruit bodies after exposure of 1.5 to 3 years. Cold temperatures ranging from 2.5 to 4°C may have been responsible for slow growth. At one 722-m exposure site, low oxygen tension may have resulted in a lack of fungi. If shipworms depend on cellulolytic organisms to prepare the wood for their settlement, probably any organism, bacterium or fungus, will serve the purpose.

B. Based on Nitrogen Requirements and Nutrition

Tabak and Cooke[1130] have reviewed the ability of fungi to grow in the presence of N_2 and CO_2 gases.

Whitaker[1227] recognized that the movement of amino acids into fungi is controlled by a series of active transport systems of varying degrees of specificity. Their presence and level of activity depend on growth conditions of the fungus prior to the assay. These transport systems have properties similar to those described in other microorganisms, plants, and animals.

Millbank[793] checked the ability of 14 strains of fungi, including strains of species of *Rhodotorula, Bullera, Torulopsis,* and *Aureobasidium* to fix nitrogen. He used both heavy nitrogen, $^{15}N_2$, and methylene reduction techniques. No nitrogen fixation was obtained, and the ability of an eukaryotic cell to fix nitrogen was doubted.

At first called proteophilous fungi, then ammonogenous fungi, and finally "ammonia fungi", Sagara[992-995] worked with a group of fungi defined as "a chemoecological group of fungi which sequentially develop reproductive structures exclusively or relatively luxuriantly on the soil after a sudden addition of ammonia, some other nitrogenous materials which react as bases by themselves or on decomposition, or alkalis." These fungi differ in habitat requirements from other known groups, although they were dependent to some extent on the vegetation type with which they occurred.

While urea was the principal treatment, aqua ammonia, aliphatic amines, and some other nitrogenous materials which react as bases in their natural form or on decomposition (ammonia-related materials) were generally as effective as urea. Nonbasic NH_4 - N, NO_3 - N, and N-free compounds were not effective. Alkalis were somewhat effective, probably because of their ability to liberate ammonia from the soil. The occurrence of NH_3 - N, together with an alkali condition, or NH_4OH seems to be the essential factor for this effect. Some of the nitrogenous materials used are more complex than urea or ammonia. In nature, most of these fungi appeared on the ground in forests where human urine or feces or dead bodies of mammals were placed by accident and decomposed in place; these are also ammonia-related materials.

Characteristic changes in the treated soil included a color change to black, increase in pH to over 7, increase in water content, enhancement of organic matter decomposition, ammonia or compost odor in the initial or early stages, and stimulated root growth in the subsequent stages. A single treatment was essential.

IV. EFFECT OF C TO N RATIO ON FUNGUS GROWTH

It has been pointed out[240] that in nature the C to N ratio of the substratum may vary from 9 and 14 to 1. In culture lower rates may result in too rich growth and poor fruiting, although some groups such as dermatophytes seem to do well on low ratio media.

V. OXYGEN REQUIREMENTS

It is generally assumed that all fungi are aerobic organisms which, under certain conditions, may produce anaerobic fermentation. Very few reports[201,516,698] concerning requirements have been published. The threshold point of lowest oxygen tension required for growth is still uncertain.

According to an experiment with *Penicillium roquefortii,*[795] growth is inhibited when oxygen in the air is reduced from the normal 21% to 2.1% by addition of nitrogen gas to the incubation chamber. Later work has dealt with six species of filamentous fungi. Growth was completely inhibited at pressures to produce 0.015 to 0.118 ppm of oxygen in the medium.

It has been found that the aquatic fungus *Blastocladia pringsheinii*[76] could grow as well under the partial pressure of oxygen of less than 1.0 mm Hg as in air. Large numbers of resistant mature sporangia were produced under 99.5% carbon dioxide, while few, if any, were produced in air without the addition of special nutrients in the medium.

Using manometric techniques[291] and pressures of less than 1 atm of oxygen, the effect of reduced oxygen pressure on *Myrothecium verrucaria* has been studied. Hydrogen gas and oxygen-free nitrogen gas were used in experiments where, under less than 1 atm, the rate of utilization of oxygen was a function of the oxygen pressure. At pressures between 0.21 and 0.40 atm of oxygen, consumption was increased while no further increase was noted above 0.40 atm of oxygen, and there was not any inhibition in pure oxygen. Under anaerobic conditions, no matter how obtained, no carbon dioxide was produced, no fermentation was observed, and organic acids were not produced.

Mycological physiology texts[201,516,698] are not in disagreement with the following statement from a tabular summary of nutritional requirements of organisms:[23] "Fungi require oxygen, obtainable from gas dissolved in the culture medium, or, in the case of surface growth, from the atmosphere." In human or animal pathogenic fungi it is indicated that under certain conditions such fungi, especially yeast-like species or species with yeast-like phases, can produce anaerobic fermentation in culture under varying conditions of nutrition, temperature, and other factors. It has been stated[395] in part: "One of the major metabolic differences between molds and bacteria is that there are no anaerobic molds ... Indeed, there is general concurrence with the idea that molds are highly oxidative organisms. This is not to say that molds will not metabolize carbohydrates anaerobically (fermentation), but rather that this is accomplished by preformed cell material and growth at the expense of fermentative metabolism exclusively does not occur."

Heavier mushrooms can be produced in excesses of oxygen,[516] while certain types of wood decay[958] proceed more rapidly in excesses of oxygen. It has been indicated[698] that certain types of fungi can survive as long as 13 weeks in water-logged soils, while in Honduras, Stover[1118] has shown that some fungi survive as long as 18 months in water-logged soils under flood-fallowing. Even when fungi, including yeasts, are used in anaerobic fermentations, they cannot survive without introduction of oxygen at some stage of their life cycle. Evidence has been cited[1042] to indicate that fungi will, at some point in their life history, need oxygen for continuing growth and activity. Studies in England[163] indicate that, in different stages of development in the Mucoraceae, oxygen supplies are critical, especially during sexual reproduction. Spores in a deep liquid culture take a long time to germinate and grow to the air, after which they develop rapidly.

In classifying the yeasts, fungi used most extensively in anaerobic fermentation processes in industry, no species are based[719] on anaerobic conditions alone, although the ability to ferment certain sugars anaerobically is used to distinguish between certain species within a genus.

Growth of fungi has been obtained in dilute media[272] in which the only oxygen present was dissolved in the aqueous medium (Figures 1 to 4). The oxygen was dissolved in the medium without the use of pressure. At the beginning of the experiments as much as 8 ppm oxygen was present in the solution; at the end of 10 days, 3 to 5 ppm oxygen were still present and growth had not ceased.

Since fungi, including filamentous species as well as yeasts, are relatively abundant in digesting sludges and other areas in sewage treatment processes where oxygen supplies are at a minimum or theoretically absent, it was of interest to discover whether these organisms are capable of metabolizing some of the organic components of sewage and industrial wastes under anaerobic or microaerophilic conditions.[1131] Some 13

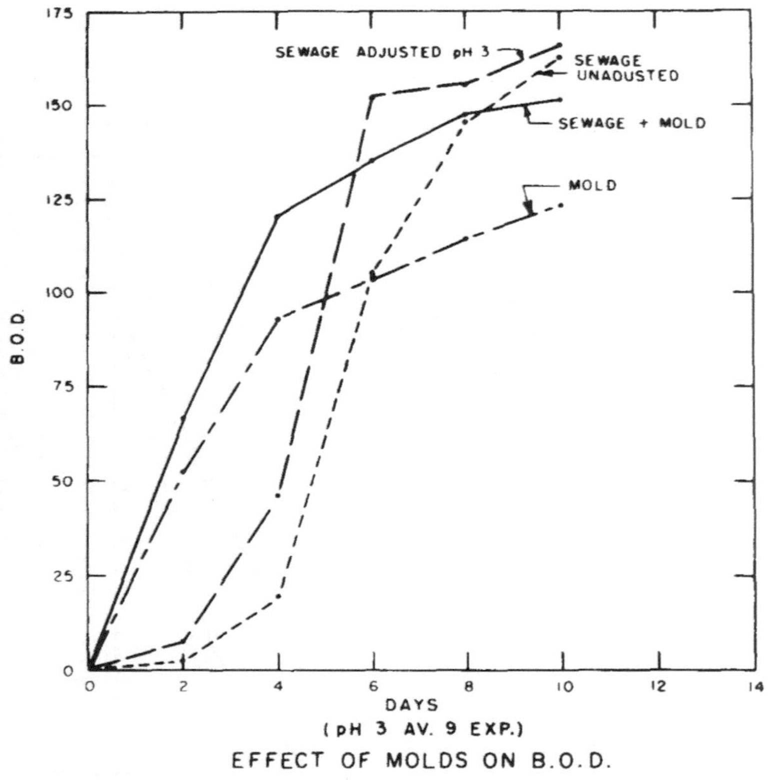

FIGURE 1. Effect of fungi on BOD. Averages of nine experiments, each using a different species of fungus.

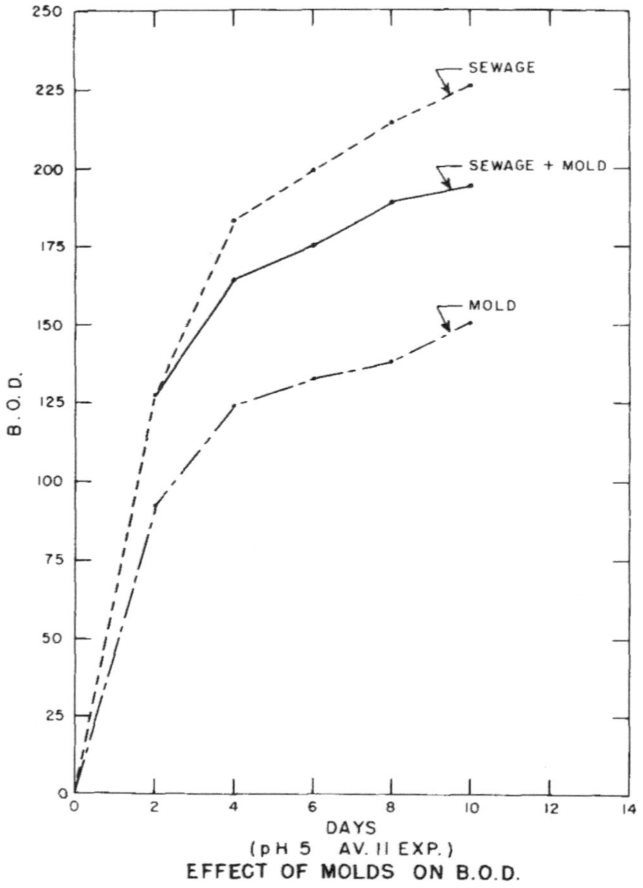

FIGURE 2. Effect of fungi on BOD. Averages of 11 experiments, each using a different species of fungus.

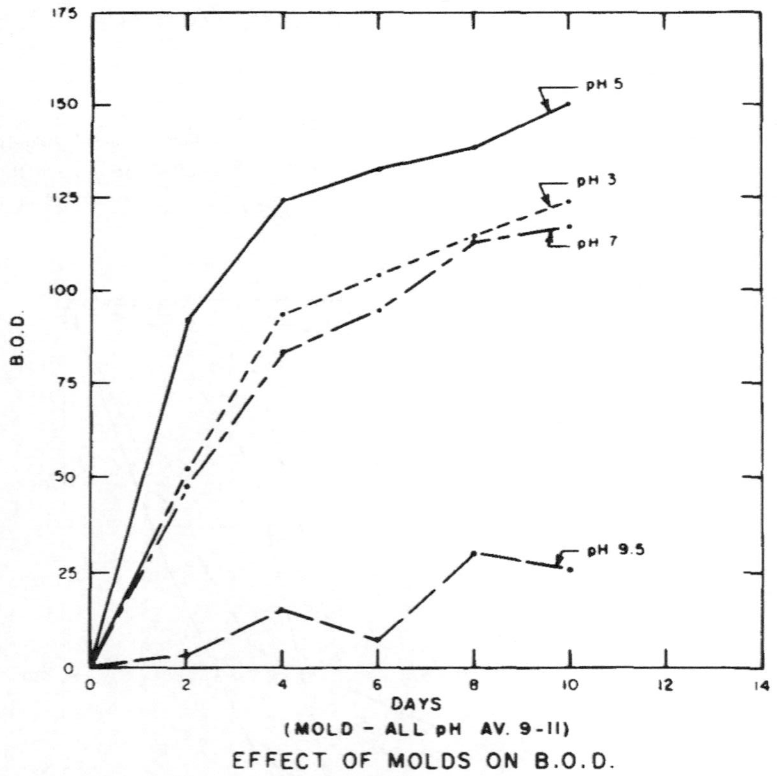

FIGURE 3. Effect of fungi on BOD. Averages of 9 to 11 experiments, each using a different species of fungus, with two exceptions.

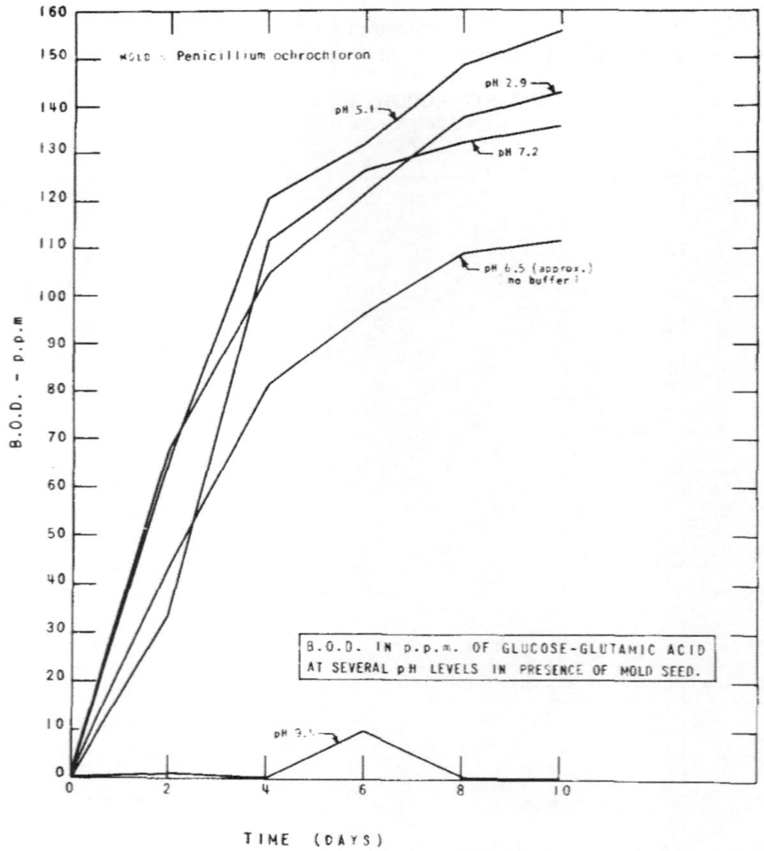

FIGURE 4. Effect of *Penicillium ochrochloron* on BOD at four pH levels.

species of fungi were chosen to determine their ability to grow, multiply, and develop anaerobically in an atmosphere of 100% nitrogen gas.

To establish whether fungi grow as well as perform metabolic activities in such habitats, liquid synthetic media containing either inorganic or organic nitrogen sources, with glucose as the carbon source and with or without the reducing agent sodium formaldehyde sulfoxylate, were used. Incubation was effected in 25-mℓ cotton-stoppered conical flasks, inoculated with shaker-grown, blender-treated, washed cells, stored in anero-jars under 100% cylinder, prepurified, nitrogen gas, monitored with leuco-methylene blue. Growth and metabolic activities were measured at weekly intervals for 4 weeks, by oven-dry weight determinations of cells collected on prewashed membrane filters, by analysis of the combustible residual total organic carbon in the culture filtrate, and by determinations of economic efficiency values.

The experimental results demonstrated that the organisms could tolerate and survive the apparently oxygen-free atmosphere. The organisms were capable of appreciable growth and development under 100% N_2 gas (Table 2) and exhibited significant metabolic activity under complete or nearly complete anaerobiasis. The species allowing the highest growth activity in a nitrogen gas atmosphere were *Geotrichum candidum, Fusarium oxysporum, F. solani,* and *Muncor hiemalis.* Those with intermediate growth and development were *Aspergillus niger, A. fumigatus, Aureobasidium pullulans,* (Figures 5 and 6) *Phialophora jeanselmei, Penicillium* spp., and *Trichoderma viride.* The fungi most sensitive to a 100% N_2 gas atmosphere but still capable of limited growth were *Phoma herbarum, Rhodotorula mucilaginosa,* and *Candida parapsilosis.*

Special emphasis has been placed on oxygen requirements because the literature appears to be weak on this phase of fungus nutrition, since the fact that fungi are aerobic is taken for granted or treated as dogmatic fact.

On the other hand, *Aqualinderella fermentans*[359,360] is anaerobic, growing in complex media in atmospheres established in sealed jars with a burning candle. It will grow in 5 to 20% carbon dioxide as well as under 99% CO_2. It also grows under CO_2-supplemented hydrogen essentially devoid of oxygen as indicated by methylene blue in the leuco phase.

According to work in Australia with rhizomorphs of *Armillariella elegans,*[1070] oxygen pressures greater than 0.04 atm in contact with the outside of the rhizomorph are inhibitory. Such partial pressures can stimulate the activity of a diphenol oxidase which catalyses a brown pigment which could overlay growing cells, preventing either uptake of nutrients or disposal of waste products by cells of the rhizomorph. In soil, such partial pressures may be normal and can be characterized by the presence of water films which can be controlled by correcting moisture content, texture, and bulk density of the soil.

VI. WATER REQUIREMENTS

The problem of the relation of moisture to the production of fungus fruit bodies and the development of mycelium has not yet been satisfactorily studied. It is well known that various types of plant disease fungi respond differently to the moisture content of the air (atmospheric humidity), but the nature of the reaction to soil moisture in soil fungi, or of aquatic fungi in the soil, is as yet incompletely understood. In Scotland,[456] moisture was considered one of the important aspects of the environment which controlled fungus fruiting. In eastern Washington and adjacent Idaho,[215] fungus fruiting was correlated as much with precipitation as with any other phase of the environmental conditions studied. In Austria,[553] fungus fruiting was correlated with the

TABLE 2

Comparison of Oven-Dry Weights of Cell Mass of 13 Fungus Species Grown in Still Culture, Anaerobically in Prepurified 100% Nitrogen Gas and Aerobically in Media With Inorganic or Organic Nitrogen Sources and With or Without the Reductant Sodium Formaldehyde Sulfoxylate

	Average weight of cell mass (in mg) after 4 weeks incubation					
	Anaerobic growth		Aerobic growth		Ratio × 100 Anaerobic to aerobic growth	
Fungus	Inorganic N	Organic N	Inorganic N	Organic N	Inorganic N	Organic N
Geotrichum	20.34[a]	20.15	25.64	43.14	79.3	51.4
candidum	19.92[b]	20.85	24.95	41.86	79.8	49.8
Fusarium	17.32	17.45	23.24	30.29	74.5	57.6
oxysporum	17.24	17.48	22.15	20.07	77.8	57.8
Fusarium	16.89	16.96	25.60	43.85	66.0	38.7
solani	16.78	16.85	28.16	39.14	59.6	43.1
Aspergillus	14.07	14.32	46.12	49.76	30.5	28.8
niger	13.32	13.90	38.15	46.09	34.9	30.2
Aspergillus	13.65	13.86	35.72	38.26	38.2	51.4
fumigatus	12.96	13.08	34.71	38.02	37.3	34.4
Aureobasidium	13.26	13.38	37.84	31.69	35.0	42.2
pullulans	13.21	13.28	39.65	34.12	33.3	38.9
Penicillium	12.79	12.85	29.26	32.82	43.7	39.5
spp.	12.62	12.71	29.81	35.47	42.3	35.8
Phialophora	12.61	12.76	39.65	26.42	31.8	48.3
jeanselmei	12.59	12.65	37.05	28.19	34.0	44.9
Mucor	12.36	12.74	21.05	23.12	58.7	55.3
hiemalis	12.21	12.20	18.91	25.18	64.6	48.5
Phoma	10.73	10.86	28.25	38.14	38.0	28.5
herbarum	10.65	10.72	29.27	37.61	36.4	28.5
Trichoderma	9.47	10.16	19.26	38.05	51.8	26.7
viride	9.38	10.00	17.64	28.96	55.7	35.6
Rhodotorula	6.84	6.95	34.56	28.54	19.9	24.4
mucilaginosa	6.75	6.73	33.04	30.15	20.4	22.7
Candida	6.02	6.25	25.72	30.89	23.4	20.2
parapsilosis	5.04	6.03	25.42	31.48	23.4	19.2

[a] Media without reductant.
[b] Media with reductant.

From Tabak, H. and Cooke, W. B., *Mycologia*, 60, 122, 1968. With permission.

moisture content of the soil. On Mount Shasta, Calif.[216] and in subalpine western North America,[217] it was noted that the fruiting of certain groups of fungi was correlated with moisture from melting snow. From various observations it may be assumed that development of fungus activities of the soil is correlated as much with soil moisture as with any other factors. When soil moisture drops below the wilting coefficient for higher plants, i. e., when the humidity of the air spaces becomes less than 100%, the soil may not be water-logged, but sufficient moisture is present to satisfy the moisture requirements of fungi. Such requirements are high, although there are few, if any, checks on water loss by fungi through evaporation. Aquatic fungi can also become active in soils at such high humidities, since they require films of water for movement of zoospores and for other activities.

Half of Griffin's volume[482] on the ecology of soil fungi is devoted to techniques for

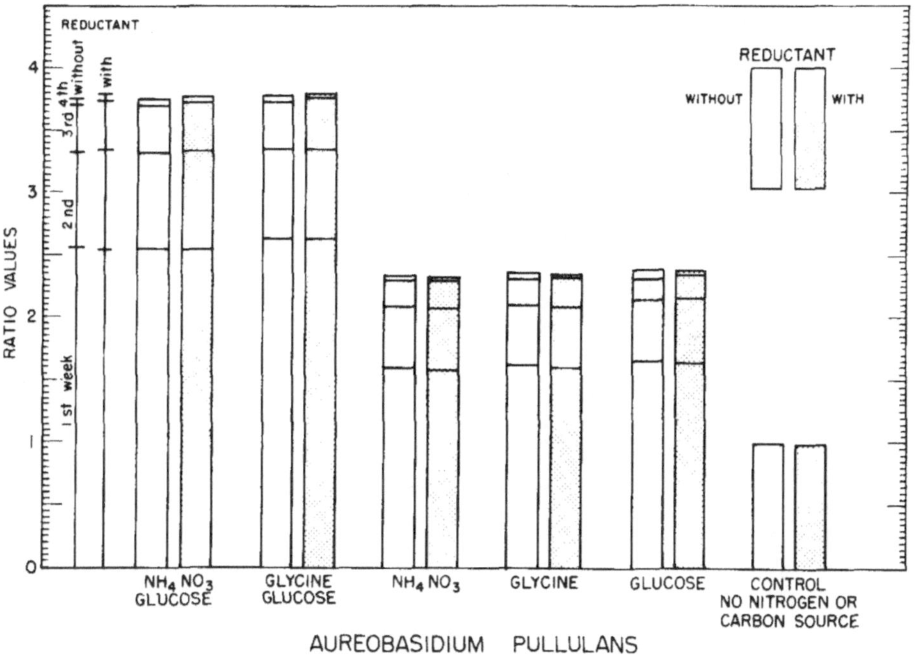

FIGURE 5. *Aureobasidium pullulans.* Comparison of the ratios of anaerobic growth increase in complete synthetic medium, carbonless medium, and nitrogenless medium, to growth in the control medium, with and without the reductant sodium formaldehyde sulfoxalate, in an atmosphere of prepurified 100% nitrogen gas at the end of each of 1, 2, 3, and 4 weeks of incubation.

the study of the physics of soil-water relations in respect to the requirements of the fungi living in the soil for water. Ingold[586] considered the water requirements of fungi with special reference to the aquatic fungi and the aquatic Hyphomycetes.

VII. WATER IN FUNGUS SPORES

Studies on conidia of powdery mildews[1282] and other fungi have shown that, while the conidia of powdery mildews contain 52 to 75% of their fresh weight in water, spores of six miscellaneous airborne fungi contain only 6 to 25% of their fresh weight in water. Correcting for hygroscopic water, these figures indicate that the physiologically active water content of *Erysiphe polygoni* conidia may be 15 times that of *Uromyces phaseoli* uredospores. The high water content may explain the ability of such spores to germinate in areas of low atmospheric humidity. Calculations based on *Monilinia fructicola* and the increase in spore volume during germination indicate that the water content of these spores at germination may be about the same as for powdery mildew conidia. These data may apply to other fungi requiring free water for germination.

VIII. CERTAIN PHYSICAL FACTORS IN THE ENVIRONMENT

A. Temperature

Temperature is an important factor in the growth of fungi. Much of the work up to 1965 has been reviewed by Deverall.[311] In relation to temperature, psychrophilic fungi may tolerate temperatures as low as −5°C, while thermophilic fungi will tolerate temperatures of as much as 60°C. However, most fungi are mesophilic. All fungi have

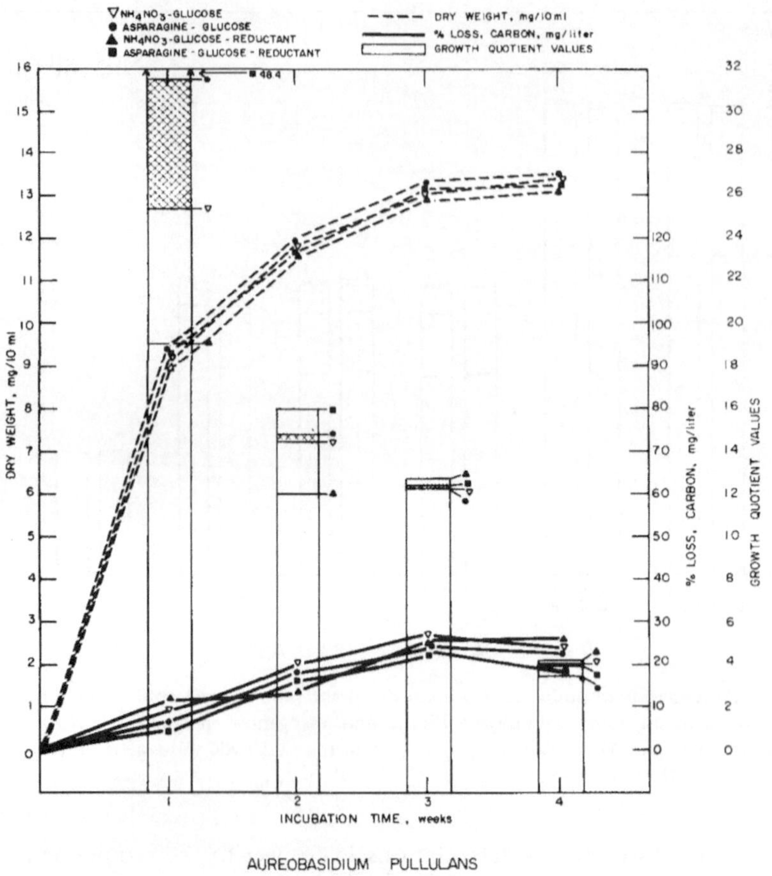

FIGURE 6. *Aureobasidium pullulans.* Correlations of the increase in growth, based on oven dry weight of mycelium in milligrams, percent loss of the total soluble organic carbon per liter of the culture filtrate, and growth quotient values at the end of 1, 2, 3, and 4 weeks incubation in an atmosphere of 100% prepurified nitrogen.

minimum temperatures below which they will not grow and maximum temperatures above which they are inactivated. They make their best growth at optimal temperatures, and these three temperature levels are referred to as the cardinal points. However, during different stages in the life cycle the optimal temperature may be different, and for the production of certain growth products the optimal temperature may be different from that of mycelial production in the same stage of the life cycle.

B. Pressure

The effects of hydrostatic pressure on fungus growth have been reviewed by Morita.[812] Fungi have been recovered from great depths[652] in the ocean, depths as great as 4610 m. Less ethanol is produced by *Saccharomyces cerevisiae* when subjected to pressures of 200 to 1000 atm than is produced at 1 atm. Kohlmeyer[652] obtained fungi from wood on panels placed in or as wood trawled from greater depths from 1615 to 5315 m in Atlantic and Pacific Ocean sites. More work has been done in this area on bacteria, and more information is needed in reference to fungi.

C. Irradiation

To test for effectivity of irradiation in preservation of fresh fruits and vegetables,

Skou[1056,1057] found that with irradiation carried out at 4°C in an indoor ^{60}Cs irradiation facility a decimation required 500 krad in the case of chlamydospores and 450 krad for older blastospores. In one experiment, surface sterilized carrots were placed in sterilized plastic bags and half of these were sealed. In the sealed bags inoculated with 4×10^7 spores before irradiation, *Aureobasidium pullulans* produced destructive attack during the 12 days of incubation after irradiation with 1000 krad of gamma radiation. The same results were obtained in the unsealed bags. In sealed bags inoculated after radiation, *A. pullulans* and *Fusarium* sp. produced destructive attack during the 12 days incubation after irradiation with 750 krad of gamma radiation.

Durrell and Shields[347] and Hunt and Durrell[581] recovered a number of fungi from soils from the Nevada atomic bomb test site and from soils from the Death Valley National Monument, Calif. It was noted that the larger number of species had spores incorporating melanin pigments in the walls. The black spores of *Stemphylium ilicis* were found to survive 60-min exposures to ultraviolet light of the wavelength 2573 Å. It was suggested that the melanin pigment may serve a protective function in highly insolated desert regions, although not all species producing dark spores are especially resistant to ultraviolet light.

D. Antibiotics and Inhibiting Substances

Concerning the existence of inhibitory substances such as antibiotics in the soil, Park[863] said,

> There is no conclusive evidence that normal mature soils contain effective concentrations of inhibiting substances. Much of the evidence suggests that such substances do exist and are effective, and this is supported by the very low level of activity of microorganisms in the normal mature soils, but none of this evidence is sufficiently critical to be regarded as adequate proof. What is known is that inhibiting substances may locally exist in effective concentrations in special situations, and that in the soil more generally inhibiting substances may at least develop quickly under suitable conditions of stimulation.

Concerning the origin of such substances in the soil, Park[863] said,

> There are several possible origins for substances in soil inhibitory to microorganisms, and in different soils different sets of factors may operate; but while specific substances may be produced by particular organisms, and may be effective under some conditions, the evidence suggests that there is a common inhibitory phenomenon that is a function of the previous general microbial activity in the soil, and which may be produced by several or many species inhabiting the soil.

E. Biochemical Differentiation of Taxa

As yet the only group of fungi in which it is practical to use biochemistry in the differentiation of taxa is the yeasts.[719] Because of morphological characteristics which are similar for large segments of this population, morphology is of little use in species identification or recognition, in contrast to other groups of fungi where morphology is the principal method of distinguishig species and other categories. Exception has been taken[74] to the use of the large numbers of carbohydrates required for the complete test for each species. However, it has been noted that the tests are for utilization, not for screening of enzymatic pathways which can be a byproduct of the identification procedures.

F. Serological Techniques

One of the earlier applications of serology to the fungi was that of Beck[89] in studies of species of smut fungi. That this type of technique is no longer considered important in plant pathology may be surmised by the brief mention given it in the definitive monograph on the biology of the smut fungi by Fischer and Holton.[383]

In a search for techniques for identification of fungi and other organisms causing deep mycoses in man and conditions resembling them, a number of techniques have been developed. These are described and discussed by Seeliger[1023] and include: the agglutination test, capsular reaction, precipitation reaction, immunoelectrophoresis, agglutination with sensitized particles, complement-fixation test, Castellani's absorption test, immunofluorescence, determination of fungicidal activity of serum, and skin tests. Five years later,[148] information on these tests was brought up to date giving details for these tests for the more important fungus-caused diseases of man, including aspergillosis, blastomycosis, candidiasis, coccidioidomycosis, cryptococcosis, histoplasmosis, nocardiosis, paracoccidioidomycosis, and sporotrichosis. Seeliger expressed the hope that serologic techniques, as refined diagnostic tools, could also be used for identification of fungi and for indicating phylogenetic relationships.

Using serological techniques, Hasenclever and Kocan[510] have shown that *Saccharomyces tellularis,* found in nature as a member of the normal yeast biota of at least four columbid species (pigeons and doves), can be serotyped. Of the 141 strains isolated, 98 strains assigned to serotype B were from *Columba livida,* the feral pigeon. A reference strain of the imperfect state, *Torulopsis bovina,* was also assigned to serotype B. While serotype A was not isolated among the original 141 strains, serotype C was harbored by *Columba leucocephala,* the white-crowned pigeon, *C. fasciata,* the band-tailed pigeon, and *Zenaidura macroura,* the mourning dove. It was noted that in only two other species of yeasts have serotype groups been established: *Cryptococcus neoformans,* the etiologic agent of cryptococcosis, and *Candida albicans,* of candidiasis. As necessity or other incentives arise, possibly one or more of the techniques of serology will become adapted to the study of yeasts and filamentous fungi.

Chapter 6

THE ORGANISM

A large number of topics under which are described a wide variety of the aspects of development and growth of the fungi have been described under the cell,[13] the organism,[14] and the population.[15] In the preceding and following pages certain aspects of these subjects are considered.

I. THE SPORE

A. Spore Dormancy

Within spores two types of dormancy have been recognized by Sussman.[1122] These are constitutive dormancy — including such processes as permeability, anhydrobiosis, self inhibitors, and metabolic inhibitors — and exogenous dormancy — including soil fungistasis, plant inhibitors, and inhibitors in marine environments. Means of activation of germination include temperature extremes, light, and chemicals, among others. Changes during germination are morphological, biological, and biochemical, the latter including respiratory changes, substrates for germination, respiratory enzymes, and the synthesis of macromolecules. Criteria for germination and stages in the germination process were outlined. Spores play an important role in nature through enhancement of survivability and disseminability of the organism producing them; they also act as timing devices in sporulation and germination.

B. Spore Release

Two basic types of spore discharge are recognized by Ingold,[588] active and passive. Among active discharge mechanisms are the breaking of turgid cells, the rounding off of turgid cells, hygroscopic mechanisms, and ballistospore discharge. Certain external conditions are necessary to violent spore discharge. Among passive spore liberation are blowoff of spores, liberation by falling drops, and mist pickup. Certain insects are involved in spore liberation, and mammals may be involved as in puffing of puff balls.

In ballistospore, or active basidiospore, discharge, the hymenial surfaces covering smooth surfaces, lining inverted cups or tubules, covering club-like structures or coralloid branches, covering teeth, lining tubes, or covering gill surfaces are oriented so that the basidiospores produced on sterigmata (Figure 1) are shot into spaces between teeth, in the center of tubes, or between gills as in *Coprinus atramentarius* (Figure 2). Inverted cups or tubules heve smooth or wrinkled hymenia from which spores are discharged directly into the air without obstructions such as opposing teeth or opposing surfaces of tubes or gills, but spore discharge is just as rigidly controlled. When the gill surfaces are strictly parallel or aequihymeniferous,[150] they usually autodigest at the margin so that succeeding basidia may be brought into discharge position; when the gills are placed at a slight angle, the hymenial surfaces are inaequihymeniferous,[151] and spores may be discharged from any position on the surface without danger of reaching the opposite gill surface before the force of gravity carries them below the tooth, tube, or gill edges when air currents take over and carry them away.

On the basis of a 2-year observation of spore release from 86 fruit bodies belonging to 27 polypore species, Nuss[838] found that

1. The sporulation of all species studied was determined by seasonal rhythms differing from species to species and which may vary to some extent based on climatic conditions.

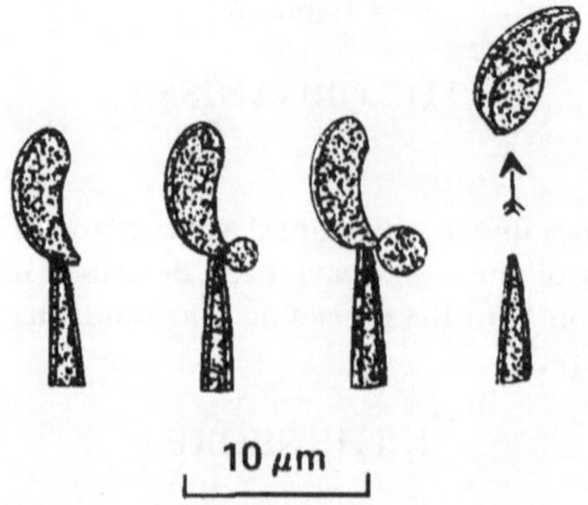

FIGURE 1. *Calocera cornea*. Discharge of the basidiospore (ballistospore) from its sterigma. (From Buller, A. R. H., *Reseachers on Fungi*, Vol. 2, Longmans, Green, & Company, 1922. With permission.)

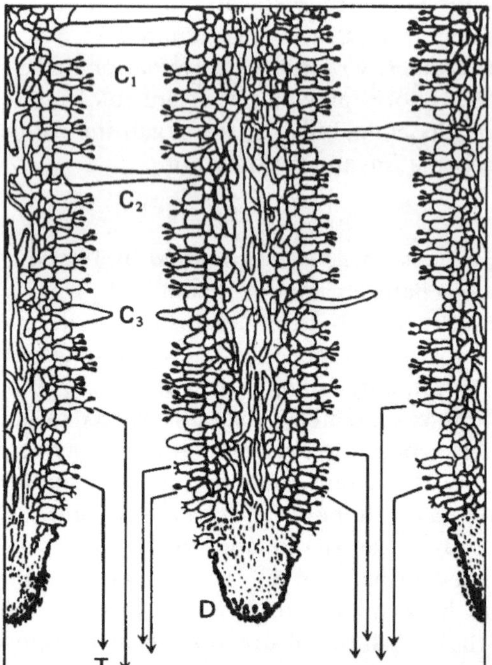

FIGURE 2. *Coprinus atramentarius*. Tangential vertical section of cap, showing gills in sectional view. C_1, unaltered cystidium; C_2 and C_3, stages in autodigestion of cystidia; D, autodigesting edge of gill with a few spores trapped in the resulting slime; T, trajectories of spores being discharged from the basidia. Note the short distance of spore fall between opposing gill surfaces. (Magnification × 136.) (From Buller, A. R. H., *Researchers on Fungi*, Vol. 2, Longmans, Green, & Company, 1922. With permission.)

2. Reduction of relative humidity to an average of 70% reduced discharge rates for most species studied.
3. Average temperatures of about 0°C, and more than 20°C, respectively, and maximum values of 27°C inhibit or prevent sporulation.
4. Many species interrupt or even stop sporulation at 0°C.
5. For some species only temperatures below 0°C seem to interrupt sporulation.

A model for the classification of the studied polypores into sporulation-type classes is suggested based on three criteria: (1) the life span of the fruit bodies, (2) the duration of spore-release from fruit bodies, and (3) the time at which spore discharge begins. In the group of species studied, 24 theoretical sporulation-types are suggested.

C. Spore Preservation and Longevity

Several methods have been devised for the maintenance of fungal spores and cultures in as compact and viable condition as possible. Such methods arose primarily from a desire to retain the large numbers of fungus cultures obtained from a variety of sources and will be described later.

The longevity of fungus spores in nature is a problem incompletely solved and rarely mentioned. It has been indicated[427] that fungus spores in the soil remain dormant until a piece of material carrying nutrients favorable to its development is deposited beside it. In most cases, this would be rarely longer than the interval between the growing seasons of a crop in agricultural soils. It has been shown[149] that collections of *Schizophyllum commune* can produce turgid spores after more than a year of complete desiccation. Assuming that a turgid spore is a viable one, spores of this species can withstand considerable desiccation.

Collections of spores representing species of Ustilaginales have been surveyed[381] for their longevity in herbarium collections. As many as 79 collections of 23 species from South America and Europe were tested on potato-dextrose agar; species from the Ustilaginaceae were held as long as 12 days, from the Tilletiaceae, 30 days. Collections ranging in age from 2 to 25 years were used. Thirteen-year-old spores of *Sphacelotheca sorghi* germinated to the extent of 40% after 2 days of incubation. Of a 10-year-old collection of *Ustilago bromivora* there was 95% germination after 2 days, and of an 11-year-old collection of *U. hordei* 90% germination occurred after 1 day incubation. In a 16-year-old collection of *Tilletia tritici* there was 90% germination after 12 days, and a 4-year-old collection of *T. levis* gave 95% germination after 7 days.

D. Air Spora

To determine the size and extent of the air spora, the spores must be trapped preferably in a viable condition or in an otherwise identifiable state. The efficiencies of suction and impact systems[767] have been considered, following discussions of isokinetic sampling, anisokinetic sampling, turbulence, intake efficiency curves, the effect of yaw, the avoidance of yaw and turbulence effects, sampling in still air and at the stagnation point, wind tunnel sampling, and sampling with very thin wires or threads. Sampling instruments include electrostatic precipitators, cyclones, cascade impactors, the Andersen sampler, the liquid impinger, and sampling instruments operated from aircraft.

The physical principles of deposition and a discussion of various types of particles, spores, and deposition efficiencies have been considered by Chamberlain.[187]

Gregory[479] said,

> Assuming that it is advantageous to a species to occupy the maximum amount of habitat and store material

for genetic variation we can now contrast the effectiveness of various dispersal methods. Autonomous dispersal (including mycelial extension) is extremely limited in range. Splash dispersal is less restricted and occurs in weather often favorable for spore germination. Dispersal by insect vectors has a precision other methods lack, and can cover long distances, although comparatively slowly. Wind transport can cover long distances rapidly, but landing is hazardous and must often occur under conditions unfavorable for further growth; for success spores must be produced in vast numbers. The most effective extensions of geographic range occur when an efficiently wind-dispersed fungus is carried across its natural barriers by Man's power-driven transport.

It has been indicated[550] that knowledge of large-scale particulate atmospheric transport is probably necessary to the understanding of local aerobiology as it is to instant dispersal, because of the fluctuation in content resulting from a variety of factors depending on local populations, sources of air masses, speed of passage of air masses, and loss from and gain to these air masses of spore loads. Descriptions are given of the nature and causes of air movement with particular reference to the distribution of fungus spores by Gregory.[475]

E. Distribution in Soil

Air dried samples of earthworm casts (*Lumbricus terrestris* L.), robin nests (*Turdus migratorius* L.), and barn swallow nests (*Hirundo erythrogaster* Bodd.) were added by McIlveen and Cole[775] to steamed soil, and in the mixture were planted soy beans (*Glycine max* [L.] Merr.), in the roots of which vescicular-arbuscular mycorrhizae developed indicating that these are organisms which may transport spores of species of Endogonaceae. By extrapolation it is indicated that earthworms may play a direct part in the distribution of spores in soil profiles, and through runoff waters they may also play an indirect role in lateral distribution.

F. Survival after Freezing and Desiccation

According to Mazur:[770]

The relatively high resistance of fungi to low temperatures in nature and in the laboratory is due to two factors: 1. intracellular ice generally does not form within them, and 2. either they are relatively resistant to dehydration, or else they form resistant bodies. The physical and biological factors that determine the likelihood of intracellular freezing are reasonably well understood. Basically the likelihood depends upon the degree to which the cells become supercooled during cooling, and this in turn depends on the rate of cooling and the rate at which water can flow out of the cell in response to the vapor pressure differential built up during cooling.

The basis of resistance to dehydration is much more poorly understood, partly because dehydrating actually involves three simultaneous processes: the removal of water per se, the concentration of ions and other low molecular weight solutes, and a decrease in the distance separating macromolecules. To this complex situation must be added the possible influence of temperature. Fungi are particularly attractive experimental cells for the investigation of the factors underlying resistance since the same organism can change from being susceptible to resistant and back again in a matter of minutes or hours. Any knowledge gained would not only be of interest to understanding mechanisms of freezing and drying injury in fungi, but ought also to be germane to understanding the ecological success of fungi and the nature of dormancy, and to understanding the causes of injury in the cells of higher plants and animals.

G. Spore Dispersal

The fungi are highly ubiquitous organisms.[254,1175] This is more true of those species whose spores are carried by air. Mechanisms of liberating spores into the air, of discharging zoospores into the water, and many highly specialized processes of spore liberation have been described in an excellent monograph by Ingold.[589] The splash-cup mechanism of spore discharge has been described by Brodie[142,143] in many groups of plants in addition to the fungi. The fact that spores can be washed down through layers of relatively coarse materials, such as sand which forms soils in some areas, has been studied by Burges.[158] It was found that "dry" spores, those produced in dry chains or clusters instead of moist spore balls, can be carried downward for considerable distances in sandy soils. It has been noted that spores of both wet and dry spored species have been found at depths of 20 ft in a coarse sand in northern Michigan.[231,235] Should human and animal disease fungi produce spores in such a way that they can be carried in water, the fact that they can get into municipal water supplies has been shown in experiments[781] with slow and rapid sand filters in Kansas. It was demonstrated that tuberculate macroconidia of *Histoplasma capsulatum* could pass through a 30-in. rapid sand filter. Removal of 80 to 99% of such spores was effected by plain sedimentation or by alum flocculation and settling. On the other hand, in other experiments, the same workers[782] have shown that spores of this fungus can be removed by using a combination of several standard water treatment methods. If 413,000 spores per 100 mℓ of raw water are passed through plain sedimentation beds, 16,900 per milliliter are left. After passing these through coagulation and settling, 205 spores per 100 mℓ are left. After filtration and before chlorination, none is left, as is true after chlorination. These values were obtained with a membrane filter. If the agar plate technique for recovery of spores were utilized, no significant difference in the number of spores was noted except that no spores were found with this technique after coagulation and settling. High counts of spores were obtained with either technique from sediment and sludge samples.

Natural spore dispersal was reviewed by Gregory[476] who considered a wide variety of conditions and spore types.

Fruit bodies of *Stereum purpureum* were produced in autumn and winter on recently dead branches on apple trees in New Zealand.[348] Individual fruit bodies shed spores in suitable weather for at least 8 months. Observations were made with a Hirst spore trap for a year. Spores were produced only when the humidity rose above 90° after rain, and fell to zero after the relative humidity fell below that level. Spores were not shed after dew. Significant numbers of spores were caught on 90.6% of the days when rain fell and on 62.2% of the days the trap was operated. Spores were shed during the day and at night, although more frequently at night. Duration of continuous spore showers ranged from less than 1 to 29 hr.

Two methods were combined for measuring the terminal velocity of the fall of basidiospores of the giant puffball, *Calvatia gigantea*, by Gregory and Henden.[477] These spores approximate smooth spheres under the light microscope and their terminal velocity(v_g) should be calculable, using (1) decay or die-away of concentration of a spore cloud during stirred sedimentation and (2) inertial partitioning of spores between the four jets of the Cascade impactor, the sampling device used to measure concentration in the decay method. Both methods gave consistent results which agreed well with calculated terminal velocities. There is no single value for v_g, but the data suggest an average of 0.6 mm sec^{-1}.

Possibly activating the rupture of the exoascus of bitunicate asci in three species of *Venturia*, Brook[144] found that red light in particular was stimulating, while blue and green light were also effective. In these three species, *Venturia* sp., *V. pyrina*, and *V. rumicis*, as well as in *V. inaequalis*, there was a sharp decline in discharge rate after perithecia were plunged from light into darkness. In the latter species, there was a high discharge rate for several minutes after the change to darkness. On the other hand, *Coleroa circinans* discharged its ascospores equally readily in light and darkness.

H. Dispersal in Water Habitats

A variety of information on spore dispersal in aquatic environments has been developed under the editorship of G. Jones.[613] Specialized cells for dispersion include flagellated zoospores and tetraradiate spores.

The tetraradiate spore (Figure 3) is characteristic of aquatic Hyphomycetes which grow on dead leaves in well-aerated water in streams and pools. The conidium is usually hyaline, and the four arms originate simultaneously or successively from a central point on a structure which may originate as a phialoconidiophore or a thalloconidiophore, according to Ingold.[591] The variety of developmental geometry suggests convergent evolution. These species are worldwide in occurrence and as many as 1000 to 10,000 spores per liter of water can be recovered from streams in the autumn. On the basis of observations and experiments, it is suggested that the biological value of the tetraradiate spore is connected with the problem of initial anchorage in the turbulent conditions of a stream. Webster[1214] has shown that on a smooth cylinder the trapping efficiency of the tetraradiate spore was higher than that of spores of different shapes which is attributed to the fact that such spores form contact at three points.

Bandoni,[71] in British Columbia, found the tetraradiate spore as one of the spore forms in litter. It was suggested that, because of the greater surface to volume ratio, advantage is given in aiding in conidial release, especially in the case of thallospores or aleuriospores. The water film which could aid in spore release would also aid in spore dispersal.

Routine isolations of fungi from streams of various types, including clean and polluted water-carrying streams, have yielded significant populations of soil fungi, fungi with no particular type of adaptation to dispersal in aquatic habitats. Such fungi could have reached the stream from soil runoff since the spores are probably carried passively.[230]

Potter and Baker[921] found an examination of 60 fish caught in three areas in Flathead Lake, Montana that 48 species of nonaquatic fungi could be isolated from swab samples from the surfaces of the fish and eight additional species from the contents of the stomach and esophagus. None of these fungi were involved in diseased conditions of the fish; all were merely being transported on or in their bodies.

I. Spore Size and Volume

Ingold[589] (Figure 4) has illustrated a series of fungal spores, showing their comparative sizes and in comparison with a small seed.

McCallan[772] determined individual fungus spore volumes together with their size distributions. Spore volumes were determined from linear measurement by assuming an appropriate geometric shape such as prolate sphaeroid. Measurements were made of 40 to 500 spores from each of 24 species of fungi. The mean spore volumes in cubic microns are given in Table 1.

The coefficients of variation in these volume values varied from 14 to 100%. The size distribution was determined by X^2 and t tests for symmetry and kurtosis. The distribution of seven species was normal, of 12 log-normal, of two intermediate, and for three (because of limited data) both types of distribution were indicated. The species with smaller mean values, i.e., less than 500, tended to log-normal distribution, and species of the same genus tended to have the same kind of distribution. The stability of mean spore volume appeared to be constant for some species but varied significantly for others, depending on the particular lot collected and time maintained in culture. Different mechanical methods of collecting conidia of *Monilinia fructicola* had no effect on spore volume. Comparisons of spore weights derived from volume measurements (density 1.1) with direct weighings for six species gave good agreement for two but were somewhat higher for four species.

J. Germination of the Spore

Sussman and Halvorson[1124] have reviewed spore germination thoroughly. Maximum

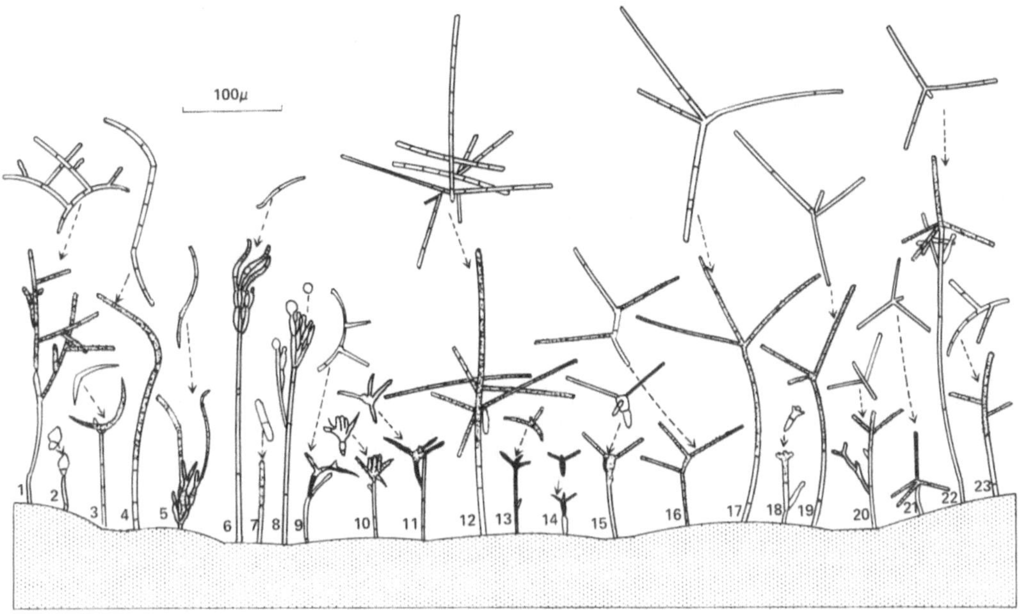

FIGURE 3. Aquatic Hyphomycetes of decaying leaves. Substratum of leaf tissue is indicated by stippling. For each species a single conidiophore is shown, the ripe, attached conidia being stippled. In addition, a liberated spore of each species is illustrated. All are drawn to the same scale. (1) *Varicosporium elodeae;* (2) *Piricularia aquatica;* (3) *Lunulospora curvula;* (4) *Anguillospora longissima;* (5) *Flagellospora curvula;* (6) *F. penicilloides;* (7) *Piricularia submersa;* (8) *Margaritispora aquatica;* (9) *Tricladium angulatum;* (10) *Tetracladium setigerum;* (11) *T. marchalianum;* (12) *Dendrospora erecta;* (13) *Alatospora acuminata;* (14) *Heliscus longibrachiatus;* (15) *Clavariopsis aquatica;* (16) *Tricladium gracile;* (17) *Tetrachaetum elegans;* (18) *Heliscus aquaticus;* (19) *Articulospora inflata;* (20) *A. tetracladia;* (21) *Tricelophorus monosporus;* (22) *Lemonniera aquatica;* (23) *Tricladium splendens.* (From Ingold, C. T., *Dispersal in Fungi,* Clarendon Press, Oxford, 1953, 166. With permission.)

germination by *Aspergillus flavus* conidia occurred at 35°C in two axenic nutritional systems.[886,887] However, the rate of germination in glucose plus amino acid nitrogen, supplied as peptone, was much more rapid than in glucose plus ammonia nitrogen. Percentage germination at the reduced temperature of 25°C in glucose plus peptone was more than double that in glucose plus ammonium. Using both citrate-phosphate and phosphate buffers, a broad pH optimum range (pH 3.0 to 7.5) for germination was observed in glucose plus peptone, while a narrowed optimum range of pH 4.5 to 6.0 was observed in glucose plus ammonium. Low soil temperature and high soil pH may restrict *A. flavus* conidial germination less in soil environments where amino acid nitrogen is the principal nitrogen source.

Trappe and Maser[1153] found that the spores of *Glomus macrocarpus* removed from fecal pellets of a vole could germinate in distilled water. Germ tubes 100 to 200 μm long were produced before they died for a lack of a suitable mycorrhizal partner with which to form a union.

The germinating fungus spore provides excellent material for detailed studies[27] of the emergence of a cell from dormancy and of the events involved in the resumption of protein synthesis. Synchronization of events is good in a spore population and can be still further improved by proper manipulation. The unicellular spores of fungi should prove to be rewarding subjects for studying the metabolic basis of the self-regulation of development and perhaps also for examining the mechanics of gene expression as they operate in a simple morphogenetic process, germination. The knowledge obtained on these fundamental problems will also contribute to an understanding

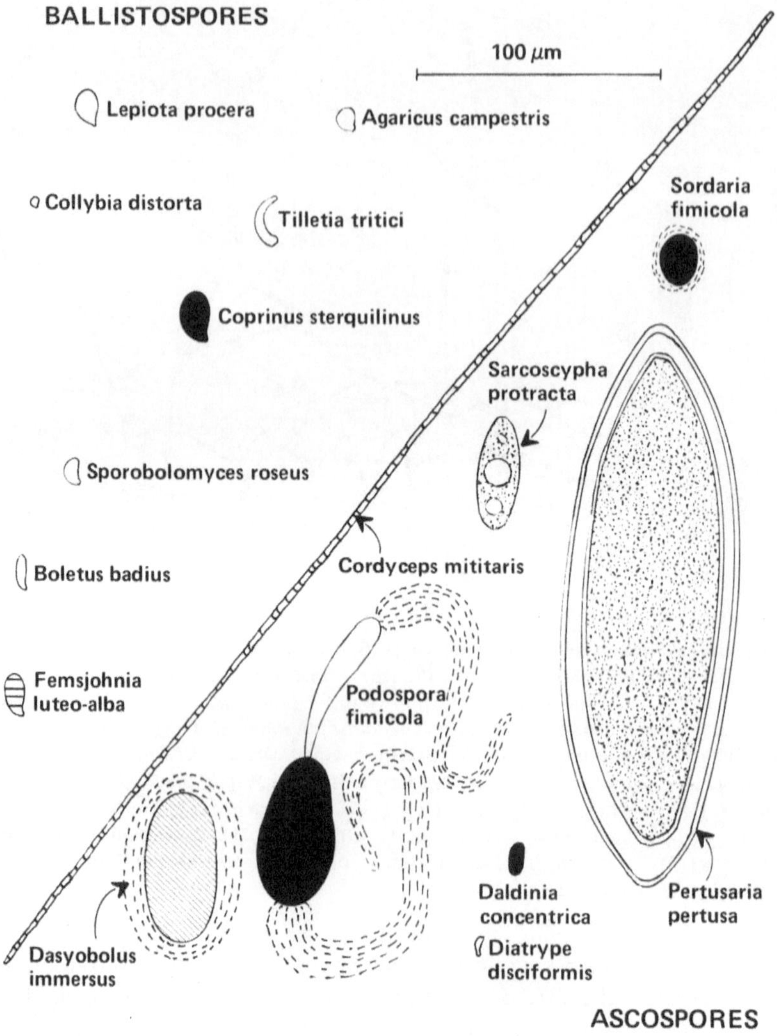

FIGURE 4. Range of form and size in two types of violently discharged spores: basidiomycete ballistospores (to left of the long *Cordyceps* ascospore) and violently discharged ascospores. (From Ingold, C. T., *Fungal Spores, Their Liberation and Dispersal*, Clarendon Press, Oxford, 1971, 12. With permission.)

of the phenomena underlying the initiation of plant infection by fungus spores and the induction of toxic responses by the uncongenial hosts.

II. THE MYCELIUM

Within the group of organisms commonly referred to as fungi, there are two principal types of growth rates and patterns as demonstrated in laboratory culture. The first of these is demonstrated by the yeasts. Here cell increase is by budding whether this is phialidic or annelidic. In shaken liquid culture probably all significant yeast-type growth is by budding. In still liquid culture and in agar culture, growth in some species can be, in addition, by mycelium production, and those cells demonstrating this type of growth would come under the second category. The second category is demonstrated by the filamentous fungi. Here growth occurs in the terminal portion of a

TABLE 1

Size Distribution of Spore Volumes of 24 Species of Fungi Based on Arithmetic Means

Species	Mean (μm^3)
Uromyces caryophilinus	15,600
Alternaria solani	9,960
Stemphylium sarcinaeforme	8,680
Erysiphe polygoni	7,550
Puccinia antirrhini	6,180
Uromyces phaseoli	6,110
Phytophthora infestans	5,360
Tilletia caries	3,730
Venturia inaequalis	616
Botrytis cinerea	556
Venturia pyrina	546
Monilinia fructicola	523
Alternaria oleracea	497
Rhizopus nigricans	323
Pestalotia stellata	321
Glomerella cingulata	292
Neurospora sitophila	283
Ustilago maydis	259
Neurospora crassa	99.3
Aspergillus niger	29.2
Myrothecium verrucaria	26.8
Penicillium expansum	16.6
Phyllosticta sp.	11.2
Cephalosporium acremonium	5.06

From McCallan, S. E. A., *Contrib. Boyce Thompson Inst.*, 19, 314, 1958. With permission.

hypha and its branches and results in elongation of a tube-like structure with occasional partial cross walls.

In shaken liquid culture, the growth habit is in the form of regular balls or irregular chunks of mycelium. In either case, colony increase could be in the form of increasing diameter as each hypha tends to elongate, growth still being confined to the terminal portion of the hypha and its branches. In still liquid culture, growth would be in the form of a mycelial pad floating on the surface of the medium. The mycelium would tend to elongate on the surface of the medium with a few cells penetrating down into the medium. Growth would probably occur only at the aerated terminal portion of each hypha. In agar culture, the mycelium would remain at or near the surface, only a few hyphae penetrating into the medium as a response to food supplies, and growth would be confined to the elongation of the terminal portion of the hypha and its branches.

Among filamentous fungi there is one exception to this growth pattern. *Geotrichum candidum,* and other fungi with arthric cell division habits, produces arthrospores which appear to be physiological equivalents of budding cells during growth phases in liquid culture.

The yeasts and *Geotrichum* appear to follow a logarithmic growth pattern in shaken liquid culture, just as the bacteria do. However, the filamentous fungi follow a growth pattern referred to as "cubic" because of their habit of cell elongation rather than

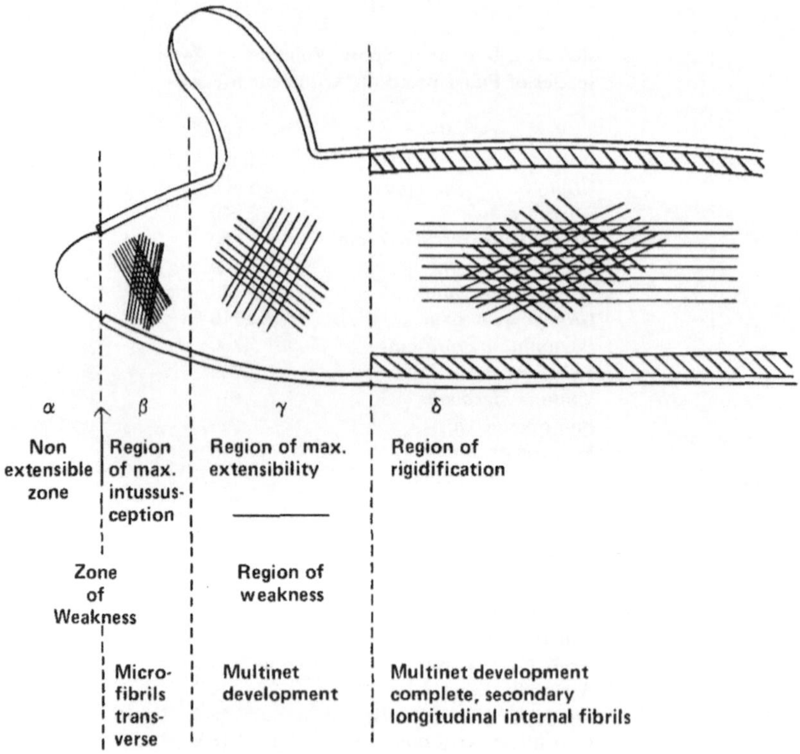

FIGURE 5. Diagram to illustrate the organization of the hyphal apex proposed by Burnett. (From Burnett, J. H., *Fundamentals of Mycology*, Edward Arnold, London, 1968, 53. With permission.) Note: A second edition is quoted by its author as being in press.)

rapid cell division. This cubic growth pattern was first noted in a study by Emerson[361] based on *Neurospora.* It was found that the cube root of the volume of the mass of the mycelial sphere increased lineally with time as long as conditions supporting unlimited growth were maintained. It was suggested that the same conclusions hold for other organisms of similar growth habit.

Hyphal growth usually occurs at the tip of the cell (Figure 5) where cytoplasm pushing against the terminal wall in the direction of growth continually forces a convexity in the membrane enclosing the protoplasm. As this area is passed, it is never static or fixed except in death, under certain conditions of dormancy or histological fixation procedures, the cytoplasm lays down a layer of molecules within the tube which becomes the hyphal wall that at first is probably only one molecule thick. The end of the tube is elastic enough to permit constant forward movement and growth of the hypha. As growth continues, additional layers of molecules are added to those already in place so that the lumen becomes slightly narrower and the cell wall becomes slightly, although finally measurably, thicker.

Burkholder and Sinnott[162] showed that 150 species of fungi grown in shaken culture produce masses of mycelium which are uniform for each species as long as conditions of culture are maintained so that optimum development could take place. The extent of the balling or the chunking of the mycelium was, thus, taken to be genetically controlled.

Going beyond filamentous fungi, Marshall and Alexander[752] went to such filamentous bacteria as members of the Actinomycetales including *Streptomyces* and *Nocardia*

TABLE 2

Major Cell Wall Constituents of Fungi Related to Taxo-nomic Groups

Major cell wall components	Taxonomic groups
Cellulose-glucan	Oomycetes
Cellulose- chitin	Hyphochytridiomycetes
Chitosan-chitin	Zygomycetes (Mucorales)
Chitin-glucose	Chytridiomycetes
	Ascomycetes
	Basidiomycetes
	Deuteromycetes
Mannan-glucan	Saccharomycetes
	Cryptococcus
Mannan-chitin	*Sporobolomyces*
	Rhodotorula
Polygalactosamine-galactan	Trichomycetes (Amoebidales)

From Bartnicki-Garcia, S., *Annu. Rev. Microbiol.*, 22, 87, 1968. Reproduced with permission from the *Annu. Rev. Microbial.* 22© 1968 by Annual Reviews Inc.

and showed that throughout a group of filamentous fungi, including *Fusarium oxysporum, Aspergillus niger, A. flavus,* and *Cunninghamella elegans* and two filamentous bacteria, cubic growth could be demonstrated for all species. The observations of Emerson[361] and earlier workers were extended to include oxygen uptake which they found to also have a cubic relationship in shaken culture.

In a review[745] of the kinetics of fungal growth, the work mentioned above was summarized, and it was indicated that the conclusions of Emerson[361] and of Marshall and Alexander[752] applied only to those organisms producing mycelial spheres. These conclusions would not necessarily apply to those fungi producing chunks of mycelium of various forms in liquid culture. Mandels[745] also noted that "no one appears to have proved that only apical growth occurs in liquid culture."

Thus, unicellular organisms — including the yeast-like fungi, which may produce pseudomycelium in some culture conditions as well as true mycelium, and *Geotrichum,* which can grow in a yeast-like fashion in liquid culture — probably produce logarithmic-type growth, while filamentous organisms, bacterial or fungal, produce cubic growth. A search of the literature has yielded no studies in which the two types of growth have been compared in one study, nor has such a search yielded studies which can be compared meaningfully with respect to growth rates of the two types of organisms in the same or parallel culture studies.

Bartnicki-Garcia[79] has summarized the chemistry and morphogenesis of the cell wall of fungi in relation to their taxonomy. The distribution of cellulose, glucan, chitin, chitosan, mannan, and polygalactosamine components is related to taxonomy as shown in Table 2.

For Austwick,[60] fungal thalli are among the more plastic forms of life. Morphological variation is a result of a response to physical stimuli resulting in changes in directions of growth, size of cells, and other factors. Physiological variation is the result of changes in the nutrient constituents of the substratum and of the parent materials of the soils concerned. Stimuli may be received and reacted upon within a very short period so that when the result becomes apparent, the basis for the modification may be no longer apparent.

Under certain circumstances, a mycelium may produce yeast-like cells rather than an elongate, septate or not, tube-like growth. Such yeast-like growth has been attributed by Nickerson and Edwards[830] to the suppression of a postulated enzyme system leading to the production of the tube-like growth, allowing the dominance of an enzyme system producing the development of short, budding, yeast-like cells which may remain in pseudomycelial chains or which may occur in clusters of separate cells. For R. Cooke,[209] the yeast habit merely indicates that the high temperature and other factors of the habitat preclude mycelial formation in host tissue.

Two principal types of development in the mycelium result in arthric or blastic cells. In the former, the cross wall is double and may divide without much apparent cause. Portions of a mycelium, or a whole mycelium, may break up into separate cells. In the latter type, buds may be produced which may become branches or sexual reproductive cells. These buds may be of the phialidic or the annellidic type.

As the hypha elongates, additional layers are applied to the wall until, by the time a millimeter of growth (1000 μm) has developed, the older portion may be a mature hypha. Secretion of exoenzymes and absorption of nutrients probably takes place mostly in that portion of that terminal section of the hypha with the thinner wall.

While concern was with the bacterial ecology of trickling filters, during the course of a study by Isaac and James[595] it was found that some significant relationships between bacteria and fungi in this habitat occurred. The lower bacterial counts in the surface layers of the trickling filters studied was related to a higher interspecific competition with fungi which were present in larger numbers. On a volume basis these fungi occupied 15% of the film, but when their greater metabolic rate was considered — bacteria 33 cal/kg hr, contrasted with fungi 50 to 250 cal/kg hr — it follows that the fungi are probably making a significant contribution to the stabilization process. It was suggested that the fungi seem particularly adapted to compete in this habitat with a high concentration of complex organic compounds and where the flushing action of jets is operative.

Microbial cells behave as macroscopic ions and quantitatively absorb onto, and desorb from, specific surfaces such as those of synthetic ion exchange resins, according to Daniels.[290] Exchange of cells is a diffusion controlled reaction. Rates of adsorption and desorption in batch systems are functions of the time of agitated contact, the concentration of the reactant species, pH, and ionic strength. The nature of this adsorptive phenomenon and the characteristics of the microorganisms, the adsorbants, and their actual environment are discussed in this review. Applications of specific adsorption include metabolic studies of adsorbed vs. unadsorbed cells, and the separation, concentration, and/or resolution of microbial mixtures.

A. Fine Structure of Cells

Some aspects of cell and nuclear division, lomasomes, sporogenesis, and organelles have been reviewed by Moore.[807] In the fungi, cell division, including cross-wall formation, is a product of plasma membrane activity. In progressing from Phycomycetes to Carpomycetes, a shift is noted from sparse septation, which is primarily abscissional, to regular septation with central pores; in the transition from Ascomycetes to Basidiomycetes, the septum becomes highly elaborated, and the shift is from functional coenocytes to functional diploids. The plasma membrane appears to play an active role in the formation of lomasomes — structures limited in 1965, with two exceptions, to the true fungi. The electron microscopy of dividing nuclei offered support to the hypothesis that fungal nuclei could divide by a nonmitotic process, a process described by the term karyochorisis.

Aronson[50] described the cell wall of fungi, including the microscopic structures of the walls, and the physical conformation of the walls, including their fibrous structure, lamellae, septations, and crystalline components. Cell walls of various groups of fungi include substances noted in Table 2 as well as proteins and several other substances.

The characteristics and structure of flagellae were described by Kole.[655] Included were data on the structure and organization of the shaft, the whiplash, and the tinsel types of flagellae and their function. Additional work in this area has been summarized by Heath.[525] Fuller[406] has given an excellent survey of the gross and fine structure of zoospores in a number of species of Mastigomycotina.

Ehrlich and Ehrlich[352] have reviewed the fine structure of the host-parasite interface in mycoparasitism. Four types of haustoria were recognized, and for the first of these are described the formation of infection structures, intercellular mycelium, host cell penetration, haustorial formation, and the mature haustorium, encapsulation, the response of the host, and compatible and incompatible interactions. Various pertinent characteristics are described for the remaining three types of haustorium; for type III haustoria, the thallus-host interface was described. Nonhaustorium-forming pathogens are also described.

A variety of types of conidiogenesis was described using both light and electron microscopy by contributors to a symposium the papers for which were edited by Kendrick.[637]

B. Translocation

The phenomenon of protoplasmic streaming has been known for some time.[153] In those fungi for which this phenomenon can be demonstrated it can be seen readily, especially where cytoplasm is granular. Because of the ability of the protoplasm to stream readily between cells of the mycelium of the Ascomycotina and the Basidiomycotina, it has been thought[674] that not only the Mastigomycotina and the Zygomycotina but all fungi are essentially unicellular and coenocytic.

An elaborate study on translocation[1017] of nutrients and dyes through fungus mycelia has been made with species selected from the Zygomycotina, Deuteromycotina, and the Basidiomycotina. No translocation occurred in species of mold fungi and, thus, no protoplasmic streaming. In all agarics studied there was translocation in certain undifferentiated series of cells which could not be distinguished from other nonfunctional cells. Translocation and protoplasmic streaming were closely linked and always occurred together. Efficient bidirectional streaming was absent.

Transpiration was found to be a vital function in the fruit bodies of agarics. It occurred under conditions of complete saturation, although at a reduced rate. The rates of transpiration and of translocation were similar. The behavior of fairy rings appears to bear out the theory that transpiration and translocation are closely connected.

It is apparent that translocation is very important to the fleshy fungi and to some mold fungi. While the morphology of the fungi may appear to be rather simple, the physiological processes within the cells are highly complex.

Wilcoxson and Sadia[1244] have reviewed translocation in fungi emphasizing translocation of viruses, of miscellaneous substances, and of dyes, and its relation to growth. Factors which influence translocation include age of the culture, light, temperature, and transpiration. It is particularly important in the formation of fruit bodies and of sclerotia. The velocity of protoplasm in the translocation stream varies from 0.5 to 10.5 mm/sec.

C. Pigment Production

In observing a series of isolates from soils or other substrata where mixed populations occur, it is readily noted that different fungi produce various colored pigments, some of which are confined to the mycelium, while others are soluble and diffused through the agar. An intensive study[542] of one species of *Coccosporium* presents a review of pigment formation and shows that temperature, aeration, pH, and carbon source are important variables in the development of the red pigment in that species. In yeasts and other fungi the presence of carotene pigments has been reviewed.[450]

D. Melanin

According to Durrell,[345] the dark pigment in many fungus spore walls is a melanin pigment. These dark spores resist ultraviolet radiation of 2537 Å. The degree of opacity depends on the amount and location of the melanin. In the presence of hexachloracetone, about half of the spores produced in culture are light colored and easily killed in ultraviolet light. Durrell's conclusions were based on field experiences with soil fungi in desert soils from Nevada and from the Death Valley National Monument, Calif. He concluded that more species were able to survive in these areas of high light intensity because of the melanin pigments in the spore walls.

Lingappa and colleagues[708,1125] studied the formation of melanins in *Aureobasidium pullulans* and *Cladosporium mansonii* in relation to light and type of medium used.

Ellis and Griffiths[353] studied melanins in the cell walls of hyphal cells of *Amorphotheca resinae,* the aleuriospore walls of *Epicoccum nigrum* and *Humicola grisea,* the sclerotial walls of *Colletotrichum coccoides,* and both within and between the microsclerotial cell walls of *Verticillium dahliae.* Melanin was extracted from each source with KOH, but because of rapid oxidation in the air, extraction and analysis were carried out under nitrogen. The pigments were examined electrophotometrically under UV and IR radiation and by means of X-ray diffraction. Characterization of the melanin in terms of its degradation products indicated it to be indolic in nature. These results were compared with published data on other melanins extracted from animal and plant sources. The indole type of fungal melanins does not conform to the Nicolaus concept of allomelanine, and therefore they belong to either the eumelanins (of the animal kingdom) or to an as yet unclassified and unique group of melanins. Before this question can be resolved it will be necessary to analyze a greater range of fungal melanins.

E. Luminescence

A number of species of fungi produce luminescent mycelia or fruit bodies. Among these the most pronounced is the jack o'lantern fungus, *Omphalotus olearius (Clitocybe illudens),* which produces fruit bodies on decaying stumps and which, under certain conditions, is luminescent. These fruit bodies are heavily laden with phosphorus and mycotoxins, because of which it is one of the poisonous fleshy fungi growing on wood. *Panellus stypticus* grows in both Europe and North America, but in North America luminescence is displayed,[154] while in Europe it is not. That these two geographic races are of the same species has been shown by Canadian workers[735] using monosporous cultures.

F. Temperature Effects

Optimal temperatures for the growth of a large number of fungi in culture have been compiled in a book published in Japan.[1145] In almost all studies of plant pathogenic fungi, optimal growth temperatures are determined.[427] It has been said that in soils in cooler parts of the world, the populations of fungi are dominated by species of *Penicillium,* and those in the warmer areas by *Aspergillus* species. The extent to which temperature controls the fruiting of fungi, especially of those that produce macroscopic fruit bodies, is only surmised. In Scotland[456] it was thought that such fruiting resulted from the concomitance of such favorable factors as temperature, precipitation, light, day length, and nitrogen supplies. In eastern Washington and adjacent

Idaho, where fruiting occurred primarily in the autumn, Cooke[215] found no correlation between these phases of the habitat. In a study of the synecology of the plant pathogenic fungi,[120] correlations with moisture were more pronounced than those with temperature. It has been shown that *Oospora lactis (Geotrichum candidum)* will not survive the temperatures of pasteurization,[449] although it is a common contaminant of dairy products both before and after pasteurization.

Comparatively few fungi are thermophilic. Most of these are associated with compost heaps during active stages of decomposition or curing of the compost. It is possible that those species which tolerate high temperatures during spore germination are thermophilic or at least thermotolerant.

Certain fungi pathogenic for man and other animals have yeast-like stages in their life cycle. The mycelial phase is encountered in nature and is obtained in pure culture in the laboratory at room temperature, but in the pathogenic phase on certain media under temperatures approximating those of the human body, the mycelial phase converts to a yeast-like phase. The reversion to mycelial phase results when the temperature of incubation is reduced. Theories concerning the conversion are discussed above.

In a more recent review,[978] additional possibilities for the explanation of the dimorphism phenomenon have been suggested. It is suggested that the oxidation-reduction potential and that certain sulfhydryl compounds play a role in this process in *Histoplasma capsulataum* and *Candida albicans.* In Blastomyces the role of elevated temperatures may be related to increased generation of reducing power. The role of a thiol-disulphide equilibrium is suggested because of its occurrence in morphogenetic processes in many organisms. The mechanism by which CO_2 induces yeast-like development in *Mucor rouxii, Sporothrix schenckii,* and *Histoplasma farciminosum* is obscure, and while it appears to be independent of the thiol-disulphide equilibrium, the elucidation of its mechanism awaits investigation.

According to Brock[140] the approximate upper limits for temperature for growth of various organisms are given below.

Organism	Upper limits (°C)
Animals, including protozoa	42—51
Eukaryotic microorganisms (certain fungi and the alga *Cyanidium caldarium*)	56—60
Photosynthetic prokaryotes (blue-green algae)	73—75
Photosynthetic prokaryotes (bacteria)	Up to 90

Rippon and Anderson[966] found in working with *Aspergillus sydowii, A. fumigatus, Penicillium lilacinum, Trichophyton schoenleinii,* and *Histoplasma capsulatum* that their metabolic rate was greater at temperatures normally encountered by the organism, as was the final Eh (oxidation-reduction potential) of growth. Adaptation to higher imposed temperatures was delayed in soil fungi, less so in a dermatophyte, and intermediate in thermophilic and dimorphic pathogenic fungi. The final Eh of growth of *H. capsulatum* was similar to that of living animal tissue and may reflect its tissue tolerance.

G. Light

Based on a bibliography of 339 titles, Marsh and co-workers[751] surveyed the information available on light requirements and light effects of more than 300 species of fungi. The earliest reference was to a study by Brefeld in 1881. Most of the references were in the field of plant pathology; medical mycology was not represented. Qualitative light sources varied widely from daylight and incandescent light to fluorescent light. Wattage was given as indicated on the bulb used and rarely according to more sophisticated methods of light measurement. Wavelength was given according to the color of filter used which was described with varying degrees of sophistication.

Light, as one of the aspects of the physical environment of fungi, has been reviewed by Page.[859] Both oriented and nonoriented types of response were recognized. Aspects of assimilative structure growth and of the development of sporangiophores of *Phycomyces* and *Thamnidium* are cited as examples of nonoriented response. The basic types of oriented response included phototaxis and phototropism. Few studies have been reported in both areas, but the orientation of several types of fruit bodies in the major classes of fungi appears to be directly affected by light, although the location of photoreceptors in the cells of many of these is uncertain.

Leach,[683] using cultures of *Helminthosporium oryzae* on potato dextrose agar, tested sporulation using near-ultraviolet light (3100 to 4000 Å) at 70 to 75°C at intensities of 76 to 370 $\mu W/cm^2$. it was found that near-ultraviolet light stimulated sporulation while visible light did not; conidia developed only when a period of irradiation was followed by a period of darkness; the minimum period of darkness following irradiation was 4.5 hr; under a cycle treatment, the minimum period of irradiation required for profuse sporulation was 4 hr under alternating periods of irradiation and darkness, under the 8- and 12-hr cycles, and under shorter cycles of irradiation when these are followed by 12-hr periods of darkness; continuous irradiation of nutrient broth cultures resulted in significant increase in growth over cultures grown in darkness.

This work was extended by Leach[684] to a survey of 34 species of fungi in relation to effects of near-ultraviolet light (3100 to 4000 Å) at 76 $mWcm^2$ at 70°F. Exposure was continuous for 3 to 10 days. Conidial production was initiated or increased in 29 species. *H. oryzae* required a period of darkness after irradiation for conidial production. *Kabatiella caulivora* and *H. sativum* sporulated as well in darkness as under irradiation. Sclerotia of *Typhula* sp. produced sporophores after a cycle of 12 hr of irradiation followed by 12 hr of darkness at 41°F.

A practical guide to the effects of visible and ultraviolet light on fungi is given by Leach.[685] Of the three major areas considered, the first deals with general effects of light on fungi which are both direct and indirect. The direct effects include mutagenic effects, lethal effects, photoreactivation, nonoriented and oriented effects on growth, effects on spore germination and spore liberation, morphogenetic effects, and such cytological and biochemical effects as effects on translocation, protoplasmic streaming and viscosity, permeability effects, luminescence, and several miscellaneous biochemical effects. Among the indirect effects considered were ozone formation, formation of hydrogen peroxide and related substances, decomposition of carbohydrates and other substances, and host susceptibility. The second major area considered was the effect and use of light in the identification of fungi. Among the several areas considered here are induction of sporulation in relation to species in the major classes of fungi, light intensity and dosage, "diurnal sporulation", the interaction of light and other factors such as temperature, pH, aeration, nutrition, depth of medium, age of colony, and surface vs. submerged cultures; under inhibition of sporulation are considered inhibition by high dosages, dark phase inhibition, and true inhibition; pigmentation; and morphological effects, including spore and sporocarp morphology and gross changes. The third major area considered involves practical considerations in the study of light effects on fungi. Here are considered light sources such as solar radiation and artificial light sources such as simple sources of far-ultraviolet (200 to 300 nm), simple sources of near-ultraviolet (300 to 380 nm), simple sources of visible light (380 to 750 nm), and complex sources of UV and visible light; light quality including isolation of broad and narrow spectral bands, transmission of plastics, glassware, and media; light measurements including nonselective radiometric devices, photoelectric detectors, and chemical methods; uniformity of illumination; temperature control, including ambient temperatures, interaction of light and temperature, and the problem of radiant heat; microirradiation; age and physiological condition of mycelium; action spectra; terminology and dosimetry; and health hazards of irradiation studies including radiation, ozone, explosions, and electrical shocks.

The influence of light on the metabolism of fungi has been demonstrated[960] in work with *Penicillium chrysogenum* and *Cephalosporium acremonium*. Growth in liquid culture was measured by optical density and dry mycelial mass. In the case of both fungi, continuous exposure to broad spectrum fluorescent light at an intensity of 1800 $\mu W/cm^2$ resulted in a 24-hr lag in the initiation of growth and a 20 to 26% decrease in the final mycelial mass in the stationary phase. Inhibition of growth was shown to be dependent on the intensity and wavelength of incident light. In both organisms, inhibition of growth was associated with irradiation in the near-UV and blue regions of the spectrum. Since the responses of the two fungi to light were similar, a common mechanism for inhibition of growth in beta-lactam antibiotic-producing fungi may occur.

H. Amounts of Space Occupied

Fungus mycelium is hyaline, colorless, and formed of many hyphae rarely more than 10 μm in diameter, usually 3 to 5 μm. It would be a rare occurrence in which only one species of a fungus would be represented in a single soil or litter habitat. In the case of fungi with macroscopic fruit bodies, the location of the mushroom on the soil or litter, or the bracket on the log, can be readily determined, but which mushroom or bracket belongs to which mycelium, and how extensive that mycelium is in a particular habitat are not so readily discovered. At the time of an observer's visit to a plot, a mycelium may be producing only one fruit body, another 10, still another 100. The comparative importance of these fungi cannot be determined readily although there are several techniques for the assignment of importance values. These include determination of the amount of space occupied, the fresh weight of fruit bodies or the dry weight of the fruit bodies.

Colonies of microfungi or soil fungi can be cultured in a variety of ways on a number of media. Colony diameter or circumference can be measured, or the amount of mycelium or yeast cells produced in liquid culture can be determined, usually on a dry weight basis. These kinds of data may not be as significant as colony counts obtained from different types of plating techniques or data obtained from mycelium grown out of the soil or other habitat on such selective media as cellophane or nylon strips. It has been estimated on the basis of colony counts that there are 1000 to 1500 lb of fungus growth in an acre of agricultural soil 7 in. deep.[794]

I. Nature of Food and Water Transporting Organs — Rhizomorphs

Certain fungi, during their growth processes, develop accumulations of mycelium which permit the actively growing and assimilating mycelium to grow where the best nutrients are available but to fruit in much different areas where the best fruiting conditions occur. Such structures are termed "rhizomorphs". In structure, they range from simple aggregations of hyphae — few to many in number, elongating together in one direction, usually called "hyphal strands" — to complex structures whose growing points superficially resemble a meristematic tissue and whose internal makeup presents a picture of fused hyphae with spiral and scalariform parts which remind one of water-conducting elements in vascular plants. These structures are best developed in

Armillariella mellea, A. elegans, Serpula lacrimans var. *lacrimans*, and *Meruliporia incrassata*. Other fungi have also developed rhizomorph-like structures, but these are of more primitive nature. A summary description of these structures has been presented in a text on fungal anatomy,[721] and recent papers have developed concepts concerning the ability of fungi to use such structures to colonize new areas of nutrient supply.[426]

Mycelial strands are found[426] in a number of fungi which produce mushroom- or puffball-type fruit bodies. Through mycelial strands such fungi as *Psalliota hortensis* colonize suitable prepared compost but not soil already occupied by older strands. For successful colonization, a certain inoculum potential is necessary, although if inoculum potential is only adequate, colonization will take longer. Inoculum potential at the growing tips of the mycelial strands appears to vary directly with the size and vigor of the inoculum food base and inversely with distance from this food base.

In the rhizomorphs of *A. elegans* in Australia, Smith and Griffin[1070] have found that oxygen supplies are critical for good development, since oxygen may affect both form and rate of development. In long rhizomorphs, the rate of oxygen diffusion is maintained because of repeated connections that are made between the central canals and the substrate-air interface. In the absence of such interfaces, rhizomorphs grow for about 20 cm when the rate of diffusion of oxygen to the apices becomes limiting. When the rhizomorphs flatten or become plaque-like, their growth slows. Such rhizomorphs are well adapted for absorbing nutrients from a host but not for spreading from host to host.

J. Nature of Food and Storage Organs — Sclerotia

One type of response to an unfavorable habitat is the development of small to large agglomerations of cells, with or without structure, which have been referred to as "sclerotia." Several types of such bodies have been described by Lohwag.[721] Chlamydospores and *Papulospora*-type cell agglomerations may be primitive developments in the pathway toward sclerotium formation. In some organisms, *Aspergillus* and *Penicillium*, for example, the sclerotium may be a precursor or a replacement for the cleistothecium.[1142] In other fungi, such as the Sclerotiniaceae, the fruiting bodies may arise from sclerotia which are considered as stromata[1226] and which serve as overwintering storage organs. In *Cordyceps, Claviceps*, the Sclerotiniaceae, *Pachyma (Poria) cocos*, and other fungi, there are small to large sclerotia of varying types, although the function may be food storage in all cases. The small sclerotia of *Typhula* are quite characteristic and have been described in detail by Remsberg.[957] In certain wood decay fungi such as species of *Lentinus, Coriolus*, and *Hypoxylon*, a black line is formed around areas of greater activity in the wood. The periphery and materials within this barrier have been thought of as sclerotial in nature.

The survival and germination of sclerotia have been reviewed by Coley-Smith and Cooke.[204] Concerning survival, structural and behavioral factors that influence survival include anatomical structure and certain features of the germination process; environmental factors affecting survival include moisture, temperature, aeration, soil organic matter, parasites and predators, and fungicides. Factors affecting germination include the mode of germination which could be that producing mycelium, spores, or fruiting bodies; physiological changes related to carbohydrate-rich and lipid-rich sclerotia; the relation of dormancy to germination, including constitutive dormancy involving low-temperature activation and light-dark period activation, and exogenous dormancy involving diffusates and mycostasis, temperature, light, aeration, and other factors.

III. THE FRUCTIFICATION

A. Fruiting Habit

The importance of position of fruit body with reference to gravity in relation to fall of spores after discharge has been emphasized[150,589] in the Hymenomycetes. This importance was demonstrated by experimental evidence as well as by evidence based on the use of physical laws relating to various phases of ballistics. In the Homobasidiomycetes, where the hymenium is not amphigenous, a fruit body must be produced in such a way that, for effective liberation, the basidia could discharge their spores into air spaces large enough so that the spores would not become lodged against opposing sides of the spaces. Thus, the tubes of a polypore or a bolete, the gills of an agaric, and the teeth of the fruit body of a hydnaceous fungus must be arranged in position on a surface perpendicular to the surface of the earth for maximum efficiency of spore discharge from a particular fruit body. Accidental change in such a position, deliberate variations as may be induced in laboratory experiments, or distortion in the production of a pileus result in loss of spores for effective wind distribution when the fruit body is reoriented. In the case of some species with perennial fruit bodies and a few fruit bodies which are annual or biennial, when a log on which they are growing is moved, the fruit body will naturally reorient itself and produce spore-bearing tissue which again is able to discharge spores in the normal manner. The fruit body arrangements have been considered geotropic. It has been shown[589,721] that in the Ascomycetes, certain structures are light sensitive or phototropic. The asci in a Discomycete fruit body may be oriented in such a way that they are directed toward the light. In this way, no spores are lost in the violent discharge process by having them aimed at the wall of the cup. Correct positions of fruit bodies for spore discharge are best shown by stereophotos of living material as used by Smith.[1065]

The cupulate type of fruit body is found in both the Ascomycotina and the Basidiomycotina. While in the Ascomycotina the cups always face upward and the asci in some, if not all, are phototropic, in the Basidiomycotina the cups face downward, and the basidia may be considered to be arranged in such a way that upon discharge the spores fall away from the hymenium by gravity. The theory[1047] has been proposed that in cupulate Basidiomycotina of the cyphellaceous type, at least in some species, the cupulate habit is a response to regressive evolution. According to this theory, what may be presumed to have been an agaric developed a resupinate habit, as in certain extant genera. Degeneration progressed beyond this resupination to a breaking up of a pileus into numerous pilei both along and across gill lines resulting in numerous, more or less irregularly arranged cupules.

Club-shaped, coral-shaped, and gelatinous fruit bodies of various shapes are, in general, amphigenous with the hymenium borne on all exposed surfaces. Here the spores are shot from basidia or asci into the air currents which always surround such fungus fruit bodies.

In relation to the origin of the cupulate Basidiomycetes, Cooke[229] has indicated that the point of origin was probably a corticiaceous fungus whose resupinate margins have raised up, the structure becoming cupulate.

In a number of genera of Agaricales, the development of the fruit body in a gastroid manner has resulted in a number of species of secotiaceous fungi now assigned to a number of genera. Since the pileus remains closed or is rarely fully open, basidiospores have lost the capacity of being discharged violently in most cases and are usually, although not always, symmetrical as in the Gasteromycetes, rather than asymmetrical as in the Homobasidiomycetes. It has been suggested that the violent spore discharge mechanism has arisen in several lines of development (resulting in a considerable diversity in these fungi) rather than arising in one level of development, following which, at more than one point, the capacity was lost in response to such possible factors as environmental changes or changes in the site of fruit body development.

According to Lange[673] in Denmark a population of *Kuehneromyces mutabilis* was found to have developed morchelloid fruit bodies. Rather than an indication of a secotiaceous habit, it was suggested that the condition was a response to a viral disease condition.

In a culture assigned to *Psilocybe mutans*, McKnight[776] obtained agaricoid, poroid, and gastroid fruit bodies that were all fertile from cow manure from Idaho. The pattern of tetrapolar sexuality was modified by high frequencies of secondary homothallism.

Studying the development of basidiocarps in the natural environment, States[1099] found that basidiocarps of *Gloeophyllum sepiarium* are initiated with emergence of assimilative hyphae from the wood, their exposure to the relatively severe, fluctuating conditions of light and moisture, and their loss of direct contact with the substratum. The hyphae become organized into hyphal strands composed of generative and skeletal hyphae. Hyphal strands direct the growth of the advancing margin and are involved in the development of a tomentum, context, and hymenophore. Their marked sensitivity to light and humidity accounts for the presence of growth zones in the sporocarp. Atmospheric humidity is considered to be the factor of primary importance in influencing the growth differentiation and orientation of the hyphae in the perennial basidiocarp during the growing season. The close association of skeletal and generative hyphae allows the fruit body to withstand the extended dry periods characteristic of its xeric habitat and to develop over a long period of time. Liquid and agar culture failed to provide all necessary conditions for fruiting.[1100] Periodic desiccation of mycelium in naturally infected and inoculated wood blocks was found necessary for normal basidiocarp development.

B. Adjusted Productivity

In a study of the productivity of fungi in forest plantations in the vicinity of Reading, England, Hora[563] used direct counts of fruit bodies of fleshy fruited fungi in untreated and treated areas. The definition of productivity was fresh weight of agaric fruit bodies per hectare per toadstool season. He concerned himself only with the 12 species common to all study areas. It was found that some species were inhibited by ammonia nitrogen but enhanced by nitrate nitrogen, and vice versa. In an experimental plot treated with hydrated lime, *Myxomphalia maura* increased from 2.33 g/ha in the control plot to 1974.39 g/ha in the treated plot. When the complete fertilizer "Growmore" was used, productivity went up from 550.42 kg/ha on a control plot to 1002.65 kg/ha on an experimental plot in one area, and in a second area it went from 177.59 kg/ha to 277.50 kg/ha.

Richardson[964] has shown that, in studying the fungal populations of an area, almost daily observations are necessary to determine the total population. In the period of peak production, a forest plantation at Selm Muir, Midlothian, Scotland, August to October, when 90% of the total fungi were found, only three fourths of the total would be observed if visits occur at weekly intervals; fortnightly visits would yield only one half of the total. These approximations would depend on climate and the specific composition of the population.

The fresh weight productivity of 265 to 460 kg/ha/year (2.1 to 3.6 cwt/A/year) compares with 1.5 to 2 million kg/ha/year (13,000 to 16,000 cwt/A/year) for intensive mushroom cultivation, and 3,000 to 6,000 kg/ha/year (25 to 40 cwt/A/year) for intensive cereal crops in the U.K. Using the figure of 6.36% for dry matter content obtained from about 2.25 kg of representative species, dry matter production ranged from 16 to 30 kg/ha (0.15 to 0.23 cwt/A).

Sporophore longevity was determined by both field observations and a model based on population decay determined by observations in a series of visits to the established plots with special reference to such fungi as *Russula emetica.* Histograms showing the length of the fruiting period and monthly proportions of single species in the population, and the population as a whole, are used to show the development of the population for the period of observation. While the mid-date of fruiting of single species varied from year to year, the development of the total population was relatively uniform.

C. The Hypogaeous Fruiting Habit

A number of species of fungi fruit beneath the surface of the soil. Most, if not all, species of endomycorrhizal fungi fruit in this way, and many species of fungi producing ectomycorrhizae also fruit in this way. Most species of agarics and boletes fruit above the surface of the soil when they are reported to "grow on the ground". In the Cascade Mountains and other western North America mountain regions, some of these fungi are to be found fruiting under the litter and duff. That these are normal fruit bodies is shown by the fact that spore deposits are found on the soil, litter, and duff under the sporulating fruit bodies. It is thought that the hypogaeous habit could have developed from such fruit bodies through eventual loss of ability to discharge spores, loss of sterigmatic function, loss of the stalk, and eventual production of spores not only on irregularly arranged remnants of gills or tubes, but in cavities in a globose or ovate completely enclosed carpophore. Many of these fungi emit an odor by means of which an animal can smell their presence, dig them out, and eat them. Many rodents[1153] use these fleshy fruit bodies as food and disperse them since their spores are passed in the faeces. In the case of cup fungi which have adopted the hypogaeous habit and lost the capacity for violent spore discharge, man has taken advantage of the odor characteristic by training dogs and pigs to hunt truffles. Other techniques used in the search include such digging instruments as garden rakes, child's garden rakes, and forked grubbing hoes. That the disturbance of the soil does not contribute significantly to the elimination of a population is shown by the study made by Fogel[390] who returned to the same woodland periodically over a period of 3 years, finding a consistently regular supply of fruit bodies throughout the period, indicating that fruit body production is more related to mycorrhizal habit than to any indication that the fungi obtain their nutrients more or less exclusively from the organic matter in the soil.

D. Transpiration in Fleshy Fungi

To help explain the distribution of fleshy fungi in Hungary, Ubrizsy[1167] studied transpiration in *Agaricus (Psalliota) campestris* and other fungi. In such fungi, the process is a completely physical one and is mostly carried out in the cuticular cells of the pileus. It occurs equally in night and day time. Such microclimatic factors as temperature, humidity, and wind strongly affect transpiration which is slower from compact, large fruit bodies than from thinner, more loosely constructed fruit bodies. The development of productive fruit bodies of fleshy fungi requires optimum conditions including 80 to 98% relative humidity and temperatures of 10 to 30°C. In artificial culture these conditions are 98% relative humidity and 20 to 32°C.

E. Drought Resistance

Bohus[112] has found in Hungary that fungi are highly drought resistant. The mycelia of 100% of the species tested survived 2 months of drought, 86% survived 4 months of drought, 71% survived 6 months, and 53% survived 8 months of drought. When subjected to drought conditions, mycelium shrank in size and assumed altered form. Mycelium in its natural substratum was revived on malt agar. Where the mycelium was not in direct contact with the moisture source, it did not survive. In the case of *Morchella conica*, after 8 months of desiccation, fruit body initials developed after a week on malt agar under 20°C. The development of certain wood decay species was rapid. *Poria vaporaria* and *Stereum purpureum*, after a desiccation period of 6 months, were in full growth 3 days after being placed on malt agar.

F. Force

Data have been presented to show that fruit bodies of fungi develop under considerable hygroscopic pressure.[152] As a mushroom is elaborated rapidly, it can raise a piece of sidewalk. In *Coprinus* an actively developing colony can destroy an asphalt tennis court covering by pushing its fruit bodies up through the asphalt.[1022] Fungus spores can be discharged under pressure from both asci and basidia. The pathways of the projectiles thus produced correspond with those of projectiles shot from guns. The fungus gun, the trajectories, or sporabolas, and related phenomena have been carefully studied by Buller[149-151] and Ingold.[589] In the Ascomycotina, such violent discharges, more of the squirt type, are accompanied by audible sound.[149]

G. Size

Most of the mold fungi are microscopic. An indication of this can be gained from data presented in an elementary text on soil science.[794] A comparison is given between numbers of colonies of individual organisms of bacteria, Actinomycetes, and fungi with their equivalent weights in an acre of good agricultural soil 7 in. deep. While 0.3 to 95 million colonies of bacteria per gram of soil are present, representing more than 500 lb live weight, 0.1 to 36 million Actinomycetes or 700 lb and only 8000 to 1 million colonies of fungi or 1000 to 1500 lb are present. Spore volumes for such fungi are given in Table 1 and comparative spore sizes in Figure 2.

Of macrofungi there are wide variations in size. Among the smaller are species of hypogaeous fungi, almost microscopic, which are uniloculate[986] in contrast to the larger hypogaeous species and the puffballs among the multiloculate Gasteromycetes. Among the larger Gasteromycete fruit bodies are the puffballs of *Calvatia gigantea* which may exceed 1 m in diameter.[101] A specimen ½ m in diameter was estimated to have produced more than 7 trillion spores.[149] Bracket fungi produce fruit bodies which vary in size between species. In many cases the size is fairly consistent, but in some species, under supraoptimum growth conditions, the fruit body may exceed the average for the species. *Ganoderma applanatum* has been found in Ohio[538] to produce conks 3 ft across, while this size has been exceeded in Alaska.[370] The Alaskan conk measured 111 cm long, 311 cm in circumference, 75 cm wide, and 81 cm high. It weighed 52.1 kg and had at least 64 layers of tubes. It has been estimated[149] that a specimen of *"Fomes" applanatus*, having a tube-mouth surface of 1 ft^2, can produce 5.5 trillion spores in the 6-month period during which it is active. In the Cascade Mountains of southern Washington and northern Oregon, a large fungus fruit body has been found of which one specimen measured at least 150 cm in height and 90 cm across.[211] *Oxyporus nobilissimus* has been located on rotting hemlock (*Tsuga heterophylla*) stumps and on the ground near noble fir (*Abies procera*) trees, but no connection with living trees has been established. The large conk weighed over 136 kg when fresh, and a "chip" from it had 35 layers of tubes.

H. Diurnalism and Day Length

In the development of certain plant diseases caused by species of downy mildews and powdery mildews, it has been shown that day length and diurnalism are important phases of the life of a fungus in its habitat. It is a difficult matter to control light, temperature, and moisture conditions in pure culture in the laboratory to simulate those conditions to which the fungus has become adapted in nature. The culturing of fungi in cases outside the laboratory window, on media containing various plant parts (natural media),[1082] and similar techniques used in the laboratory of the Department of Plant Pathology at Berkeley, Calif. are steps in this direction. Use of culture chambers with too rigidly controlled conditions of either single temperatures and humidities or of alternatively applied temperatures and humidities may not give any more natural conditions than use of simple laboratory room conditions.

For the study of rhythmic development of any fungus populations or any segment of such a population, the ideal condition would be one in which a location would be subject to the same climatic conditions year after year during the study period, and the fungi would be growing on litter, humicolous materials in the soil, and other nutrients, of the same age and stage of decay each year during the study period. Because of the vagaries of the weather and of the other members of the population in any given area, such rigidly controlled conditions are rarely, if ever, found. Thus, the same species groups are never found in a study site year after year on the same date, but within certain limits some of the species may be found regularly, others within certain limits more or less irregularly, and others in cycles of time more or less related to periods in which the area receives more or less erratic phenomena.

Kalamees[615] in eastern Estonia has noted:

> In their year's rhythm fungi, like higher plants, are subject to regular seasonal changes. Such seasonal changes in the development of fungi are connected with the alternation of two stages of development — the assimilative stage and the stage of development of fruit bodies. Seasonal changes in the development of fungus cover can be termed aspects. This is due to the extremely close dependence of the appearance of fruit bodies of fungi on weather conditions. Besides, aspects of the fungal cover in forests are not so distinctly pronounced as aspects of the plant cover — the fungal fruit bodies are always met with in forests in fewer numbers than higher plants; moreover, many of them are small and not easily discernible, e. g., *Marasmius, Mycena*, etc.

In Estonia[615] it is possible to distinguish, on the basis of seasonal changes, seven periods of development of fleshy fruit bodies of fungi: the early spring aspect, the spring, early summer, summer, autumn, late autumn, and the winter aspect. The summer and autumn aspects in these forests were periods richest in fungi. Each aspect was characterized by listed species.

Kalamees[617] determined development and sporulation in several fleshy-fruited fungi. Glass slides were arranged under agarics in place in the field and were examined daily at 9 a.m. Intensity of sporulation was recorded using a three-point scale:

+	Weak sporulation, spores observed under the microscope
+ +	Moderate sporulation, powder film visible on glass surface
+ + +	Strong sporulation, thick layer of spores visible to the naked eye

The majority of observed fruit bodies were damaged by slugs and snails. One species was apparently damaged by rodents. Apparently fruit bodies start to sporulate before they reach their maximum size, and sporulation may continue for several days or weeks depending on species and climatic conditions.

Matseeb[764] suggested techniques for predicting periods of fruiting of edible fleshy fungi in Russia. Critical factors were thought to be soil moisture and air temperature. A formula was given for calculating values on which predictions could be made. Other habitat factors were also given. It was also necessary to know the past history of mass fruiting of fleshy fungi in concerned areas.

Each fungus may have its own rhythm. Since man's rhythm is circadian, or about a day, on a 365-day-year basis, we tend to assume that this applies to all organisms; however, this is not necessarily true. Aside from this, various factors may interrupt or speed up whatever rhythm exists within an organism.

In Lassen Volcanic National Park, Calif., weather seems to be a prime factor to which the fugus organism must be tuned for successful growth and reproduction. This factor may be surveyed briefly as follows. Following the melting of the snowpack, providing irrigation of the soil and rotting wood, there is a dry period. The length of this dry period varies depending on the arrival of autumn rains which may start as early as mid-August. Occasional summer showers are not significant. With the start of autumn rains, the soil is irrigated to a greater or less extent. Between this time and the natural lowering of temperature to a level bringing freezing conditions to the soil moisture, there is an active period of the production of fruit bodies. This may be interrupted by freezing temperatures which may not reach significant depths in the soil or in rotting wood. Growth, at the expense of soil organic matter or wood components, continues, although slowly, and food may be stored for the production in the next season of fruit bodies. With the coming of the snowpack, freezing temperatures in the soil and the wood may be alleviated somewhat. The snow forms an insulating layer, and temperatures beneath it, at least at the melting edge in the spring, rarely reach down to the freezing level.

A number of species of fungi fruit only under the snowpack or at the edge of it as it melts. These include smothering fungi of conifers, sooty molds of herbaceous plants, most fungi of rotting herbaceous debris, and certain fleshy "snowbank" fungi. Other fungi, especially those fruiting on the soil under litter and duff or in rotting wood, depend on moisture left by, or provided by, the melting snow. Species depending on vascular plants for nutrients, such as plant parasitic fungi and certain mycorrhizal species, may fruit later in the season as the soil or rotting wood are drying out. Finally there is a massive flush of fruiting of species producing fleshy fruit bodies following the arrival of late summer or early autumn rains. This is halted by freezing conditions produced by autumn temperature declines.

If conditions are such that any portion of the above described cycle is disrupted, the entire cycle may be disrupted for certain species.

I. Space Occupied

A mycelium may produce one, two, or many fruit bodies at a time or in a season. The size of the fruit body is usually genetically controlled, but certain environmental controls may also be operative. The numbers of fruit bodies may also be genetically controlled, and some environmental controls may also be in operation. The amount of nutrients required to produce each fruit body is unknown. One technique for[215] determining the space occupied by a fruit body is the determination of the amount of space a cylinder of the same dimension (diameter of cap and length of stipe) would occupy.

J. Types Produced

The only types of fruit bodies with which it has been practical to work in mycocoenological studies are those of the macromycetes — cupulate, club-shaped, mushroom-like, and puffballs. Micromycetes, as well as fruit bodies less than 1 mm in diameter or even 1 cm in diameter, are considered in the same studies, but their size, arrangement, and substratum make them less satisfactory subjects for mycocoenological studies, except in certain limited types of studies. Some workers have stated categorically that soil fungi are unsuited for mycocoenological research.

K. Spore Production

Once a fructification is initiated, the production of spores, assuming conditions continue to be favorable, proceeds on a genetically and probably an environmentally controlled plan, at least in nature. In the laboratory, the adjustment of nutrient solutions, physical nature of the medium, other environmental factors, and other habitat factor adjustments may modify the ''normal'' spore production process.

Chapter 7

FUNGI IN THE ECOSYSTEM

I. THE ECOSYSTEM (INCLUDING THE BIOCOENOSIS)

Billings[104] briefly defined the ecosystem as the "energy-driven complex of a community of organisms and its controlling environment." Milne and Milne[804] amplified this slightly: "an ecological community and its physical environment of extent such that the cycling of nutrient materials and energy form a closed system." However, studies of higher plant and animal ecosystems rarely include data concerning fungi in that ecosystem, although fungi form an integral part of that ecosystem.

II. THE ENVIRONMENTAL COMPLEX

That the environment in which a fungus grows is exceedingly complex[103] has been demonstrated in a series of proposed outlines to be used in the study of ecological life histories of fungi.[212] This is also indicated by the efforts of numerous researches to indicate one or more phases of the life cycle of fungi being studied specifically as plant or forest tree pathogens, such as saprobes in the deterioration of various types of material, and as organisms accountable for the reduction of dead organic matter to simple elements or compounds in forms assimilable by the plants or other organisms filling the ecological niche once inhabited by the decaying organisms.

In the study of the ecology of the higher plants, a number of concepts have arisen which may, in the development of a more nearly complete picture of the ecological life history of fungus species, become adopted by the mycologist. One such study presents a functional, factorial approach to the study of plant ecology.[742] The functional equation

$$V = f(cl, p, r, o, t)$$

where V is a plant community, a distinctive organized grouping of the plants living together in a specific area, which is a function of the regional climate (cl), the parent material (p), the topography or relief (r—slope, exposure, and certain ground water conditions), the biota (o), and time (t). The description is qualitative rather than quantitative. That such information can be applied to the development of a fungal population was demonstrated by Cooke[228] in a description of the biota of a trickling filter in the Dayton, Ohio sewage treatment plant. He substituted b, the biota of an area, for V, and used cl, the regional climate; p, the physical and chemical quality of the habitat; r, the fixed operational conditions; o, the biotic effects of the comunity; and t, time.

In a further extension of the environmental complex to include all factors operative on the organism,[103] it has been shown that a very wide range of habitat and environmental influences is operative during the development of an individual, a species, a community, or a population of organisms. Such factors are operative on all organisms in the population complex, and this complex is only one of the factors in the environmental complex.

The relationships between fungi in relation to mutual pathogenicity, i.e., symbiosis, have been critically reviewed.[310] It has been found that some fungi are parasitic on others to the extent that one species may be restricted to one strain of another species or may have a wide range of selectivity of species within an order. Pathogenicity of one fungus on another may be expressed in reduction of size of a fruit body, or in the elimination of the ability by the host to sporulate in favor of the parasite. The whole question of the relation of one fungus to another was developed on the basis of the MacDougall pattern of symbiotic relationships of which Daubenmire's modification may be cited.[293]

III. THE CONCEPT "ENVIRONMENT"

In an extensive language analysis of the concept "environment" the following definition has been developed by Mason and Langenheim:[762] "The environment of any organism is the class (of phenomena) composed of the sum of those phenomena that enter a reaction system of the organism or otherwise directly impinge upon it to affect its mode of life at any time throughout its life cycle as ordered by any other condition of the organism that alters its environmental demands."

IV. FUNGI AND ENVIRONMENTAL EXTREMES

It has been noted that environments which may seem extreme and inhospitable to man may be quite normal to other organisms. The extremes of high and low temperatures, high or low salts or sugar content of the substrate, oxygen levels, various types of fumigants, heavy metal ions, indicative isotopes, aquatic or xeric conditions, and other factors, may not only be tolerated by fungi, but preferred by certain species.

V. FUNGAL ADJUSTMENT TO THE ENVIRONMENT

Cantino[177] has said,

> While no single organism or group of organisms has provided sufficient evidence for establishing the exact methods involved through ontogeny and phylogeny in the manifestation of living systems, evidence for establishing the exact methods involved has been derived from a study of the Blastocladiales and other aquatic fungi. In particular, conspicuous differences in metabolism exist among the undifferentiated, differentiating, and differentiated cells of *Blastocladiella emersonii,* the existing evidence pointing sharply to a manifold induction and repression of enzyme systems as a partial basis for these changes. Just as interaction between nucleus and cytoplasm can be approached most directly by studying synthetic capabilities in a cell deprived of its nucleus, an equally direct approach would be to study a "living" nucleus deprived of its cytoplasm, a system approximated by the spore of *B. emersonii.* For the investigation of the way in which visible light exerts control over growth, development, and differentiation, both *B. emersonii* and *B. britannica* demonstrate their utility. In the last analysis, the intellectual profit to be gained from working with water fungi such as the Blastocladiales stems from the fact that they can be observed and manipulated under controlled conditions as *intact, living, growing, differentiating systems in operation.*

For Waid[1192] five techniques are useful for determining mechanisms of physiological and biochemical adjustment of fungi to their environment: (1) mixed cultures in a natural environment, (2) pure growing cultures growing in undefined laboratory media, (3) nonproliferating cells in pure culture on chemically defined substrates, (4) pure growing cultures in highly defined media, and (5) cellfree enzymes and coenzymes on pure substrates. Among the mechanisms of fungal adjustment considered were the nature of the viable material and the establishment of the assimilative phase, self-inhibition of spore germination, substances influencing germination, including fungistasis in the soil and rate of germination; assimilative growth and utilization of the environment; hyphal structure and utilization of the environment; extracellular and surface localized enzymes; hyphae and mechanical forces, fungi in ecological succession; influence of hyphae on their environment; interactions among microorganisms in ecological succession; mechanisms of antagonism and tolerance; parasites and predators of fungi; antibiotics; competition for space, water, nutrients, oxygen, and carbon dioxide; tolerance of suboptimal conditions; tolerance of toxic materials; and propagation, dissemination, and survival.

Two questions arose out of the discussion based on these points. Do fungi possess means of adjustment to their environment that are different from those of other organisms? Failing this, do some or all fungi make greater use of certain mechanisms of adjustment than do other organisms?

To these questions Waid answered that "fungi do not appear to possess unique biochemical or physiological mechanisms of adjustment to their environments" but that out of a wide selection of mechanisms which are widespread or have become highly developed among fungi — including adjustments involving activities such as nutrition absorption and assimilation, aerobic metabolism, propagation, survival, and tolerance to toxic materials — fungi are able to make adjustments. In the case of a fungal parasite or symbiont, survival must depend on adjustment throughout the assimilative phase of the life cycle so that the functioning harmonizes with that of its host or symbiont and the partnership is prolonged. The ability of many fungi to form such associations is, perhaps, one of the most striking features of their biology and one which must involve intricate, delicate, and little understood mechanisms of adjustment.

VI. TEMPERATURE ADJUSTMENTS

Soil temperatures measured by thermocouples on several soil surfaces at Moscow, Idaho were reported by Daubenmire.[292] These included:

Soils	5mm below surface	Surface	5 cm above surface	10 cm above surface
Black, dry, exposed	32	47	42.5	40.5
Gray, dry, exposed	32	42.5	38.5	38.5
White, dry, exposed	33.5	37.3	34.5	32.3
Gray, wet, exposed	27.8	34.5	32.5	31.8
Gray, dry, shaded	27.5	31.5	29.5	29.7

Fowler[396] used a U-shaped spring held in place by temperature-sensitive wax and placed at 0 to 1 cm above the soil surface in a climax *Artemisia tridentata-Agropyron spicatum* association. The indicators used released at 100, 109, 119, 131, and 144°F.

With the increasing importance of thermal pollution, the release into natural waters of warmer cooling water from power plants and from other types of natural cooling processes, the annotated bibliography by Mackenthun and collaborators[733] could be a useful reference. The effect of temperature on all types of aquatic life except the fungi forms the basis of this report. The fungi are excepted merely because no studies were available in which they have been used in this connection.

Feyret and Parguey-Leduc[377] have found that in the development of the perithecia of *Gnomonia leptostyla* various stages was temperature-sensitive. In culture at 29°C perithecia remained immature, only developing to the stage of ascogenous hyphae produced from an ascogonium. At 10°C the ascogenous hyphae completed their development.

VII. FAIRY RINGS AND CLANS

Shantz and Piemeisel[1027] developed a classic study on fairy rings which are known under a variety of names in many languages and which are formed in grasslands and forests by mycelia of fungi growing outward from a single point, the point of spore germination, synchronously in all directions. Outward growth may be interrupted by a number of obstacles, thus rarely resulting in true rings but forming numerous arcs and partial rings, which appear to continue growth in the original direction of that segment of the ring. Growth proceeds at the average rate of 12 cm per year with surges to 60 cm in favorable years. The age of the largest rings observed, which reach 200 m in diameter, may be as much as 420 years. In the development of a ring, several stages were noted. In the center of young rings is an apparently dead area in which the mycelial mat is impenetrable. As this mat breaks down, moisture gets through, and conditions are favorable for development of a succession of plants starting with an early weed stage, progressing to a late weed stage, a short-lived grass stage, and finally returning to the original short-grass cover in the short-grass prairie. Fungal and bacterial degradation of dead grass tissue yields ammonia and nitrate nitrogen which enable the grasses to grow more vigorously producing darker green leaves and larger growth. The fairy ring fungi, of which 63 species were reported in 1917, do not infect the grasses or other associated plants. It should be noted that in order to study these fungi a relatively large area of undisturbed prairie or forest would be required for the length of the study period.

Transeau[1146] contrasted fairy ring and "clan" patterns of growth as demonstrated in the latter case by *Coprinus variegatus* growing on a stump of *Ulmus americana* in Columbus, Ohio. Periodicity and yield of fruit bodies was observed for 7 successive years, and data for local weather conditions during that time were also given. Temperature and available moisture were considered the more significant aspects of these data. A second series of charts giving the distribution of fruit bodies in space appears to indicate the development of a number of independent mycelia and the patterns of the clumps of fruit bodies that developed from these. Considering each independent mycelium a clan, the clan pattern was contrasted with the fairy ring pattern of development which represented the expanding growth in a widely and evenly distributed medium. Fairy ring patterns may continue to develop indefinitely, probably showing a continuous and steady source of nutrients; clan patterns show a relatively limited developmental pattern in time, probably based on the elimination by the fungus of its food base.

VIII. SPORE SLIMES

Bandoni[72] studied the surface activity of spore slimes in 25 species of fungi. Four principal types of slimes were found, including slimes highly surface active and spores dispersed primarily at the air-water interface; slime rapidly spreading in the bulk phase and spores spread throughout; slime weakly spreading or insoluble in water; and slime insoluble in water but showing some activity in oil or lactophenol. The significance of this activity with respect to spore dispersal could include the possibility that surface active slimes may aid in tension movements of spores on litter surfaces; they may be involved in insect dispersal mechanisms, although in litter fungus spore production may take place in cool months when insect activity is low or nonexistent; they may be involved in protecting spores against arthropod predation; and they may result in other factors in spore dispersal.

IX. OSMOPHILY

Concerned with the ability of certain organisms[593] to survive and grow in habitats with high concentrations of sugars, studies were made on them in solutions with high concentrations of sugars. Osmophily is best known in certain bacteria, yeasts, and the *Aspergillus glaucus* group. A number of observations lead to the conclusion that the internal solute concentration of their cells is roughly compatible with that of the medium in which they grow. While most work has been done with bacteria, there is no evidence of any significant difference in this respect with organisms with different types of reaction, such as intolerant, facultative, or obligate.

X. SALINE SOIL

James[598] was interested in the recovery of fungi from a saline soil in Manitoba compared with that from a Red River Valley clay soil. One fifth as many fungal colonies were recovered from the saline soil. In a comparison of different media, it was found that soil extract agar with 0.02% dipotassium phosphate, 0.1% glucose, and 0.1% peptone gave higher yields than Waksman's modified medium, Martin's medium, or Smith and Dawson's medium. Also, when soil extract was replaced with water in the above medium with the three chemicals, counts remained as high. When different levels of potassium phosphate were tested, counts varied inversely as the amounts of potassium phosphate; the same was true when sodium phosphate was used instead of potassium phosphate.

In a test of Death Valley National Monument soils for microfungi, Hunt and Durrell[581] found that numbers of fungi and fungus colonies isolated decreased with increasing salinity of the soil samples. Apparently organic matter also decreased in the more saline soils. In the laboratory, the two most common species of fungi, *Stemphylium* sp. and *Sphaeronema* sp., recovered from saline soil grew well in media containing up to 6% NaCl.

XI. MUTUALISM AND ANTAGONISM IN FOREST SOILS

A survey of mutualism and antagonism among isolates of fungi made from tree seedlings, nursery soils, and forests was described by Vaartaja and Salisbury.[1173] Selected isolates from tree seedlings, forest nursery soils, and fruit bodies of Basidiomycetes were paired on cornmeal agar and the mutual effects recorded. In the majority of 1205 different combinations, antagonism was present, but in many there were no mutual effects. Stimulation took place rarely. The most common effect was the inhibition of growth for a distance of 0 to 20 mm. Very often the growth of the colony was inhibited on contact or soon thereafter. Lysis was observed in several cases. Although antagonism or sensitivity to antagonism was typical of many isolates, it was lacking in certain combinations of each isolate. Isolates of the following genera were among those exhibiting strong antagonism: *Armillaria, Cylindrocarpon, Gliocladium, Penicillium, Russula, Suillus,* and *Trichoderma*. Many species of *Fusarium* and *Gliocladium* were the more insensitive isolates, and several basidiomycete and phycomycete species were the more sensitive ones.

XII. PESTICIDE EFFECTS

Three methods to investigate the effects of certain herbicides on the growth of fungi were used by Wilkinson and Lucas:[1249] measurements of hyphal extensions across agar plates, measurements of hyphal extension across sterilized plant material, and manometric techniques. Three points, in particular, emerged from these studies. First, there was no stimulation of fungal growth. Herbicide interference in fungal growth included suppression of spore germination, inhibition of the rate of the linear extension of the mycelium, and abnormalities in growth habit and in patterns of spore production. Second, some herbicides, e.g., linuron and paraquat, were more fungitoxic than others, e.g., MPCA and simazine, to a range of organisms. Third, there were differences between fungi in their sensitivity to different herbicides. All three methods have shown consistent differences between fungi in their ability to tolerate paraquat. *Trichoderma viride,* in particular, has been found to be sensitive to paraquat. The inhibitory effects were observed at concentrations well within the range likely to be expected in the field.

In considering the relationship of soil microorganisms and plant protection chemicals, Walker[1201] said,

> The above survey of the available (up to 1964) information on the effects of microorganisms in detoxifying plant protection chemicals illustrates the relevance of microbiological studies to this problem. Understanding of these effects is still fragmentary; only a few bacterial species have been studied and even fewer fungi. Interactions in soil are complex and metabolic pathways often differ in different microorganisms. At present little if anything is known of the possible role of other small organisms such as algae, protozoa, the host of other microbial species, microscopic soil fauna, etc., in the disposal of chemicals added to the soil. Compounds that are very persistent or insoluble in water may also be affected by slow chemical reaction with soil constituents and by irradiation by visible and ultraviolet light. These effects complicate microbiological studies and much more work is clearly desirable.

Passage of the Federal Environmental Pesticide Control Act in 1972 has resulted in a number of changes in herbicide, insecticide, miticide, and fungicide formulation and usage. A certain amount of information on such chemicals published prior to 1972 is now of historical interest only.

Precursors to that Act resulted in the following statement by Brandes[125] in the introduction to his review on advances in fungicide utilization.

> Present concern over pollution and the declining quality of the environment is justified, and so the contribution to pollution of agriculture pest control chemicals must be accurately and dispassionately appraised. Abuses must be corrected and prevented. At stake are the health of the world's crops and the well-being of its people. Fungicides will make a significant contribution to this well-being if the prediction is correct that we are entering a new era of fungicide utilization. This will be achieved through the development of more efficient application equipment and methods that result in reduced spray volume, lower costs, and faster, more timely delivery as well as improvements in formulations, adjuvants, and the fungicides themselves. Such advances will make the use of fungicides practical and economical even on our vital basic crops.

XIII. EFFECT OF OR ON CHEMICALS

Trinitrotoluene (TNT) at concentrations of more than 50 mg/ℓ inhibits the growth of most fungi tested, according to Klausmeier et al.,[648] although at concentrations below 20 mg/ℓ most test organisms grow. Growth of an organism in media containing TNT reduced the amount of TNT present, although no organism was found that could utilize TNT as a sole carbon source. In all cases examined, regardless of the type of organism used or the initial concentration of TNT present, if growth was good, the TNT disappeared. This may have significance in the case of landfill and other solid waste disposal projects involving explosives.

Over 100 isolates of representative estuarine bacteria and fungi were tested by Bourquin et al.[121] for sensitivity to polychlorinated biphenyls (PCB). Super absorbent sensitivity discs impregnated with up to 0.5 mg PCB were placed on freshly inoculated agar plates. PCB had negligible effect on the growth of those fungi tested.

Strains of the fungus then referred to as *Phialophora jeanselmei,* now probably more correctly called *Exophiala jeanselmei* (Lager.) McGinnis and Padhye, are common in polluted habitats. A strain isolated from sewage pollution was used by Cooke and

Matsuura[269] to determine its ability, if any, to utilize the hard detergent alkyl benzyl sulfonate (ABS). In solutions with or without glucose, it was found that the fungus could utilize this substance. That it absorbed the ABS rather than adsorbing it, as in the case of lignosulfonic acid, was indicated by the fact that washings of harvested mycelium yielded insignificant amounts, if any, of it.

Chapter 8

TOTAL ECOSYSTEM VS. INDIVIDUAL SEGMENTS

I. INTRODUCTION

Especially in relation to forest pathology, but with application to other areas of mycology, Ehrlich[351] considered three principal groups of fungi: parasites producing pathogenic conditions in living tissue, pathogens producing decay in dead tissues in living plants, and saprobes or saprophytes producing saprobic, or saprogenic, or decay conditions in dead tissues. Basically, all fungal activities can be fitted into these patterns.

The fungi occupy a great variety of ecological niches, occurring almost wherever there is organic matter acceptable to fungus enzyme systems. On first glance it could appear possible to arrange the fungi present in any location in synusiae, but the bases of such distribution patterns have been considered by Cooke[252] to be ephemeral. However, a number of habitat groups have been delimited, and these will be indicated in the following discussion. The yeasts will be considered separately although they occur with filamentous fungi in almost every habitat group to be considered.

For Garrett[427] there are four principal groups of soil fungi. These include (1) those fungi living saprobically in the soil where they develop only on dead organic matter; (2) weak plant pathogens living in the soil, capable of using soil organic matter, but developing as occasion arises on young roots or on senescent roots; (3) ecologically obligate plant pathogens, including those which can parasitize the actively growing roots, produce diseases like vascular wilts, and, following the death of the host, utilize its tissues as nutrients until their activity is arrested by more actively competitive saprobic fungi; and (4) obligate plant parasites such as endotrophic mycorrhizae, rusts, powdery mildews, and downy mildews, which apparently require living plant materials for their substrates.

II. YEASTS

As a group, yeasts have received little attention ecologically. In an early report, Lund[727] surveyed habitats in Scandinavia for yeast populations. Since 1940 several reviews and books have been written concerning their classification[719] and biology.[904] With intensive study, information about the ecology of yeasts can be gathered from the various topics which have been presented by Lodder[719] and from several chapters in the book by Phaff and colleagues.[904] An early survey of the sexuality, development cycle, and phylogeny of yeasts was produced by Guilliermond.[486] Among more recent general works are those of Ingram[592] and of Phaff and colleagues.[904] The genetics and cytology of the yeast cell have occupied workers in St. Louis,[706] especially in relation to improvement of beer yeast strains. The taxonomy, ecology, and distribution of *Hansenula* have occupied researchers in Peoria,[719,1239] and the most comprehensive manual of the taxonomy of perfect and imperfect genera of yeasts and yeast-like fungi has been edited in Holland.[719] A comprehensive volume on the biology of yeasts from England[592] indicated that little is known about their ecology, especially on a quantitative basis and in relation to other organisms in the same environment and habitat.

In recent years, studies have indicated that yeasts occur in a number of locations throughout the natural scene. One of the more comprehensive statements of this type is that of Lund.[727] Yeasts were found in large numbers in decaying fruit bodies of fleshy fungi, sweet, juicy fruits, exudates of trees, and insects vsiting flowers. Present in smaller numbers, yeasts have been found in flowers, on leaves and bark, root crops containing sugar, dung of horses and cattle, soils, and a variety of other habitats. Asporogenous yeasts were more common than those producing ascospores. Techniques of study at both the levels of isolation and laboratory investigation of habitat factor effects were autecological rather than synecological in nature.

Yeasts have been found in increasing numbers in fleshy fungus fruit bodies (mushrooms) as they mature and start to decompose.[30] Different mushrooms harbored different species of yeasts. In France, yeasts were detected[115] in liquor from plant material used for tanning. In studies on the surface populations of shrimp obtained in the Gulf of Mexico off Texas,[906] yeasts were obtained in considerable numbers. During the ripening of apples in Quebec, yeast populations changed[1250] so that low numbers occurred in August, high numbers in October, and different species prospered in the change of seasons. No fermenting species were common. Davenport[295] has described the yeast populations on surfaces of developing apples and grapes in England. The human body has been found[208,1054] to be a habitat for yeasts. *Lipomyces starkeyi* was found to be present on many areas of the skin, and populations were found to develop within 3 hr after bathing. The pigmented genus *Rhodotorula* was of considerable trouble to this group of workers. A system has been developed in which five carbohydrates, three nitrogen sources, and other tests were employed to separate species. This system has revealed that at least five species of *Rhodotorula* occur on the human body. However, if a larger number of carbohydrate sources were used, there was no special pattern of utilization, and it was concluded that the genus was monotypic, consisting of only *Rhodotorula glutinis* which was made up of a number of strains, each able to use varying types of carbohydrates. On the other hand, the recent monograph[903] of red yeasts has shown that, using a larger variety of assimilation and other tests, several species are involved, including two now assignable to the basidiomycete genus *Rhodospirillum*.

In New Zealand,[778] a search has been made for yeasts in soil samples, especially for those which might be of interest as human pathogens. In Canada, yeasts have been isolated from soils by various techniques. To reduce the numbers of bacteria and filamentous fungi, the workers[797] used acid, rose bengal, and oxgall. Of these, the latter was preferred.

In the Yosemite region of California an extensive study was made on the occurrence of yeasts in the same habitats as species of the fruit fly *Drosophila*.[181,325,908] A number of species have been found in the alimentary canal of the flies and their maggots, suggesting that the yeasts may form a part of the food supplies of these insects. The natural breeding sites of the flies have been studied intensively for their yeast populations. Factors of the microclimates of the habitats and the nutrition supplies were considered. Finally, in this series of studies attention was given to the differential attraction of species of fruit fly to various species of yeasts.

Work done in St. Louis indicated[704] that if assimilative stage yeast cells can be induced to store excess materials of fats and carbohydrates, they can be induced to become dormant. Conditions for reactivation such as starvation were also discussed. The same workers[705] analyzed the mechanism of budding in yeast cells based on structural considerations. Lindegren[703] has shown that it is practical to breed yeasts for use as food supplements; and Schmidt[1014] has described the use of yeast as foods and feeds in Germany during World War II.

In their volume on the biology of yeast, Phaff et al. have considered some aspects of the ecology of yeasts. While yeasts do not break down complex natural substrates such as cellulose and starches, they may occur in habitats where such compounds are known to occur, depending on other organisms for the initial breakdown process. Yeasts are inhibited by antibiotics produced by species of Actinomycetes, probably in nature as in the laboratory. Acetic acid produced by bacteria from alcohol formed by yeasts is inhibitory to yeasts. Ingestion of yeasts as food by insects, protozoa, and other animals occurs commonly. Yeasts are associated with plants in a number of ways, occurring on leaves, flowers, and fruits, in tree exudates and as plant pathogens; they are associated with animals on surfaces, in respiratory, digestive, and reproductive tracts, and as pathogens; they occur in soils and in fresh and salt waters. Yeasts are associated with food spoilage and occur in fermentation processes; they are found in substrates with a high sugar content and are associated with products of low sugar content; and they are associated with milk products and with miscellaneous human infections. Yeasts are used in a number of industrial processes.

In studies of yeasts in Great Lakes waters and sediments, Hedrick and his colleagues have sampled at a number of sites and depths in lake waters and bottom sediments in Lakes Ontario, Erie, Michigan, and Superior.[526,527,529,530] While yeasts were found at all depths in the waters as well as in bottom sediments, larger numbers were found in subsurface waters than in surface waters. Species and genera present varied among the different lakes sampled. Organisms in the water were recovered by membrane filtration followed by incubation of the membrane on a defined agar medium. Sediment samples were suspended in a defined nutrient broth, membrane filtered, and the membrane incubated on agar prepared with the defined medium. In Lake Michigan the numbers of yeasts in surface waters are less than in subsurface waters. In deep open waters of the lake, the density of the yeasts was higher at 100 m than at either 0- or 200-m depths; in waters of intermediate depths, the number of yeasts at 40 or 50 m was greater than the number at 75 to 90 m. In more shallow waters with a maximum depth of 15 to 20 m, the greater density of yeast cells was just above the bottom. In mud samples, the variety of different kinds of yeasts was greater than that in open water. This was also true for samples of beach water. In Lake Erie, the density of yeasts per unit volume was less than in Lake Michigan, but the variety of yeasts was larger. In the case of Lake Ontario, for the 27 stations tested, the average number of yeasts isolated per 100 mℓ of water was 10 for water at 1 m, 130 for water at mid-depth, and 460 for water near the bottom; sediment yielded 36 colonies. For the total yeast population there was a delicate relationship between the numbers of cells at the different water depths and the organic nitrogen and nitrate nitrogen concentration of the corresponding depths. In Lake Superior, there was no apparent difference between the yeast populations of the north- and southside tributaries. The population of any yeast species isolated from different stations could be related, in general, to the amount of organic nitrogen reported for each habitat. The number of microorganisms recovered and the organic nitrogen content of the different ecological niches may be arranged in the following progressive series: (1) open waters, (2) areas of the lake near harbor and river entries, and (3) streams entering the lake.

In Lake Michigan[528] a series of cores from the lake bottom was taken at various locations. Yeasts were recovered from some sections of all types of cores regardless of the depth of the lake from which they were secured. The variety and the incidence of the yeasts in the cores were related more to the type of sediment in the core than to the lake depth from which the core originated. The cores with the largest variety of yeasts were the black-viscid types which came from the basins in the lake. The cores with the next larger variety of yeasts were the red-clay cores which were secured from plateaus in the lake. These latter core sediments did not permit the growth of many yeasts on the membranes; however in the jar incubation series, these samples gave rise to several different yeast species. Only a few species of yeasts were isolated from the grey-black cores. The cores which gave the lowest number of species were the black and red banded and the black mottled cores.

Fresh waters at or near the Indiana-Illinois line[1278] were sampled to compare yeast populations in mildly polluted water with those in waters with heavy loads of industrial wastes and those with heavy loads of sanitary sewage. Generally, large yeast populations were found ranging as high as 37,000 cells per 100 mℓ and averaging 3000 cells per 100 mℓ. In a statistical study[1279] to determine the ecological relation between these yeasts and their habitat, a computerized stepwise regression analysis was used. Some 15 different pollution parameters were correlated with the total yeast population, six genera, and the species of yeasts from the three locations. Higher percentage correlations were generally obtained when more narrowly defined taxa were used in the regression analyses. The sampling location which was polluted with industrial wastes was considered to be a highly complex ecosystem in which such large numbers of factors influenced the yeast population concomitantly that significant percent correlations were not often obtained.

In a discussion of the ecology of yeasts, Windisch[1261] pointed out that, from his point of view, the yeast association is similar to that of the higher plants, in that as the population matures, the nutrients are depleted. Artificial renewal of resources may result in symbiotic associations with other organisms such as those between yeasts and insects. A comparison is also given between the principles of the ecology of yeasts and bacteria, and certain similarities were pointed out.

A selective medium developed by Rao and coworkers[938] for yeast isolation made use of a proteose peptone yeast extract dextrose agar to which had been added tellurite at 150 mg/ℓ and streptomycin at 150 μg/ℓ.

Slime fluxes on the bark of certain trees furnish excellent habitats for certain yeasts. Yoneyama[1283] studied 268 isolates from slime fluxes on *Quercus* and *Pinus* species in Japan and from soils beneath them. It was found that, from the morphological point of view, strains of yeasts in nature differ from those of the same species in culture. Noteworthy characteristics of bark-inhabiting yeasts were correlated with properties of the bark.

From the slime flux on the bark of a tree of *Ulmus carpinifolia*, during an annual cycle at Davis, California,[907] *Pichia pastoris* was most abundant and was found in all seasons. *Trichosporon penicillatum* was also found in all samples but was less abundant. *Prototheca moriformis* was abundant in the summer but absent in the winter. Other species of yeasts and *Prototheca* were present but occurred more rarely. It appeared that these species were probably introduced to this habitat by insects. Isolations of 107 species in 21 genera of yeasts from slime fluxes, activities of bark beetles, or exudates from stumps of recently cut trees (from 90 species of trees from Japan and western North America between California and Alaska), were recovered in the spring of 1967 and the summer of 1968.[905] The study was designed to expand knowledge of yeast species associated with tree bark. Ubiquity of some species of yeasts was demonstrated.

Standard and Ahearn[1097] reported that species of *Hansenula* and *Candida* tested are resistant to high concentrations of linear (LAS) and branched chain (ABS) anionic alkyl benzylsulfonates. Subinhibitory concentrations of these detergents were degraded in the experiment.

The South Saskatchewan River flows through the city of Saskatoon. In a survey of its yeast population, Spencer and colleagues[1091] found that yeast populations increased as the effluent from the sewage system of that city was received. In the summers of 1964 and 1965 upstream waters yielded from 400 to 500 cells per liter, while in the water immediately downstream the numbers increased to 4600 cells per liter. In the summer of 1968, when the flow of the river was very low, the upstream water yielded 150 cells per liter while the waters immediately downstream yielded 30,000 cells per liter.

A survey of yeasts in the Pacific Ocean by Yamasato et al.[1281] was made along longitude 150° E from latitude 44° N to the equator. In December 1967, 184 samples were taken of which 27.7% yielded yeasts averaging 56.7 cells per liter in the yeast-positive samples. Yeasts were found from the surface to a depth of 4000 m. Experiments to determine the effects of salt concentration, temperature, pH, and hydrostatic pressure on *in situ* growth and reproduction of yeasts were carried out. Maximum NaCl tolerance was between 9 and 21%, but test strains showed fairly good growth at the NaCl concentration of sea water. The pH threshhold for growth in the alkaline range overlapped with that of sea water. The pH 8.1, widely observed in the ocean, was deleterious for most of the test strains. It is suggested that marine yeasts may be subjected to extensive pH effects by different bodies of sea water. The maximum tolerance to increased hydrostatic pressure among the test strain atm, was 400 to 500 corresponding to a depth of 4000 to 5000 m. Marine yeasts appear to be more tolerant to hydrostatic pressure than terrestrial yeasts. Adverse effects were not noted above 2000 m and were more important below 4000 m. Fell[379] has surveyed the occurrence of yeasts in oceanic regions including marine waters.

An interesting association has been found by Meyers and colleagues[786] between the estuarine angiosperm *Spartina alterniflora* and the ascosporogenous yeast *Pichia spartinae.* The yeast occurs in concentrations as high as 9×10^7 cells per gram in cell liquid within the culm and viable cell tissues. The yeast assimilates lipids of the host, plant hydrolysates, various amines, and both alpha and beta glucosides.

That yeasts are an important part of the population of sewage, sewage treatment systems, and polluted waters has been demonstrated by Cooke.[236,257,273] A modified indicated-number technique has been used for a semiquantitative estimation of the numbers of cells present. In this technique, 250-mℓ Ehrlenmeyer flasks are used. To each of a pair of flasks is added 25 mℓ YNB (yeast nitrogen base, Difco). To one flask is added 25 mℓ 2× 1% dextrose, to the other 25 mℓ 2× 20% dextrose. Each of the three solutions is prepared in sufficient quantity for use with the number of samples to be tested. These solutions are filter sterilized, not autoclaved, before being added to the smaller flasks. In blending the two solutions, the final product becomes 1 or 20% glucose — YNB solution. To each of one pair of flasks is added 1 mℓ of the original sample, or a 1:10 dilution of the sample. As many samples may be used as equipment is available to handle them and time is available for processing. The flasks are placed on a rotary shaker for 64 to 72 hr. Upon removal, the flasks are placed on the laboratory bench for about 4 hr or until the yeast cells which have developed settle to the bottom, leaving a clear solution or one clouded with bacteria. In either case, some filamentous fungal colonies will remain on the surface, float in the medium, or settle to the bottom. After settling, the flask may be tilted and a loop of yeast cells removed from the meniscus and streak smeared on two or three plates of yeast extract — malt extract — dextrose agar or Diamalt agar. After 2 or 3 days, the resulting growth should be restreaked on Diamalt agar plates. This process is repeated until discrete colonies can be recovered for identification. The growth in the flasks, as they are removed from the shaker, can be evaluated for visual turbidity. For more critical evaluation, an indicated-number technique can be used. In the form in which this technique was used during its development, an 80-place shaker was used, 40 places for 1% glucose-YNB, 40 for 20% glucose-YNB media. In all cases, 250-mℓ Ehrlenmeyer flasks were used. Eight samples were used, each in five dilutions approaching extinction: 1:10, 1:100, 1:1000, 1:10,000, and 1:100,000. The flasks were handled in the same way as for the simple dilutions, and isolations were made in the same way. Prior to settling, records of visual turbidity were taken. Until identifications have been completed, white yeast growth may be recorded as 1, 2 if more than one type can be differentiated, and, in addition, red yeasts can be recorded as 1. Thus, the first record of minimal number of cells in 1 mℓ of sample incubated at a dilution of 1:10 and for which a series of flasks approaching extinction was incubated could be 111:110 if only white yeasts were present. The values presented in Table 1 for three series of isolations were obtained as described above. Certain apparent anomalies may be noted. The same species of yeast may have developed in both 1 and 20% glucose, or the species may be different, so that the values obtained from one sample in both media can be neither added nor equated without results of final identification. Enough work has not yet been completed to explain the apparent discrepancy in which a value of 200, 010 is obtained from one sample in one medium, why no yeasts were obtained in three of the flasks, but why one obvious type in the lowest and two in the highest dilutions are presently considered as an accident of the occurrence of viable cells in the milliliter of sample with which the medium was inoculated.

TABLE 1

Minimal Number of Yeast Cells per Milliliter of Original Sample Based on Number of Species Isolated (Total Species = 29)

		Sampled habitat							
		1	2	3	4	5	6	7	8
Date	Medium	Raw sewage	Settled sewage	Filtered sewage	Final sewage	Settled sludge	Digested sludge	Drying sludge	Trickling filter slime
1962									
May	YNB-1% glucose	10,210	110,210	1,010	110	110,310	110,010	110	202,040
June	YNB-1% glucose	100,230	102,110	200,010	210,110	103,110	1,210	10,100	100,310
May	YNB-20% glucose	12,410	1,110	100	120	110,220	111,120	10	11,020
June	YNB-20% glucose	201,020	200,110	100,110	201,120	112,200	10,020	1,210	121,110
1963									
April	YNB-1% glucose	110,220	10,310	11,130	120	120	10	1,020	133,100
	YNB-20% glucose	101,220	1,320	1,520	200,220	1,010	210	110	132,320

Note: Test Conducted at Dayton, Ohio Sewage Treatment Plant.

From Cooke, W. B., *Mycologia*, 57, 702, 1965. With permission.

Chapter 9

POPULATION GROUPS — SOIL

I. INTRODUCTION

As a natural habitat for fungus populations, the soil has probably been studied more intensively than any other habitat. This is probably partially because of the importance of fungi as crop pathogens and partly because of their importance in the decomposition of plant and animal residues and, thus, as agents in the beneficial carbon and nitrogen cycles in the soil in relation to crop plant nutrition.

In a broad sense, the organic layers on the surface of mineral soil are included here, although the litter is treated as a separate chapter, as part of the soil complex. Such layers include litter, the fermentation layer, duff, and humus. Components of these layers include decomposing or decomposed leaves, twigs, branches, trunks of trees, stumps, dead and discarded parts of flowers and fruits, and other materials, including roots. Throughout the process of decomposition, fungi, alone or in cooperation or in antagonism with other organisms, are among the more active members of the population. In addition to fungi, of course, are bacteria, protozoans, numerous invertebrates, and some burrowing vertebrates. In ways to be mentioned below, the fungi may secrete antibiotics or toxins which may eliminate competitors and, in some instances, in turn be eliminated.

Depending on the parent material and the plant cover of a wild soil, the soil reaction may be acid or alkaline in forest, meadow, or prairie. This may be changed as the vegetation cover changes or as the soil is put to different crop uses when recovered from its primitive condition by man. When such changes occur, the fungal population tends to change. The fungal population may be different for a forest, a meadow, or a prairie,[215,354,437] and as the plant cover changes in any one location, the associated fungi will tend to change.[437]

Doeksen and van der Drift[326] edited the papers presented at a symposium based on soil organisms. These papers include much information on the occurrence and activities of soil fungi. Garrett[432] has shown how fungi are useful in the development of soil fertility. Gray and Williams[460] have given a detailed account of the biology of soil microorganisms. Panasenko[862] reviewed the ecology of microfungi considering temperature, humidity, light, radiation, pH, and aeration in relation to growth of fungi in general. He also described such substrate groups as fungi of grain and grain products, sugar and sugar products, meat, fish and eggs, butter and fats, and cellulose.

Observations of fungi in meadows and prairies have been largely confined to surface areas, although some depth studies have been made. The classic study of fairy ring fungi was made in the short grass prairies of eastern Colorado.[1027] It has been said in France[437] that long after the forest has been removed from an area which has become meadow, fungi usually associated with forest types continue to produce fruit bodies in the vicinity of places where trees grew with which they were presumed to be mycorrhizal. Studies in eastern Washington and adjacent Idaho[215] have shown that there is considerable difference between populations associated with grasslands and those associated with forests. Because of the economic importance in the life of crop plants of organisms which produce decay in mulches, fertilizers, crop residues, used trash, etc., the fungi of farm soils have received a great deal of attention. Ever since the question "Is there a fungus population in the soil"[1195] was answered in the affirmative, work has been directed toward answering questions concerning the members of that population, the activity of certain single members,[446,1194] and the activity of the population as a whole. The work of many students of this problem has been summarized by an outstanding soil microbiologist.[1196,1198] It has been shown by Harvey[508] that there is an exensive population of Oomycetes in the soils of the Chapel Hill, North Carolina area. Their contribution to the activity of the ecosystem has not been surveyed to a significant extent. It has been shown[405,446] that soil fungi cannot assimilate free nitrogen. Since 1950, directions of inquiry have changed toward discovering which nutrient portions the fungi were using, although how many and what kinds of fungi are present to use them is also considered important.

Gilman's manual of soil fungi[445] listed 756 species of fungi reported from soils up through 1956. These include members of the Mastigiomycotina, Zygomycotina, Ascomycotina, and Deuteromycotina. The species were isolated by techniques to be discussed in a later chapter. Currently, students of field soils and of wild soils are interested both in the activity of the fungus or microbiological population and in their specific and quantitative composition. Moreover, some attention is given to species which demonstrate a hazard to crop plants or which compose part of the populations of the rhizosphere and the mycorrhizae.

The activity of soil micropopulations was described by Garrett[433] as follows:

> Throughout much of the year over most of the earth's cultivated surface, the soil is both warm enough and moist enough for some microbial activity to go on. Sources of energy for this continual microbial activity are provided both by the dead remains of plants and animals and by excretion of organic nutrients by the root systems of living plants and the living bodies of animals. Thus in every soil at all times when temperature and moisture content permit, the soil microflora of fungi, bacteria, and actinomycetes is active in the decomposition of organic substrates, from which it derives energy and essential nutrients for the synthesis of microbial protoplasm.

Inoculum potential is a useful concept both in plant pathology[427,433] and in the study of saprobic fungi.[864] It is defined as the energy of growth per unit area of host or substrate surface of a parasite or saprobe available for infection of a host or substrate at the surface of the host organ or substrate to be infected.

Five methods of survival of a fungus from the exhaustion of one nutrient source to the arrival of a new nutrient source were listed by Garrett[427] and Park.[864] These include survival as competitive saprobes on dead organic substrates; saprobic survival on dead tissues of a host crop or on weeds infected during the parasitic phase; dormant survival as resting propagules, such as sexually produced oospores, ascospores, or basidiospores, asexually produced chlamydospores, and multicellular sclerotia; parasitic survival on living roots and other underground parts of weed hosts and "volunteer" susceptible crop plants; and parasitic survival on living root systems of plants that show no symptoms above ground. Substrates for soil fungi include living or dead, undecomposed or partially decomposed plant or animal tissues lying in or upon the soil and soluble products diffusing from them. In addition to the definitions given by Garrett, Park[864] gives several useful ones.

Competitive saprophytic ability, according to Garrett,[427] is "the summation of physiological characteristics that make for success in competitive colonization of substrates." It includes four general attributes:[427,864] rapid germination of spores and a high rate of hyphal growth, both favoring rapid colonization; good enzyme production, which favors rapid and extensive substrate utilization; production of substances toxic to other organisms, which may reduce competition for the available substrates; and tolerance of antibiotic substances produced by other organisms.

Substratum resistance is defined as the "resistance of a living or dead substratum to colonization." Substratum is used to designate the material medium in which the fungus occurs; substrate is used to designate the chemical substance acted upon by an enzyme. Antagonism is the factor that makes for the difference between saprophytic colonization in pure culture and saprophytic colonization in nature where other microorganisms are present and may be expressed in three ways; antibiosis, competition, and exploitation.

Saprophytic survival occurs specifically under conditions of antagonism. The term can be used for saprobic as well as parasitic fungi. Survival as inactive structures has been described as dormant survival. The term inactive survival can include both dormant survival — truly dormant in the physiological sense of being incapable of germination owing to immaturity or endogenous inhibition — and inhibited survival, that could immediately germinate given favorable conditions but that remain inhibited by environmental antagonism (antibiosis).

The half-life concept is used for viruses and plant parasites in an unfavorable environment which decline in numbers at a logarithmic rate, and in such a situation there can be no accurate statement of the direction of total longevity.

In relation to factors influencing fungal ecology, Park[864] noted that Garrett[432] stated: "In the ultimate analysis it is the physico-chemical characteristics of a habitat that determine the kind of community that will occupy it. But the species composition of that community will be the result of competition between different species of organisms." Colonization is affected by available nutrients, temperature, water, pH, the overall physicochemical nature of a soil, the biotic environment, etc.

Two types of succession may be mentioned: substratum succession, in which, when a potential substratum becomes available for colonization usually a succession of organisms colonizes it; and seral succession, in which a whole habitat starting from parent rock material passes through a series of successions before arriving at a more or less mature stage. For Garrett, a fungal succession reaches a climax with the elimination of the fungal habitat by those fungi consuming the substrate as food.

Park[864] stated that "autecology and synecology cannot at this level be maintained as separate departments of knowledge because both are elements in the relationship of fungi with their full environment." This statement can be disputed or used as an argument for the maintenance and further development of mycosynecology.

Griffin[482] has presented a thorough analysis of the relation of soil, water, oxygen, and carbon dioxide to fungus growth.

In the study of soil organisms, it must be realized that those fungi which are involved are primarily making use of the organic matter present which is available to their enzyme systems. Secondarily, the nature of the parent material of the soil can affect some species and apparently has no effect on others. The organic matter acted upon by the fungi may be present as humic acids and unrecognizable plant and animal remains incorporated in the inorganic soil by percolation, earthworms, or rodent activity. Immediately adjacent to the layers of soil there are inorganic components and layers of litter in various stages of decay from freshly fallen leaves, twigs, cones, and parts of flowers and fruits to such structures being actively decomposed eventually leading to humic acids and humus. In organic soils where these materials remain on the surface of the inorganic soils or include only a very small percentage of inorganic matter derived from sources other than parent materials, these same materials are present, but their extent of decomposition is usually less extensive because, in part at least, the soil becomes more readily water-logged, aerobic decomposition processes are less rapid under such conditions, and those decomposer organisms which may be active have a somewhat lower range of enzymatic processes on which to rely.

As Warcup[1208] has noted, while it is possible to sketch outlines, there are major gaps in our knowledge of the occurrence and growth of fungi in the soil. New techniques will undoubtedly aid acquisition of fresh data, but there are many points of information which would greatly aid our conception of fungus biology in soil which can be developed by present techniques. The species list should not be the sufficient reward for study of fungi in soils since there are many other aspects of their occurrence and activity waiting to be investigated.

II. DECOMPOSITION OF ORGANIC MATTER IN SOILS

In the study of the decomposition of organic matter in the soil, Mayaudon and Simonart[768,769] used radioactive carbon to determine decay rates of proteins, hemicellulose, cellulose, and lignin. In the same period of time, 80% of cellulose was decomposed while only 30% of lignin was reduced. At the same time, 5.9% of the radioactivity from cellulose and 34.2% of that from lignin was recovered in humic acids.

III. SAND DUNE SOILS

Occasionally opportunity is afforded to develop a study on the fungus populations of sand dunes. In France[811] and elsewhere, reports have been published listing species from such habitats but not always correlating them with populations observed on more stable soils.

In sand dunes studied by Webley and co-workers[1212] in England, samples from the main seral stages in two dune systems, taken along transects passing inland from bare sand above the high water mark, showed a marked increase of fungal populations with the start of plant colonization. Numbers rose steadily with the development of the vegetation and increasing complexity of the plant community.

Pugh[929] in England found much larger numbers of fungi in plant colonized dune areas than in nearby salt marsh soils. The differences were quantitative more than qualitative and were probably related to growth conditions in the two habitats.

The distribution of fungi in two adjacent communities of plants representing two stages in the development of populations on sand dunes in Indiana was studied by Wohlrab and colleagues.[1273] A relatively rich fungal biota was found, containing few species apparently confined to the dune habitat. Significant differences in the composition of the mycobiota were found to exist in the two communities suggesting a succession of species comparable to that of the higher plants. The activity of those species isolated from a single sample is not understood. Further studies by Wohlrab and Tuveson[1272] confirmed the earlier study and extended it to other dune areas in Indiana and Michigan along the southern end of Lake Michigan. Five fungal species were found to occur in all communities of *Andropogon scoparius* var. *septentrionalis* studied, but in *Ammophila breviligula* communities no such characteristic fungal populations were found. Monthly samples indicated no seasonal variations in the numbers and composition of the mycobiota. It was concluded that "the presence of a characteristic fungal flora in the *Andropogon* communities and its absence in the *Ammophila* communities indicates a successional trend from a harsh, unstable environment, low in available mineral and organic nutrients, to one more stable and suitable for fungal growth."[1272]

A pine (*Pinus sylvestris*) plantation on a fixed inland sand dune in Tennessee was studied by Witkamp[1269] with emphasis on the fungal populations of the soil. Fine sand on coarse sand varied from 0 cm in thickness at the base to 150 cm at the top. Dependence of litter and humus formation on primary production and of mycelial growth on litter and humus resulted in significant correlation of these factors and moisture. Mycelium length was significantly correlated with fungal plate counts in January. The annual cycle of mycelial length did not show a peak in autumn as was found in broad-

TABLE 1

Fungal Activity Measured by Three Different Methods[160,880]

Horizon of layer	Percent organic matter	μℓ O_2/5 hr/g organic matter O_2 uptake	Dilution plate count per gram dry weight	Meters of hyphae/mℓ soil sections
L	98.47	2406.0		
F_1	98.06	1428.0		
F_2	89.30	274.6		
H	54.58	148.3	281,900	5.56
A_1	17.16	77.7	199,800	3.96
A_2	1.87	238.8	27,250	3.78
B_1	10.62	91.9	11,370	1.09
B_2	5.24	56.6	1,970	0.377
C	1.42	96.3	225	0.03

leaved forest, presumably as a result of differences in nutrient release from the litter. Reasons for the low (0.6 to 0.9%) fungal mass in the soil may be its ephemeral character, low influx of organic matter into the soil, and the presence of much mycelium on top of the soil which was not measured.

IV. SOIL PROFILE STUDIES

Using the distinct layers of a podzol profile on glacial sand under a forest of *Pinus sylvestris*,[160,880] fungal growth and activity was studied in the several layers. The value of using different techniques such as washing, soil sections, and oxygen uptake measurements was emphasized (Table 1).

Williams[1255] extended this type of study to a consideration of the vertical distribution of fungi on such substrates as dead roots, humus particles, and mineral grains.

V. MULL AND MOR SOILS

Witkamp[1265] compared the populations, especially of fungi, of mor (a thick layer of duff on raw humus on acid mineral soil) and mull (a thin layer of humus over a calcareous soil) soils. Freshly shed litter had a restricted population, quantitatively and qualitatively, of fungi which increased through the second year before starting to decline. The fungal populations of mull soils were lower than those of mor soils, probably because of water relations and the calcium content of the soils. Populations of acid mull soils were intermediate between those of mor and mull. The composition and activity of the population of the mineral soil were directly affected by the litter, the desiccation and pH of the soil, and the activity of the saprophagous soil fauna and the nonfungal microbiota. Fluctuations in the numbers of microorganisms and mycelial growth in the mineral soil, similar in the various forest floor types, were caused by fluctuations in temperature and moisture and by the addition of fresh litter. The concentration of mycelium in an oak forest was highest in fall and winter, lowest in spring and summer. This annual fluctuation was not found in a pine stand. The rate of breakdown of mycelium was almost equal in calcareous mull and mor although in the calcareous mull there were four to ten times as many chitin decomposing and mycolytic organisms as in the mor. In the mor, mycelium feeding oribatid mites prevailed; their individual consumptive capacity was three times as high in summer as in winter.

The mycelium concentration in the mineral soil of one forest floor type appeared to be positively correlated with humus and moisture content and negatively with depth. Contrasting percentage occurrence on soil suspension slides with that on Cholodny slides, the predominance of thick and brown hyphae on the former and their low percentage on the latter, indicates slower decomposition and growth of this mycelium than of slender white mycelium. Macrofungi appeared in highest numbers about 2.5 months before the maximum mycelium development. Their numbers and masses were not correlated with the mycelium concentrations in the various floor types. Their occurrence was partially correlated with moisture content of the substrate, since drought reduced the number of species of mushrooms fruiting. There were characteristic macrofungi for each forest floor type. In the calcareous mull these were mostly humus and litter fungi. Mycorrhizal fungi dominated in the acid mull and mor. Litter fungi appeared in all types. Concentrations of mycelium, formed under natural conditions, were able to inhibit the development of microorganisms.

VI. FOREST SOILS

Sample populations of the soil microfungi in five southern Wisconsin maple-elm-ash floodplain areas were isolated using dilution plates, and lists of the species present and their frequencies of occurrence were compiled by Christensen and co-workers.[196] The average number of microfungal species per community was greater here than in any vegetational unit previously surveyed in Wisconsin. A comparatively high proportion of the species encountered were of *Penicillium*; that genus accounted for 43 species represented by 493 isolates in a total of 199 species represented by 1082 isolates. There were qualitative and quantitative differences in species composition among the stands, particularly between the two stands representing the extremes in vegetational composition. A linear ordination, based on species composition in the five sample populations, resulted in an ordering of the stands in the exact sequence dictated when the stands were ordinated on the basis of dominant tree species.

The soil microfungi in five northern Wisconsin open bogs and ten conifer swamps were surveyed by Christensen and Wittingham.[197] As many as 200 isolates from eight sites in each community were identified or characterized, and frequencies for all entities were calculated. The number of entities per community ranged from 10 to 68, and the numbers of entities per unit, representing 1000 isolates, were 108 in open bogs, 104 in *Picea-Larix* swamps, and 181 in *Thuja-Abies* swamps. Only five to nine species in each unit had average frequencies of 30% or higher. Approximately one half of the identified species appear to be absent or rare in mineral soils, but several are well-known wood, wood-pulp, or litter forms. Species composition in the microfungal communities is correlated with species composition and maturity in the overlying higher plant community.

The species composition of soil microfungal populations[1240] in adjacent stands of red alder (*Alnus rubra*), conifers (*Pseudotsuga menziesii, Tsuga heterophylla*, and *Picea sitchensis*), and mixed alder-conifer correlated strongly with the dominant vascular vegetation. A total of 92 species was isolated: 55 from the alder stands, 45 from the conifers, and 46 from the mixed alder-conifer, with few species (16, 7, and 5 in the three plots, respectively) reaching average frequencies of 50% or higher. *Penicillium deleae*, which occurred with a frequency of 83% in alder soil, appeared to be a rare fungus elsewhere. There was little difference in species composition among soil horizons within a stand.

The populations of higher fungi of two oak woods on slate soils and two mixed woodlands on limestone soils were studied in England by surveying 12 permanent

TABLE 2

Estimates of Weight Production of Higher Fungi in Forest Stands

Reporter	Forest	Number of visits annually	Production (kg/ha)
Bohus and Babos (1960)	Oak and Beech, Hungary	4—5	7—160
Hofler (1937)	Beech, near Vienna	1	18—170
Rautavaara (1947)	Conifers, Finland	[a]	32—302
Cooke (1955)	Conifers, Washington and Idaho	3	150—270[b]
Hering (1966)	Oak on slate soil, Lake District	8	12.5—95
Hering (1966)	Mixed on limestone soil, Lake District	8	3—37

[a] Estimate for whole year.
[b] Estimated from Cooke's volume results.

Compiled by Hering, T. F., *Trans. Br. Mycol. Soc.*, 49, 375, 1966. With permission.

quadrats of 100 m^2 eight times each autumn for 3 years.[535] Lists of 40 to 50 species were obtained for each woodland, and the biotas of the two soil types were clearly demarcated. Total numbers of fruit bodies in all four woodlands were similar, but the slate woodlands had more abundant mycorrhizal species and a fresh weight production of 95 kg/ha against a maximum of 37 kg/ha in the limestone woodlands. A comparison is given of estimated production of fruit bodies of fleshy fungi reported in the synecological literature in Table 2.

VII. COASTAL GRASSLAND SOILS

Apinis and his students have made extensive studies of fungi in coastal soils in England with special reference to thermotolerant (thermophilous) species, biocoenoses with which they were associated, and mycocoenoses in which they were associated. As many as 19 species of thermophiles[36] were found to be associated with grass litter, humus, and the A horizon of the soil where five species of aging and dead grasses occurred. Seasonal variation, occurrence in various layers of particular grasslands, their relation to the mesophilous mycobiota, and their pattern of distribution in various grasslands were considered. Apinis[37] reported on studies yielding 92 species of Phycomycetes from 23 soils in five plant communities. Species in populations of these fungi may be restricted to certain locations or habitats or absent from others. Differing water relations of alluvial soils appeared to be a major factor governing the type of soil formation, the patterns of higher plant communities, and those of phycomycete populations.

In grassland soils various occasional groups of fungi produce distinct mycocoenoses dependent upon the type of vegetation, climate, and factors of the soil environment.[36,38] The continuous presence of these organisms depends on a continuous supply of organic energy sources which were produced by the roots of the vascular plants in the biocoenosis. In areas with an oceanic climate, such as western Europe, root production in a permanent grassland may reach 6000 kg/ha, but in semiarid countries in eastern Europe root yields are reported as high as 20,000 to 25,000 kg/ha annually. In a balanced environment, the amount of organic matter produced annually is more or less the same as the annual rate of decomposition by soil microorganisms, including the fungi. Various fungi of these grassland soils possess a wide ecological diversity in decomposition of organic matter and display favorable (symbiosis) or unfavorable (parasitism) relationships within the grassland symbiosis, as well as influencing in one way or another higher plants and other soil organisms by their biosynthetic activity. The rhizosphere and the root surface mycobiota of the grassland vegetation appear as an initial phase of the colonization by fungi which increases on the aging roots, indicating biological crumb formation in the turf layer of the top soil. The population of soil fungi plays a distinct part in the formation of this crumb structure from the outset, through the optimum, to the decline and disintegration of this important microniche in the soil.

Apinis[39] indicated that there are three principal mycocoenotic structural characteristics in five coastal grasslands studied in England: (1) the populations, or synusiae, of fungi on the surface (phyllosphere) of plants including the various parasitic and saprobic fungi colonizing upright, senescent, and dead parts of the plants; (2) the fungi of the litter and fermentation layers including those on seeds (spermatosphere), fruits (carposphere), and rhizomes, as well as those on culm and stem bases; and (3) the saprobic microfungi of the various soil horizons, including the parasitic, rhizospheric, root surface, and mycorrhizal fungi, as well as synusiae of the larger fungi producing their fleshy sporocarps above ground.

About 70 species of thermophilous fungi,[40] including thermophilic, psychrotolerant thermophiles, microthermophilic, mesophilic, and transitional mesophiles distributed in 40 genera, have been recorded from various grasslands in Great Britain. In topsoil thermophilous fungi are confined to large crumbs, roots, and plant debris. The frequency of occurrence of thermophilous fungi is higher in the litter layer and in senescent and dead standing grasses than in the top soil. At present, there is no conclusive evidence that thermophilous fungi constitute a significant part of the rhizosphere population. Upland barley and wheat fields possessed a considerably higher number of thermophilous fungi than alluvial and coastal grasslands. Presence of thermophilous fungi in grassland vegetation and soil including fields of cereal crops involves two general biocoenotic aspects: (1) the health hazards of man and animals because a number of thermophilous fungi are pathogens and (2) storage problems of agricultural produce, such as hay and grain in which thermophilous microorganisms initiate hot spots and general self-heating which frequently leads to combustion.

VIII. AGRICULTURAL SOILS

Under semiarid conditions in the central coastal plain of Israel, a comparative study of the soil mycobiota was carried out by Joffe[610] on the plots of a fertilizer trial that had been running continuously for 40 years. Various manure and fertilizer treatments were applied to a heavy soil in an unirrigated 5-year rotation of maize, berseem clover, wheat, purple vetch, and oats. In plots supplied annually with NPK fertilizers, or once in five years with cow manure, the number of fungal isolates greatly exceeded that on unfertilized control plots. However, while the plots under NPK treatment outyielded the manured plots in four out of five crops, fungal isolates tended to be more numerous in the manured plots. No relation was found to exist between fertility level and number of fungal species found under the various treatments.

IX. NORTHERN, ARCTIC, AND ANTARTIC SOILS

A study of forest soils collected from latitudes between Hungary and Finland[378] was extensive, covering all members of the soil population and comparing soils of forest types of numerous latitudes across Europe. In this study it was found that, as one proceeds northward, the ratio of fungi to bacteria in soils increased, there being relatively fewer fungi than bacteria in more southern latitudes and more fungi than bacteria in more northern latitudes.

A series of litter and soil samples from locations adjacent to Mendenhall Glacier near Juneau, Alaska and the Muir Glacier, Glacier Bay National Monument, Alaska[266] was processed using a standard plating technique for quantitative and qualitative evaluation of soil fungal populations. As the natural plant communities matured and as the nitrogen content of the soil increased, the total fungal populations increased both in numbers of colonies and of species (Figure 1). As the vascular plant populations matured, more and more fungi were found in deeper soil layers. The soil fungus populations in the two areas were relatively homogeneous. No single species was characteristic of any horizon or stage of development of the forests. Certain species appeared only in pioneer communities, others in transitional communities, and others in mature forests; some were present only in mineral soil, others in soils with organic matter, and others in the litter.

Records have been kept in relation to the year by year retreat of the glaciers and the advance of the adjacent plant communities. Thus, samples for the above and the following studies were taken from dated sites in the seral succession from bare ice to mature forest.

Baxter and Middleton,[88] in a study of the soils in the same area, introduced the term "geofungi" for soil fungus populations including species of *Pythium*. Fungi obtained were isolated from roots and rhizospheres of trees and shrubs. Fewer fungi appeared in colder gravels on roots of willows, alders, and poplars than in forests on older soils on roots of hemlock and spruce (Figure 2). *Pythium* species were found on roots of all arborescent trees, regardless of whether these were in gravels poor in organic matter or organic soils in peety muskegs where vegetative deposition exceeded decomposition.

Holding and colleagues[558] edited the results of a symposium dealing with soil orga-

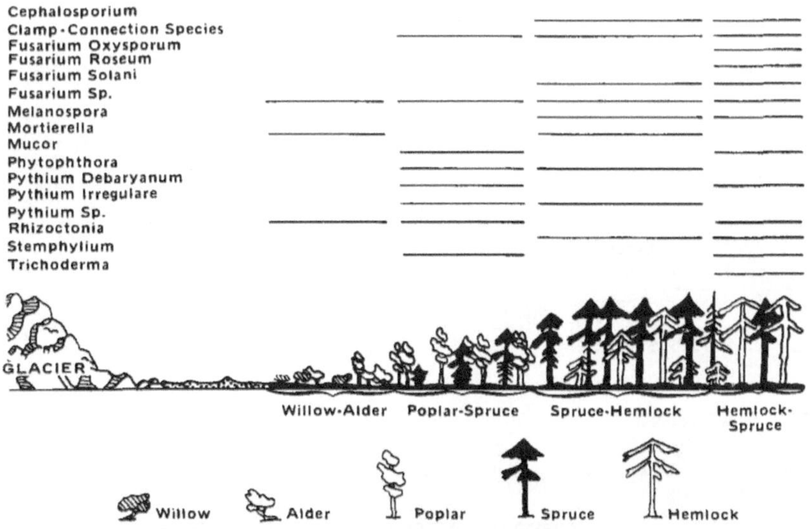

FIGURE 1. Glacier Bay, Alaska. Average numbers of colonies of fungi per gram dry weight of sample developed on five agar plates and corrected to a dilution of 1:100. Zero line separates mineral soil (below) from overlying organic layers which have accumulated largely through deposition of plant parts. Collection stations arranged from left to right in order of increasing surface age following recession of glacier ice. (From *J. Ecol.*, 47, 538, 1959. With permission.)

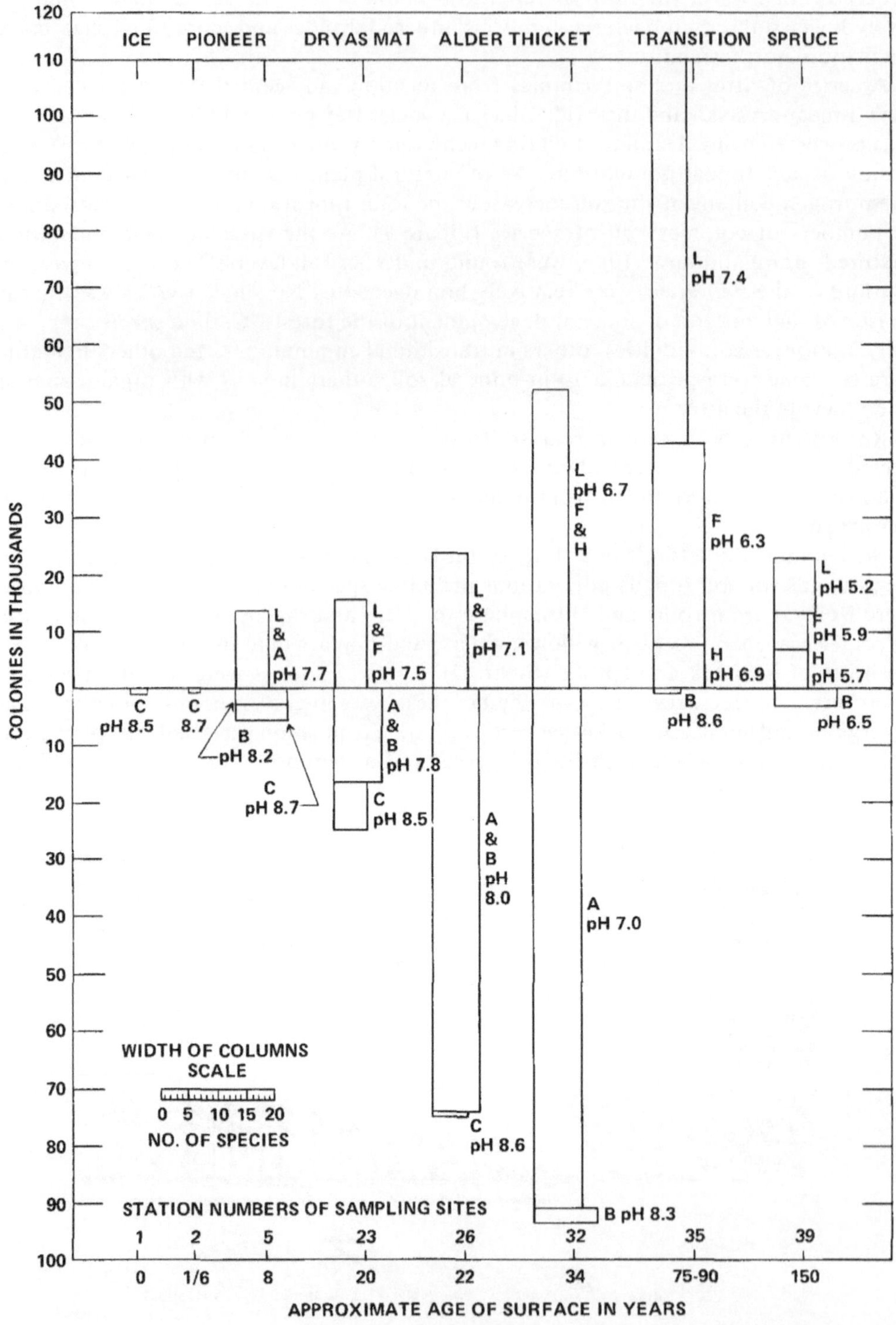

FIGURE 2. Diagrammatic representation of geofungi isolated from roots and rhizosphere in plant communities that have developed following ice retreat. (From Baxter, D. V. and Middleton, J. T., in *Recent Advances in Botany*, University of Toronto Press, Toronto, 1961, 1516. With permission.)

TABLE 3

Numbers of Microorganisms per Gram of Soil from Wheeler Dry Valley, Antarctica

		Fungi		
			Yeasts[b]	
Soil sample	Sample depth	Molds[a] + 20°C	+ 2°C	+ 20°C
609	Surface	20	0	0
	2 cm			
615	Surface	200	2	0
	2 cm			
616	2—10 cm	25	12	0
617	Surface	35	0	0
	2 cm			
618	2—10 cm	15	0	0
619	Surface	50	0	0
	2 cm			
620	2—10 cm	200	0	0
621	Surface	0	4	0
	2 cm			

[a] Rose Bengal agar.
[b] Neopeptone-dextrose agar, pH 4.5

From Cameron, R. E., King, J., and David, C. N., in *Antarctic Ecology*, Holdgate, M. W., Ed., Academic Press, London, 1970, 702. With permission.

nisms and decomposition in the tundra. In relation to fungi, some relationships between fungi and their environment in tundra regions were considered, as well as physiological groups of decomposer fungi on tundra plant remains.

Kjøller and Ødum[647] studied soil samples from a placer mine near Fairbanks, Alaska. From top soil they isolated 1000 colonies of fungi per gram of soil at 24°C. Four samples of permafrost beneath the topsoil yielded no organisms, but two samples yielded several types of bacteria on Sabouraud medium.

In studies with an acid peat soil from Signey Island, Antarctica, Baker[66] found that yeast numbers, like those of filamentous fungi, decreased rapidly going down through the profile. From the uppermost layer, 10^6 cells of yeasts per gram dry weight were isolated. At the 11- to 12-cm level there were approximately 5×10^4 yeast cells per gram dry weight. The pH levels were 4.1 and 4.4, respectively. There was no apparent seasonal variation.

Using standard and modified techniques, Benoit and Hall[96] processed 400 soil samples from the McMurdo Dry Valley, Victoria Land, Antarctica. Some soils which appeared sterile in one technique yielded interesting results in another. Filamentous fungi were detected infrequently, but in the vicinity of mosses in some areas some psychrophiles were found. The maximum temperature of growth of these fungi was 20°C; the optimum temperature was near 18°C.

In a study of soil samples from Wheeler Dry Valley, Victoria Land, Antarctica, certain Ascomycetes, including species of *Penicillium* and yeasts such as *Candida* spp. were isolated by Cameron and co-workers[170] (Table 3).

X. SOIL-BORNE PLANT PATHOGENS

Proceedings of a symposium on the ecology of soil-borne plant pathogens, with emphasis on biological control, were edited by Baker and Snyder.[67] The 40 papers covered a wide range of ecologically related subjects. Particularly related to fungi, growth and reproduction, dispersal, survival, and dormancy in the soil are considered, as well as the physical and chemical factors of the soil as they relate to the microbial populations. The anatomy and physiology of plant roots are related to the biochemistry of root exudates and its effect on the organisms in the rhizosphere. Pathogenesis by soil fungi and resistance to them by plants are important characteristics of this environment. In relation to antagonism, factors of importance are competition, antibiosis, fungistasis, parasitism, and predation. All of these factors are involved in the development of biological control technques.

XI. PLANT RESIDUE DECOMPOSITION

In a series of experiments on the biological consequences of plant residue decomposition in the soil, Snyder and colleagues[1085] found that substances possessing phytotoxic properties may be formed during decomposition of a variety of plant residues under laboratory and field conditions. The degree of phytotoxicity depended on the length of the decomposition period: little or none was present soon after the crops were turned under; levels gradually increased until 15 to 25 days decomposition, followed by a decrease until, after 30 to 45 days, little phytotoxicity remained. Green plant residues in early stages of decomposition were more toxic than mature crop residues. Temperature and moisture were the major factors determining phytotoxicity: cold, wet soils were more favorable; warm, well-aerated soils produced little, if any, phytotoxicity. Injurious effects included delay or complete inhibition of seed germination, stunted overall plant growth, injury to the root system, deranged nutrient absorption, chlorosis, wilting, and killing of plants. Roots of plants appeared to be highly sensitive. In both field and laboratory experiments, phytotoxins of significant potency were detectable only if the decomposing plant residues had been freed of most of the adhering soil prior to extraction with water. Two major components were noted in the phytotoxic extracts, a water-soluble, ether-insoluble fraction and a water-soluble, ether-soluble fraction, the latter containing the phytotoxin proper. On the basis of bioassays, paper chromatography, and gas chromatography, the more important phytotoxins were identified as benzoic acid, phenylacetic acid, 3-phenylpropionic acid, and 4-phenylbutyric acid, many of which had not previously been reported as crop residue decomposition products. Successions of microorganisms involved in decomposition of green barley and phytotoxin production revealed a marked enrichment of bacteria and the dominance of *Geotrichum* and yeasts in laboratory experiments. Yeast isolates increased the concentrations of benzoic acid and phenylacetic acid, while *Clostridium* spp. and *Geotrichum* sp. increased the concentration of hydrocinnamin acid and phenylbutyric acids. A wide range of microorganisms yielded higher amounts of phenyl acids than when only one organism was used. Root disease of bean caused by *Fusarium solani* f. sp. *phaseoli, Thielaviopsis basicola,* and *Rhizoctonia solani* appeared to be increased in the presence of phytotoxic substances. This occurred even when relatively small amounts of these substances were added in the presence of these pathogens. The phytotoxins lowered the disease resistance of otherwise resistant varieties. The important mechanisms operating in the host-pathogen interaction complex included direct injury to host cells, increased exudation from host cells, and stimulation of germination of resting propagules of the pathogen.

XII. SOIL PEZIZALES

Analyses for pH, conductivity, and organic matter content were made on soil samples collected from habitats in which ten species of Pezizales were found by Petersen[896] in Denmark. The observations were compared with results of similar investigations of habitats of Basidiomycetes. The Pezizales (Ascomycetes) differ from the majority of the agarics, which prefer acid substrates, in growing in habitats where the pH is a little below the neutral point. Concerning pH, conductivity, and organic matter content of the soil, the Pezizales habitats resemble those of a group of Gasteromycetes from the open country, although the Pezizales in Denmark mostly grow in open forests.

XIII. PESTICIDES IN SOILS

While not indicating specific agents of degradation, Norris[837] indicated that 2,4-D, 2,4,5-T, amitrole, and picloram were degraded in forest litter but at markedly different rates. When 2,4-D and DDT are applied together, the degradation rate of 2,4-D was stimulated. Although 2,4,5-T is somewhat more persistent than 2,4-D, it approaches 90% loss in four months. The degradation of 2,4,5-T is not influenced greatly by the concurrent application of 2,4-D. Amitrole has an initial rate of degradation exceeding that of 2,4-D and is rapidly lost in the forest floor, but this degradation may not be completely biological. Picloram is considerably more resistant to degradation than the other herbicides studied, but it is considered to be biodegradable.

One month after aerial application of DDT[1135] at 12 oz/acre, 3 oz/acre residues were detected in the forest floor in eastern Oregon forests; three years later the DDT content had decreased by more than 50% and had not leached into the surface mineral soil. At the time of spraying, water from two streams draining the sprayed area had a total DDT content of about 0.3 ppb. This low concentration decreased rapidly to levels below the limits of analytical detection. No effect of the spraying was noted on the soil microbial populations, nitrification rate, or amount of nitrate nitrogen in the soil. Of the 12 oz DDT applied per acre, about 26% reached the ground surface initially,and over 36 months, about 6% more was brought to the ground in litterfall. Thus, approximately one third of the sprayed chemical reached the forest floor.

XIV. USEFUL MATERIALS AND INHIBITORS

A. Cellophane

Tribe[1157] described results of a study in which cellophane strips were laid on or buried in soil. Using direct microscopic observation and direct isolation techniques, it was possible to follow the succession of microorganisms which occurred on the cellophane. After establishing a normal course of development of such populations in a chosen soil, deviations in other soils could be described. The use of thin sections of other materials such as wood shavings could stimulate the development of other communities from the soil population. By the variation of substrates, soils, and conditions of soil treatment, it was suggested that it should be possible to accumulate information on the interrelations of a number of microorganisms.

Strips of cellophane were laid[1158] on sand saturated with a mineral salts solution and inoculated with fungi which could then be studied singly or in pairs. Two species of *Pythium* which were noncellulolytic and six species of Fungi Imperfecti with varying degrees of cellulolytic activity were studied. Individual competitive characteristics led to dominance of one fungus over another in any given pair, and cellulolytic activity was considered to be secondary to this. The species of *Pythium* used may be considered

TABLE 4

Microbial Populations in Soil Uninoculated, Incubated, and Incubated with Tannin Treatment

	Fungi		Bacteria	
Treatment and days inoculated	Total (thousands/g)	*Penicillium* (%)	Total (millions/g)	*Streptomyces* (%)
Preacher Loam				
Soil alone (0)	3	68	0.05	34
Soil alone (180)	43	94	2.00	0
Soil ÷ tannin (180)	1073	98	4.80	2
Astoria Clay Loam				
Soil alone (0)	195	39	9.30	45
Soil alone (180)	240	23	20.90	69
Soil ÷ tannin (180)	2472	93	42.20	92

From Bollen, W. C. and Lu, K. C., *U.S. For. Serv. Res. Pap. PNW*, 85, 7, 1969.

as sugar fungi, or as parasitic on, or commensal with, cellulose-decomposing fungi. The species of *Pythium* used in this experiment may also be considered as parasitic sugar fungi.

Cellophane strips were used by Gams[418] in a study of cellulolytic soil fungi in Germany. "Rooting branches" of the fungi on the cellophane were found to parallel the cellulose fibers forming the film and were used to facilitate the isolation of the fungi from wooden flower boxes and from forest soil, although many of these isolates remained sterile. When media were enriched with carboxy-methyl-cellulose (CMC), additional fungi were recovered, and some of the nonsporulating mycelia sporulated. It is suggested that in each case a different cellulose-splitting enzyme is used by the fungus, and that in the case of the cellophane film, only one of these enzymes is produced.

In a series of isolations from forest soils of different types, from washed roots, soil particles, and cellophane strips, sterile dark mycelia of the *Mycelium radicis atrovirens*-type were commonly isolated.[420] A few strains of such fungi developed conidia of *Phialophora dimorphospora* Kendrick.

B. Chitin

Microorganisms involved in the breakdown of chitin in an acid soil were recovered by Gray and Bell[459] using a modification of the Tribe cellophane strip technique. Four common soil fungi were among those fungi isolated, indicating a wide distribution of chitinolytic enzymes.

C. Tannin

Bark residues form an important aftermath of logging.[116] Tannins purified from Douglas fir (*Pseudotsuga menziesii*) bark were tested in soils in the laboratory to determine decomposibility and possible toxicity of decomposition products. Amount of tannin present in a 2-in. bark mulch was used in widely different soils (Table 4). About 22% of the tannin was decomposed in 180 days. Soil microbiota was generally increased. Nitrate production was slightly decreased in the presence of the tannins. Addition of the tannins produced no apparent toxic effects. The moderate decrease in nitrification bears consideration in agricultural use and could be useful in watershed management by decreasing the hazard of nitrates in water supplies fed by forest streams.

D. Copper

From ten soil profiles in a copper-rich swamp north of Sackville, New Brunswick, 35 samples were studied by Kendrick[635] for fungal populations. The copper content of the samples ranged from less than 1 to 68,000 ppm. Fungus colony counts ranged from 0 to 8000 per gram of sample. Of the 31 species found, 13 were found only in samples containing more than 7500 ppm copper. The commonest of these was *Penicillium ochrochloron,* a fungus previously reported to be copper tolerant. Of the remaining species, none were found only in samples with less than 5 ppm copper, and the others were apparently unaffected by high or low copper contents of the samples.

E. Serpentine Soils

A collection of 279 species of fungi obtained from serpentine and nonserpentine soils in the autumn of 1963 and the spring of 1964 on sites in Kittitas County, Washington was discussed by Maas and Stuntz.[728] They made 14 trips to the plots which were 150 to 200 × 50 to 100 ft in extent. Of the 279 species of macromycetes collected, 67 were either lignicolous or parasitic and were not related to soil types. Of the remaining 212 species, 19% were found only in serpentine soil areas, 63% were found only in nonserpentine soil areas, and 18% were common to both. A higher proportion of species were mycorrhizal symbionts in serpentine than in nonserpentine areas.

XV. THE ECOPEDON

Szabo[1128] has used the term ecopedon for an arbitrarily chosen unit of soil in a soil ecosystem consisting of the upper part of unconsolidated parent material together with the inhabitants corresponding to the minimal area of activity of the most important, locally acting, biological and abiotic soil-forming factors and processes. The boundaries of ecopedons are gradual rather than abrupt. A plant association or community may occupy a number of these ecopedons, which have also been called polypedons. They are made up of a number of pedons which vary from 1 to 10 m in diameter, may be elliptical or hexagonal according to different usages, and may be as deep as the root systems penetrating the soil. Szabo has worked with a number of ecopedons and the organisms forming the ecosystems of which they are part, but except for summarizing tables and charts, the work is largely based on bacterial relationships.

XVI. A DECOMPOSITION MODEL

Without regard to kinds of organisms involved, a model has been developed by Hunt[582] to simulate the dynamics of decomposition and substrates in grasslands. Humic materials, faeces, and dead plant and animal remains were considered as substrates. The latter three materials were further subdivided into rapidly and slowly decomposing fractions. The proportion of rapidly decomposing material in a substrate was predicted from its initial nitrogen content. Temperature and soil water being the more important factors in the belowground region, it is divided into layers based on their occurrence. Decomposition rates were predicted on the basis of temperature, water content, and inorganic nitrogen content. The predictions of the model were reported to compare favorably with data on CO_2 evolution and to litter bag experiments, but not to ATP estimates of active microbial biomass. A profound influence of depth on decomposition rates and on decomposer biomass dynamics, growth yield, and secondary productivity, was indicated by the model.

XVII. SOME GENERAL CONSIDERATIONS

It may be indicated that, at some point in time, with the surveys of speciation of soil fungus populations the law of diminishing returns becomes active, and fewer and fewer species are added to the list of fungi occurring in the soil. From isolates on hand in culture collections, studies in progress, and work such as that of Domsch and Gams[328] as translated by Hudson,[569] occasional novelties may be added to the list. While the list that Gilman[445] developd is not complete and may be added to from lists pulished in the last 20 years and from studies in progress (and not yet conceived), fewer and fewer of the species appearing on such lists will be found to be new to the master list. At various points in the study of fungi associated with decaying organic matter, attempts have been made to delimit habitats. Litter is separated from soil, the phyllosphere is separated from litter, decaying wood including twigs and branches is separated from litter, goods manufactured from natural fibers, wood, etc. are separated from litter, and other habitats are segregated. From the ubiquity of many of the species involved, it is obvious that the fungus does not respect any restriction but to organic matter. While there is a fungus population of the soil, the fungi which compose it are not necessarily restricted to the soil itself as the only habitat in which to find organic matter. In the comparison of section headings in chapters or the chapter headings in texts dealing with microorganisms, especially fungi, there will be noted a high degree of parallelism. Thus the chapter by Warcup,[1208] the book by Gray and Williams,[460] by Brock,[141] and others cover the same territory, although each uses a somewhat different approach and a different technique of presentation.

Chapter 10

POPULATION GROUPS — LITTER

I. INTRODUCTION

As a specific habitat for a specific association of fungi, the litter is becoming important in ecological and other studies.[318] The ability of certain known and unknown organisms to decay portions of the litter has been studied,[138] and the ability of certain species of litter fungi to use lignin and cellulose has been considered.[701] An extension of plant pathology and of forest pathology into this field has been made by the study of the life cycle of pleomorphic fungi which attack the leaves of living plants. During the growing season, most species of woody and herbaceous plants are subject to attack by fungi causing leaf diseases. Usually the asexual state of the organism fruits in the leaf during this time, but after the leaf falls to the ground and overwinters, one or more additional asexual states in addition to the sexual state may be produced, and the spores resulting from these fruiting types can reinfect the young leaves in the spring. Some of these apparent leaf disease fungi may be associated only with senescent leaves, and these are involved in a similar cycle.

In addition to the populations of larger fungi on leaves and on the forest floor or in a similar habitats on the ground in meadows and prairies, there is a large population of fungi which grows on the totally unrecognizable remains of plants in the fermentation, duff, and humus layers under the litter. The populations of these fungi which produce mushroom-like fruit bodies have been studied by techniques described earlier.[113,215,554] In addition to these fungi, there are small Discomycetes, as well as other fungi, mostly, as far as is known, of the mold type. Until recently,[318] little interest has been taken in populations of these fungi, although work has been reported in attempts to correlate microfungi with forest types[1156] and in preparing lists of fungi from as yet little studied areas.[374] The habitats of these fungi extend to the deeper mineral soil into which humus and humic acids have been leached or originate as a result of the decay of roots. It has been found that fewer fungi are present as deeper layers of mineral soil are reached.[158,235,1204] In addition, it has been noted that, of the larger fungi, the mycorrhiza-producing species penetrate the deepest into mineral soil.

Litter can include fallen leaves, dead herbaceous material, parts of flowers and fruits, twigs, branches, and logs including bark, and, for some, even the roots which penetrate shallowly or even deeply into the soil. Litter, then, can include any organic matter which is still recognizable as having come from plant or animal sources and which is recognizable as to the organ which has died from one cause or another including pathogenesis. The fungal populations of the litter can be derived from previous generations of litter materials, the soil populations, or the phyllosphere and related areas of plant surfaces which have become colonized from their initiation by one fungus or another.[317,926] For Pugh,[930] litter also includes any man-made decomposable organic product, in place or not. A relatively extensive literature is being developed on this subject.

II. WOOD LITTER

It has long been observed that woody plant materials including logs decompose at varying rates of speed, depending on species of tree, location, fungi involved, and other factors. Studies of wood decay fungi have been related more to methods of identification of the causal organism. Few studies have been made on the succession of fungi involved or their contribution to the rate of decay. In Russia[806] a report was published on the rate of decomposition of dead pine and spruce trees. It was noted that after 3 to 6 years, symptoms and signs of rot began to appear. After 8 years, in 60% of the trees which had fallen on the study stand, 7% bore fruit bodies of *Gloeophyllum sepiarium*. After 14 years, 60 to 76% of the dead trees had sporophores of that fungus. After 16 to 26 years, sporophores of *Fomitopsis pinicola* were abundant. After 20 years, decay was effected chiefly by bacteria and fungi, the fungi including *Xeromphalina campanella* and *Tylopilus felleus*. After 20 to 25 years, sporophores of *Gloeophyllum sepiarium* ceased to appear, and after 50 to 60 years, fruit bodies of *Fomitopsis pinicola* ceased to appear. No doubt other fungi were involved in the decay of the observed trees. Chestukhni[192] analyzed ecologically the decomposition of plant residues in young pine plantations.

Texts on forest pathology[86,122,566] indicate the kinds of fungi associated with forest disease, which includes decays of wood of forest tree species. In surveys of such species, primary and secondary producers of decay are considered without reference to specific time sequences or fractions of host tissue required to produce known quantities of mycelium as assimilative or fruiting tissues.

As an introduction to the Aphyllophorales of the Muddux National Park in Sweden, Eriksson[367] presented information on the wood decay fungi found there. Wood, such as that of spruce, pine, deciduous trees, and shrubs, in the natural or burned condition, is described in relation to being habitat for the wide variety of wood decay fungi described in detail in the main part of the work. The distribution of the Aphyllophorales in Sweden was considered in general terms.

III. FOREST LEAF LITTER

In Finland, Mikola[789-791] was interested in the decomposition of litter by forest soil Basidiomycetes. Many litter-decomposing species are capable of breaking down both lignin and cellulose. The proportions of decomposition of these substances are not constant for a species but are dependent on environmental and genetic characteristics of different strains. The breakdown of litter by species of *Collybia* differs greatly between pure culture and natural environments. The natural population of litter uses water-soluble compounds first while *Collybia* attacks lignin. In culture, *Marasmius perforans* appears to prefer birch leaves; in litter it is found only on conifer needle litter. A rich population of lignin-decomposing fungi, such as *Clitocybe* species, may indicate healthy humus conditions. Mycorrhizal fungi probably participate very little in the transformation of organic matter in the soil. Mixed and deciduous forests provide better conditions for a rich fungal population and more active decomposition of litter than pure coniferous stands. However, the assessment of the importance of Basidiomycetes in the ecology of forest soil requires more information on the physiology of individual species.

Viro[1185] studied the decomposition of spruce litter and found that potassium, nitrogen, and phosphorus are returned to the trunk before the yellowing leaves fall. On the ground, the decrease of organic components varies less than that of inorganic compounds. Needles contain most nutrients, cones contain least. Soil fertility is better studied in the soil than in the litter. Decay of spruce litter increases silicon in the soil at the rate of 34.6 kg/ha annually, resulting in increasingly poor soils in spruce forests and plantations.

A number of fungi are reported as decomposers of *Pinus sylvestris* litter.[481] Including pioneers, these include primary and secondary decomposers. Needles rich in carbohydrates have a population different from those poor in carbohydrates. Microclimatic factors including water supply play a selective role. The decomposition process appears to proceed along the following lines: (1) decomposition of the tactic substances, (2) decomposition of sugars, (3) decomposition of cellulose, and (4) decomposition of lignin.

Certain biological aspects ofthe decay of *P. sylvestris* litter were investigated in England by Kendrick and Burges.[639] Several techniques were used concurrently, and a well-defined succession was worked out. Colonization of the interior and exterior of the needles was carried out by different groups of fungi. The fungi are largely responsible for the development of the subdivision of the organic horizon known as the F_1 layer. subsequently soil animals produce the condition recognized as the F_2 layer. Eventually, about 10 years after the needles have fallen from the tree, their now comminuted remains enter the H or humus layer, where biological activity is greatly reduced. On the basis of this study,[634] a method was described for determining the duration of the various stages in the decomposition of coniferous litter on mor sites, and a list[636] of the species involved was developed. It was suggested that knowledge of the time required for the decay process in different species would allow useful comparisons to be made of their inherent mor-forming tendencies.

Fungal succession[1033] on needles and twigs of *Pseudotsuga menziesii* in the Oregon Cascades exhibits a well-defined sequence which was documented by counting thalli and fruiting bodies under a dissecting microscope. As a result, 25 categories were tabulated according to young trees and old-growth trees, needles and twigs on each, and on each of these four categories, upper surface, lower surface of leaves, and nodes and bud scales of twigs. Detailed information on the distribution of fungi in this habitat suggested that their mode of nutrition has yet to be elucidated.

In a study on the decomposition rates of tree litter,[294] recently fallen leaf litter from 11 species of conifers and two species of dicot trees were allowed to decay after uniform inoculation with a complex mixture of natural saprobes. Cultures were kept at 10°C with duplicates at 25°C, and loss of air dry weight after 100 days was taken as an index of extent of decay. No relationships were found between net decay and analyses for pH, N, P, K, Ca, with hot and cold water-soluble cellulose, lignin, protein, ether-soluble, or ash fractions. The conclusion is supported that a rating of species according to the decomposability of their litter, when done in an artificial environment, is largely an artifact determined, among other things, by the temperature level chosen for the test.

IV. BROAD-LEAVED LITTER

Various aspects of the decomposition of beech litter in Japan have been studied by Saito. The litter[999] isfound in several layers from freshly fallen leaves above, to thinner, partly degraded leaves at an intermediate level, and at the bottom the leaves, as such, are unrecognizable. Bacteria, filamentous fungi, and Hymenomycetes participate in the activity resulting in litter decomposition. Carbon dioxide production is prominent in the horizon which is rich in bacteria and Basidiomycetes.

Beech litter[998] can be described more accurately as consisting of several layers of leaves which include brown, yellowish, thin yellowish, moldy (overgrown with basidiomycetous mycelium), and fibrous leaves in increasing degrees of decomposition. Early stages of decomposition are characterized by the presence of Basidiomycetes capable of using lignin, resulting in an increase in water-soluble substances and total nitrogen, leading to a vigorous development of a variety of microorganisms which participate in the subsequent processes of decomposition. Apart from grayish-brown leaves which are rich in lignin, the rate of decomposition of cellulose is slower than that of lignin.

Brown, newly fallen, apparently uninfected beech leaves[999] experimentally show a greater weight loss with basidiomycetous mycelium infection than with filamentous fungi. Inoculation of leaves undergoing decomposition is difficult unless the leaves are sterilized, following which new inocula grow well. Removal of the water- or alcohol-benzene-soluble fractions affects the growth of filamentous fungi to some extent. Basidiomycetous mycelia attacking leaves are finally broken down by bacteria associated with them, although Basidiomycetes isolated from the litter show an antibacterial activity.

Using powdered beech leaves,[1000] it was found that filamentous fungi grow rapidly but cause only a small loss of weight. Hymenomycetes decompose this vigorously in spite of their slower growth. The increase of water-soluble substances and reducing sugars accompanies the decomposition of cellulose and lignin when powdered leaves are inoculated with Hymenomycetes. In mixed cultures of filamentous fungi and Hymenomycetes, there is an initial flare-up of the former, which is soon reduced as the Hymenomycetes become predominant. As nutrients resulting from hymenomycete activity become available the filamentous fungi again develop.

While antibiotic activity[1001] was noted by *Collybia* against bacteria and weakly against microfungi in culture, this was not noted in natural litter. In the process of degradation of lignin in the *Collybia*-infected leaves, considerable amounts of vanillin and syringic acid were released, and then vanillin and syringaldehyde were detected.

In the leaf litter layer[1002] of the climax beech woodland, the decomposition process can be divided into two sequential patterns; the white rot process and the brown rot process. They are not always separated in nature but form a linked process. In the fungus succession in the white rot process, the litter-decomposing Hymenomycetes are gradually replaced by microfungi in varying degrees. In the brown rot process, sugar fungi and cellulose-decomposing fungi are relatively close groups both in time and space, as compared with the Hymenomycetes. Litter-feeding fauna showed a remarkable preference for the white-rotting leaves and no preference for the brown-rotting leaves. A diagram of these processes is given in Figure 1. Chemical changes in these processes were reviewed by Saito.[1003]

In a study of the breakdown of forest litter,[1270] it was concluded that differences in the forest floor type were primarily conditioned by edaphic factors which influence the vegetation, litter composition, air humidity, soil microbiota, and ultimately the forest floor type.

Using leaves contained in nylon net bags placed in litter at 5200-, 3400-, and 850-ft elevations in Great Smoky Mountains and Oak Ridge Forest, Shanks and Olson[1025] found weight losses during the first year of:

Fagus grandifolia	21%
Acer saccharum	32%
Quercus alba	35%
Quercus shumardii	54%
Morus rubra	64%

Average losses for all five species placed in spruce, hemlock, and pine stands were 29, 34, and 40% for leaves placed in beech, cove hardwood, and white oak stands.

The microbiota of five species of tree litter[1267] decomposing on north and south slopes of hardwood and coniferous stands at 260, 1040, and 1600 m was estimated in the four seasons of the year by the serial dilution plate method. Microbial counts from the various leaf species, stands, and altitudes differed significantly and were positively

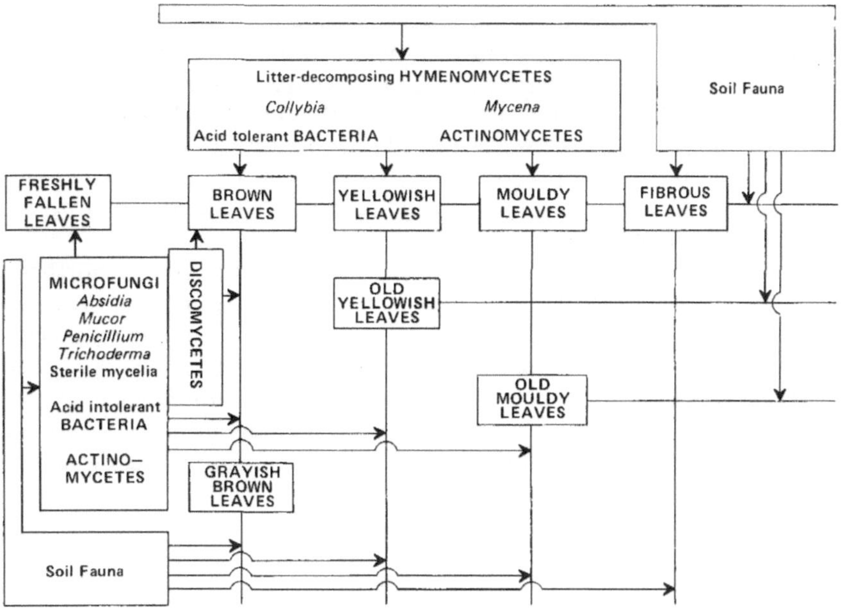

FIGURE 1. The sequential pattern of decomposition of beech litter in relation to the ecological disposition of the main microbial population. (From Saito, T., *Ecol. Rev.*, 16, 246, 1966. With permission.)

correlated with differences in litter breakdown rates. Microbial counts made in the four seasons were also significantly different. Chief factors controlling the rates of litter breakdown and the microbial populations were the stage of decomposition and moisture for the five leaf species; temperature and moisture at the three elevations; and possibly soluble inhibitors and pH in the two types of stands. Moisture, temperature, and stage of decomposition of the litter governed the microbial populations in the different seasons. Microbial respiration rates at the end of the 1-year cycle were correlated with the microbial populations, total weight loss, and the percent moisture of the leaves. Most correlations were not significant. While enumeration of the microbiota throughout the year reflected breakdown rates of the litter, one-time measurements of microbial populations and respiration did not.

In further studies of the decomposition of leaf litter in relation to the environment, microbiota, and microbial respiration,[1268] it was found that microbial respiration was controlled in decreasing order by temperature (T), bacterial density (B), moisture (M/D), and the number of weeks since leaf fall (W). An effective model for prediction of microbial respiration (C) is $C = 46.5 + 3.2T + 26.9\ M/D + 11.4 \log B - 0.6\ W$. Mean CO_2 production was 0.17 ℓ/g of substrate decomposed. Production was higher for rapidly decomposing leaf species dominated by relatively inefficient bacterial populations than for slowly decaying litter with predominantly more efficient fungi. Loss of weight and respiration were highly correlated with a microbial population estimate combining bacterial and fungal counts.

The study of the growth of litter fungi in a forest soil in England was carried out by Hering[534] by measuring mycelial extension of five litter fungi growing over sterile oak leaves buried in soil inside cellophane covers. Preliminary estimates in four soil layers showed inhibitions, compared with a sand control, varying from 0 to 50%. Material from the H layer and the A layer was inhibitory to *Trichoderma viride* and

Mucor hiemalis, while all soil layers inhibited *M. ramannianus*. These effects are tentatively referred to as diffusible inhibitors. Inhibition of the remaining two species, *Cladosporium herbarum* and *Polyscytalum fecundissimum*, was not indicated.

Fungal decomposition of oak leaf litter[536] was tested at 9 to 15°C. Decomposition by ten species of fungi from litter was 1 to 17% dry matter in 6 months. Lignin and hemicelluloses were most actively decomposed by *Mycena galopus* and *Collybia peronata*, cellulose by *M. galopus* and *P. fecundissimum*. No significant effect on the solubility of nitrogen compounds was detected.

Hering[537] studied fungal associations in broad-leaved woodlands in northern England. Species decomposing broad-leaved litter could be grouped into at least three distinct associations: those colonizing leaves on the trees, those colonizing newly fallen litter, and those colonizing litter from the soil. An overall estimate of the importance of a given species could be obtained by combining data on abundance, duration of activity, and decomposition ability of sterile litter. The populations of higher fungi in high-altitude oak woodlands was shown to have significant differences from those at low altitudes.

The succession of fungi on sterilized leaves placed in bags in the litter on forest floors in Japan was studied by Tubaki and Yokoyama.[1162] Leaves of *Castanopsis cuspidata* and *Quercus phillyraeoides* were used. Buried leaves were returned to the laboratory once a month from October 1969 to March 1970. The fungi developing on the leaves were isolated and identified. A succession of four groups was determined among the 58 genera of fungi recovered, including:

1. Transient fungi present on the leaf surface as propagules, probably transported to the leaf from the soil.
2. A group of fungi actively growing and sporulating on the leaf surface throughout the decay. These primary invaders grow better at high humidities and low temperatures. These fungi are plurivorous, tolerate a wide range of conditions; the process of colonization on the leaf is rather rapid.
3. A group found in the early decay process, which may also be among the primary invaders but which declines early. (Members of the second and third groups show an extensive growth over nearly the whole surface of the leaf and possess an efficient spore dispersal method.)
4. A group of species which do not colonize newly fallen leaves but which do invade leaves in which decomposition has been initiated by earlier colonists.

A sharp line between "soil" and "litter" fungi is not recognized, although in this study some common genera of soil fungi have not been found in the litter. There are no apparent differences between age of the leaves used of either species or between the kinds of leaves used.

Mangenot[748] in France studied the decomposition of five pure litter and one mixed litter in vitro in artificial soil and in different conditions of pH and humidity. For each, the loss of weight, development of pH, and the composition of the microbiota were determined. It was shown that each litter exercises a selective effect on certain groups of microorganisms, including certain fungal species. Those which favor the bacteria, the Actinomycetes, and the cellulolytic fungi strongly are rapidly decomposed; this is the case with apple, *Melandryum*, and, in the forest, *Festuca sylvatica*. *Calluna*, developing a brown humus, selected filamentous fungi, and myrtle occupied an intermediate position.

Macauley and Thrower[729] studied the succession of fungi which developed on fallen leaves of *Eucalyptus regnans* during a period of 60 weeks. Leaves were separated into a series of layers. It appeared that species of Coelomycetes and Moniliales were primary colonizers, particularly those species able to attack cellulose and pectin. Toward the end of the series, species of *Penicillium* and Mucorales appeared to be predominant. Intermediate stages were studied using radioactivity and other techniques.

Frankland[398] was particularly interested in determining the biomass of fungi active in decomposition of litter. She considered active fungi those with turgid cells filled with protoplasm as determined by using phase contrast microscopy, in contrast with inactive living cells and dead cells.

Studies were made in England on the production of ammonia and nitrate nitrogen, rate of decomposition, and comparative changes in microbial populations during decomposition of deciduous and coniferous litters.[596] Liberation of nitrogen in the form of ammonia occurred somewhat more rapidly in decomposing deciduous litter than in decomposing coniferous litter. Toward the end of the experiment, the nitrification process had started in the deciduous litter but was absent in the coniferous litter. The deciduous litter decomposed more rapidly than the coniferous litter. Bacteria and actinomycetes were consistently more numerous in the deciduous litter than in the coniferous; in the case of fungi, the reverse was true. Cycloheximide, an antifungal antibiotic, had little or no effect on the numbers of fungi in the decomposing coniferous litter, while streptomycin and chloramphenicol, antibacterial antibiotics, reduced slightly the numbers of bacteria and actinomycetes. Numbers of bacteria and actinomycetes increased greatly in cultures receiving the antifungal antibiotic, and fungal growth was apparently stimulated where antibacterial antibiotics were added. In spite of these observed shifts in microbial populations, the rate of decomposition of coniferous litter was apparently unaffected.

The study of decomposition products in litter was extended to amino acids and hexosamines.[1087] On an oven-dry basis, there was a gradual increase in the percentage of most of the amino acids as carbon was lost from the system. The hexosamine content also increased with time, and this increase was greater than the carbon loss. More galactosamine relative to glucosamine was found in the deciduous litter compost. The similarity in the amino acid composition of the decomposing leaf litter appears to indicate that changes in the composition of the proteins play little part in the formation of mull and mor. The relatively higher proportion of glucosamine in the total hexosamines of the coniferous litter composts, considered along with the higher relative numbers of fungi found therein, might be an indication that fungal products are of relatively greater importance in mor formation.

While no direct statement concerning fungal action was made, previous studies may be used to extrapolate an implication of fungal activity resulting in weight loss in a study by Olson and Crossley[847] by tracer studies in litter breakdown. Labeled litter was placed in nylon net of fiberglass curtain material bags which were then placed in plastic boxes with side and bottom openings, and these boxes were placed in approximately natural positions on the forest floor. These were repeatedly counted with radiation tracers in standardized counting positions over a scintillation crystal. These analyses supplemented previous studies using repeated weighing and chemical analyses. Cobalt-60, ruthenium-106, and strontium-85 and strontium-90 were released from leaf litter at rates similar to or slightly greater than the rate of weight loss, and caesium-138 was released considerably more rapidly. Much of the nuclide released was intercepted by the surface layers of humus, especially in types where mineral soil layers were included. Witcamp[1266] has discussed factors which influence the immobilization of certain radionuclides in the process of leaf litter decomposition.

Autoradiographic procedures were described by Waid and colleagues[1193] to detect metabolically active hyphae in litter habitats. Experiments with hyphae on the surface of beech leaf litter showed that active hyphae could be detected after exposure for 2 hr to 10 μci/mℓ of uniformly labeled ^{14}C-glucose followed by film exposure for 2 days. Controls showed that there was negligible background radioactvity and that there was no adsorption, or nonmetabolic or passive, uptake of radionuclide by hyphae.

The microlayers of water, less than 1 mm thick, covering leaves, etc. in litter, were described by Bandoni[70] as a habitat for fungi from which species with tetraradiate spores have been isolated, as well as helicosporous fungi, predaceous fungi, and soil fungi.

Litter decomposition in fresh water is described by Willoughby.[1259] A critical factor is the oxygen content of the water. Depending on the species of plant from which the litter originated, rate of leaching, and rate of decomposition by aquatic fungi, aquatic Hyphomycetes and soil molds vary. Microbial successions were described, as were biogeochemical developments in sediments attributable to litter decomposition.

V. ROOT LITTER

Assuming that any recognizable dead tissue is part of the litter, even though that tissue is formed in the soil, dead roots may be considered part of the litter. The distribution of fungi within the decomposing tissues of ryegrass (*Lolium perenne*) roots was studied in England by Waid.[1191] Ryegrass roots, classified according to the degree of visible decomposition of the cortex, were washed in ten changes of sterile water and plated either in tact or, after dissection, as fragments of outer cortex or of inner cortex plus stellar cylinder. Mycelial populations in the three different root habitats were compared. The most active populations originated from root surfaces of intact roots. The progress of root degradation was paralleled by an increase in fungal activity in each zone, although the extent of this increase was not so great as the gradient of activity across the cortex. These results agreed with direct observations made on a semiquantitative basis. A characteristic order of succession was found, and the composition of the hyphal population was different in each zone of the cortex. The primary colonizers were fungi with a low degree of competitive saprophytic ability; they included an endophyte which was not isolated in culture. Secondary colonizers were found commonly isolated from soil habitats and having a high degree of competitive saprophytic ability.

VI. COLONIZATION OF DEAD GRASS CULMS

Studies of the colonization of the various parts of plants of *Phragmites commnis* were initiated by Taligoola and colleagues[1132] in England. The submerged portions of the standing culms were initially colonized by Ascomycetes and Sphaeropsidales. This differed from colonization of aerial parts which was effected primarily by Hyphomycetes. Parts of the culms in the muds below the water were colonized by a different series of fungi than those in the water. Sporulation of fungi on submerged surfaces did not begin until the culms were moribund or dead, usually after flowering. It was suggested that it is the fungi on the upper portions of the submerged parts of the host which may contribute to the weakening and eventual collapse of the culms.

Apinis and co-workers[43-46] reported that the origin, character, and changes of the leaf surface microbiota on *P. communis* are complex because of interactions between the host in providing the leaf surface area and nutrients, including various excreted organic and inorganic substances; the physical environment, including moisture, temperature, and dust particles; and the various activities of the other organisms in the leaf surface biocoenosis. Developmental stages of the host also affect the dynamics and the character of the leaf microbiota which consists of bacteria, actinomycetes, fungi including yeasts, and algae. The quantitative and qualitative composition of the fungal spore load on green leaves is similar on both adaxial and abaxial leaf surfaces. Colonization by yeasts, yeast-like fungi, and Deuteromycetes begins at leaf apices and sheaths. Apices of young leaves harbor larger numbers of species compared with the basal part of the lamina. This is similar for the leaf sheath, but colonization is delayed, indicating that the spread of certain fungi proceeds from the sheath tip to the base of the sheath. The stages of leaf development, young leaves to mature leaves, advancing stages of senescence, and decay primarily influence the species composition of the leaf mycobiota. Of 49 species recorded, only six were common to the six habitats studied.

Microfungi colonizing inflorescences of *P. communis* were surveyed by Taligoola and co-workers.[1133] Relativley few fungi were found to colonize inflorescences, especially the spikelets, and then at low frequencies. The common fungi in the rhachis and spikelets were *Alternaria alternata, Cladosporium herbarum, Cephalosporium* spp., and *Epicoccum purpurascens*. In certain habitats, *Acremoniella atra* was common on inflorescences but almost absent on other parts of the host. In general, microfungi colonizing *P. communis* inflorescences are of the same species which are common on other parts of the host plant, differing to some extent in host plant habitats. *Claviceps purpurea* var. *microcephala* attacks the spikelets severely in some habitats but not in others. Dormancy of the ergots of this fungus can be broken by low temperature treatment. Prior low temperature treatment enhances germination and shortens the lag period of ergots. It is possible that attack by this parasitic fungus may severely affect seed production by this plant.

The mycobiota of nodes and internodes of culms of *P. communis* has been described.[43] Species composition differs with the habitat. Both in number of species and occurrence, nodes seem to be better sites than internodes, young culms are poorly colonized, and colonization increases as nodes and internodes age. Initial colonization is by members of the Hyphomycetes and Sphaeropsidales. Species of *Hendersonia*, Discomycetes, and other Hyphomycetes colonize dead culms. Discomycetes and certain Hyphomycetes are most frequent on old nodes and internodes of upright culms, although some fungal species consistently sporulate on the upper or lower parts of the culms. In certain species, the vertical level—in which they sporulate abundantly—varies from one habitat to another, apparently being subject to environmental conditions in a particular habitat.

Webster[1213] described the succession of fungi found on decaying culms of *Dactylis glomerata* collected at monthly intervals from three localities in England. He found five distinct patterns of the fungi: (1) a group of primary saprobes which colonize tissues as they become moribund and advance up the stem as new leaves unfold, (2) species normally confined to the damper tussock base early in the succession, (3) species which fruit in the tussock base during spring and summer following flowering and may be confined there by humidity requirements, (4) species which fruit in the tussock base in the summer following flowering and persist on old stems until the second summer, and (5) species which fruit on the upper internodes in late summer of the year following flowering.

Continuing the study of fungi on *D. glomerata*, Webster and Dix[1216] compared sporulation and growth of some primary saprobes on stems, leaf blades, and leaf sheaths. Comparative studies of spore germination on glass slides and mycelial growth on pulverized sterile grass at various relative humidities have shown that primary colonizers often show more rapid spore germination, more rapid growth, and a better ability to grow at lower relative humidities than those secondary colonizers with which they were compared. On the basis of tests using grass tissues collected from upper and lower internodes throughout the year and measuring the growth rate of primary colonizers on a medium composed of 3% dried grass in 2% tap water, it was found that upper internodes may attain a higher nutritional status than lower internodes during the early period of colonization, but this situation may be reversed later. It is suggested that upper internodes, having a higher level of nutrients, are more attractive to the more vigorous primary saprobe colonizers than the lower internode tissues with lower nutrient levels. These rapid colonizers also colonize the tissues subject to more rapid senescence such as leaves and upper sheaths.

Hudson and Webster[568] found that the succession of fungi on flowering and non-flowering tillers of *Agropyron repens* followed broadly the pattern described for *D. glomerata*. Bases for differences may be attributed to physical differences (especially differences of water content of upper and lower internodes, nutritional differences in the level of available nutrients at the time of leaf senescence, differences in host resistance resulting from rates of senescence between upper and lower areas, and competition.

VII. TUNDRA STUDIES

A variety of studies on microorganisms in relation to decomposition processes in the tundra were brought together in a symposium held at the University of Alaska.[558] Basic differences in technique and differences in approach to classification of soil organisms have resulted in inconclusive results in attempts at statistical analyses of relationships between fungi and their environments in the arctic tundra.[332] Lack of uniformity in the level of determination of taxa prevented all but the most general of comparisons. Careful work by Flanagan and Scarborough[386] has demonstrated that the physiological capabilities of fungi identified as the same species may vary considerably. In view of the relative ease of standardization of biochemical tests when compared with taxonomic identifications, such tests may be more worthwhile in comparisons of "ecosystems", especially in the case of studies, such as those in the IBP (International Biological Program), which place the elucidation of ecosystem functions as a greater goal than community structure.

Relationships between respiration, weight loss, temperature, and moisture in organic residues on tundra have been studied.[387] It was tentatively concluded that the majority of tundra microorganisms respond to moisture and temperature like microorganisms anywhere, and that in large part the processes of decomposition are basically the same in tundra as elsewhere. For the amount of organic materials produced each year, the resident microorganisms and leaching processes can recycle as much organic matter, proportionately, as is recycled in temperate or boreal regions. Respiration and, most likely, decomposition can occur at very low temperatures (−7°C), reflecting the presence of some very cold-tolerant and/or psychrophilic microorganisms in the tundra. Tundra moisture levels depress decomposition at the very low (20%) as well as at the high (400 to 2000%) range. Tundra temperature/respiration relationships are best expressed by an exponential $Q_{10} = -4$ and are better correlated than are moisture and respiration.

Physiological groups of decomposer fungi were studied in the tundra of northern Alaska.[386] Evidence strongly suggests that despite the differences in space, the pattern of substrate utilization in the typical fungus succession is very similar to that reported by Garrett.[427,433] In the phylloplane in Barrow tundra, the sugar fungi abound while the cellulolytic forms are scarcely present. The leaves fall and reach the litter zone where the cellulolytic and ligninolytic forms have increased their population size. In the soil zones, the cellulolytic forms are more numerous still, and potentially ligninolytic forms reach their greatest numbers. The taxa present in the phylloplanes at Barrow show no strong successional pattern during the first 3 years of decomposition. At Prudhoe Bay, a seasonal change was seen in the soil where the major spring fungus population was mesophilic, and that of the autumn was psychrophilic. From month to month no significant succession of physiological types was seen within the soils and litter studied at other sites. Cellulose decomposers were active down to −7°C, and 20% of the species had an optimal cellulase activity at 6 to 8°C. Thus, the low temperatures of tundra soils may make these fungi important decomposers, although very little cellulose decomposition occurs in the phylloplane before dead leaves become part of the litter studied at other sites. Cellulase decomposers were active down to −7°C, and 20% of the species had an optimal cellulase activity at 6 to 8°C. Thus, the low temperatures of tundra soils may make these fungi important decomposers, although very little cellulose decomposition occurs in the phylloplane before dead leaves become part of the litter. Commonly, cellulolytic tundra fungi produce extracellular phenol oxidases and, less commonly, oxidize humic acids. While 80 to 90% of the tundra fungi are amylolytic and pectinolytic at room temperature, at temperatures of less than 10°C this activity is lacking in most species, while cellulolytic and ligninolytic activty is not as adversely affected by low temperatures.

In an attempt to determine biomass of fungi in litter decomposition in the tundra, Swift[1127] found that of the several techniques tried none was sufficiently developed to give an adequate idea of this character.

VIII. THE USE OF LINEAR REGRESSION MODELS

On seven stands representing four mature vegetation types in western Oregon, Fogel and Cromack[391] developed a linear regression model for the decomposition of Douglas fir (*Pseudotsuga menziesii*) needles, female cones, branches, and bark. Rate constants (k) for annual weight loss for needles ranged from 0.22 to 0.31 $year^{-1}$, for cones from 0.047 to 0.083 $year^{-1}$, for branches from 0.059 to 0.089 $year^{-1}$, and for bark from 0.005 to 0.040 $year^{-1}$. The decomposition constant (k) for needles had a negative linear correlation (P −0.01) with maximum plant moisture stress and temperature growth index of the seven stands. In comparing substrate quality of needle and woody litter components, (k) was more closely correlated with lignin content than with C:N ratio.

IX. CALCIUM OXALATE

One of the commoner by-products of fungal metabolism is calcium oxalate (or possibly some other form of oxalic acid) which is found as crystals on the surface of hyphae of species in a number of widely separate genera in most classes of terrestrial fungi in both perfect and imperfect states. Its occurrence has been used as an indication of relationship in some cases.

In the litter layers of soils from which samples were collected,[457] two from western Oregon, one each from New Mexico and coastal California, and two from Connecticut, calcium oxalate was found in abundance associated with fungus mycelium. Whewellite, the monohydrate, and weddellite, the dihydrate form of calcium oxalate, were found. Although these minerals are rare in geologic environments, crystals of calcium oxalate have been found commonly attached to fungus mycelium. Mycelium in the Oregon material may have been that of *Hysterangium crassum*, a hypogaeous fungus probably producing ectomycorrhizae with *Pseudotsuga menziesii* in the sampled area.

Because of its ability to readily form chelates with iron and aluminum, calcium oxalate is an important chemical in the litter layer and, to some extent, in the A horizon. The fungi producing it are thus important members of the litter and soil populations.

X. PHYLLOPLANE, PHYLLOSPHERE, AND LITTER SURVEYS

Before leaves and other plant structures fall from the plant producing them to form litter, their surfaces become populated with a number of kinds of organisms which may react with each other and the leaf. Beyond maturity the leaves go through a period of senescence before falling as a response to certain physiological and morphological factors. Preece and Dickinson[926] edited a series of reports describing this habitat and its populations. The papers are organized in five groups including the characteristics of the leaf surface, saprobes on the leaf surface, pathogens on leaf surfaces, the microbiology of senescing leaves, and interactions on leaf surfaces. Dickinson and Preece[317] developed a second symposium on this subject in which information presented earlier was updated and new materials in each of the above mentioned areas was presented. Much of the material in these symposia is basic to an understanding of processes involved in the decomposition of litter.

Dickinson and Pugh[318] brought together information on the ecology of plant litter decomposition. Here the decomposition of litters generated by such diverse organisms as lower plants, herbs, angiosperm trees, coniferous trees, wood, roots, and digested litter are considered by people who have devoted time to the study of each area. The organisms involved in the decomposition of litter, including aquatic and terrestrial fungi, are discussed. Finally, the environments in which such decomposition takes place, such as the soil surface, in the soil, in fresh water, in marine water, agricultural crop debris, and urban wastes, are described. For the understanding of the dynamics and processes of litter decomposition, these chapters can be most useful.

XI. BURNED AREAS

The ability of fungi to grow on burned areas or areas sterilized by intense heat has been recognized for some time. Some fungi have been called "pyrophilous".[1021] More recently this term has been redefined[815] on the basis of a study carried out on fungi, bryophytes, and higher plants growing on areas burned over by forest or campfires in the Austrian Tyrol. Four types of response to such habitats were found, and organisms giving such responses were defined as "anthracobiont", "anthracophilous", "anthracoxenous", and "anthracophobous". The first two categories included pyrophilous fungi, making the term ambiguous, at least in this context. A number of anthracobiont fungi were found. These could grow only on burned areas, and this restriction was attributed to certain unexplained changes in available nitrogen. Most of the mosses, liverworts, and vascular plants could and do pass the initial stages of growth on burned areas and continue to develop on the area after the influence of the fires has been eliminated.

Studies on the effects of slash burning at three different intensities on north and south facing slopes of logged off Astoria silt loam in the Oregon coast range were initiated.[825] Samples of soil were collected 2 days after the burn and at seasonal intervals thereafter for a year. Soil moisture and water-holding capacity were reduced significantly during the year. The pH of the soil increased from 0.3 to 1.2 units and was related significantly to the intensity of the burn. Ammonium nitrogen increased significantly in burned areas up to 6 months after the burns, after which it decreased in comparison with unburned soil. Nitrate nitrogen was low at all sampling periods. Total nitrogen increased slightly, then declined. Total carbon content was generally 1 to 2% higher in burned than in unburned soils. In all instances, the C:N ratio was higher in burned than in unburned soils. The population of molds was reduced significantly by slash burning, the decrease being most noticeable in the spring.

Some of the papers dealing with the effects of fire on soil fungi have been summarized in a note dealing with fungi found in a controlled burn. In clearing chaparral brush from an area to be developed as a pine plantation at the southwest base of Mount Shasta, Calif., Cooke[251] arranged to receive a set of samples of soil for study of effects of flash burning. Litter and soil (at depths of 0 to 2 in., 2 to 6 in., and 6 to 12 in. in undisturbed chaparral and in a 30-year-old pine plantation) and soil samples from the area to be burned (both after preparation and after burning) were obtained. At the 0- to 2-in level there was a decline in the number of colonies recovered on agar plates, but at the 2- to 6-in and 6- to 12-in levels there was an increase in colony numbers recovered.

Petersen[897] studied "fireplace"fungi on burned areas in Denmark. Following clearing and slash burning, a number of habitat changes were noted: the release of nutrients in felling and clearing the area; the heating of the soil beneath the burns, hardly enough for complete sterilization but influencing the microbiota below and in a narrow zone around the burn; and the formation of an ash layer which may gradually leach into the humus layer. About 75% of the resulting fungal populations were assigned to the Pezizales. Selected groups of species were noted: those mostly associated with breakdown of such organic matter as burned roots; those associated with the humus; those associated with fireplace mosses; those at the edge of or outside of the heated soils; and those associated with an increased pH in the soil. That these fungi are associated in an environment requiring extreme specialization may refer to the fruit bodies only, although the high pH values, or some correlated factor, may be of importance to the assimilative mycelium. It is still a matter of question whether the assimilative state of the fireplace fungi takes part in the decomposition of plant roots and other remains and humus in unburned areas in nature.

Petersen[898] studied the microfungi of a burned pine plantation in Denmark and found the population of Discomycetes similar to those reported for a slash burn. There were 11 species in both areas at the same length of time after the burns. The species thought to take part in the decomposition of roots were largely common to both burned areas. The species associated with mosses were largely absent from the burned pine forest, and the pH amplitude of the species found may be broader than indicated in these two studies.

XII. AMBROSIA AND FRAS FUNGI

A variety of lower Ascomycetes is associated with habitats in which beetles have used or are in the process of utilizing wood as nutrients. These are recovered by culture techniques. They form at least part of the diet of the insects with whose activities they are associated.[80,81]

Some fungi play an important part in the food supplies of other insects. Among these may be included bark beetles and blueing fungi, possibly *Cryptoporus volvatus* and bark beetles, yeasts and yeast-like fungi, and *Drosophila* (see above), *Termitomyces*, and other fungi and termitaria, etc.

XIII. HUMUS

Humus[159] is the amorphus dark brown or black organic material in the soil distinct from partially decomposed plant and animal remains. It includes humin, humic acid, fulvic acid, and hymatomelanic acid, all of which occur in soils and composts in varying proportions depending on soil types. The humic acid fraction represents either a single chemical substance or a group of closely related substances. It is probably nonnitrogenous, the presence of nitrogen probably being due to secondary combination with amino acids or proteins. In the soil, humic acid normally occurs as a metallic humic acid complex, the metal varying with the soil conditions. Decomposition of humic acid has been discussed by Burges and Latter.[161]

Humus is slowly decomposed by soil organisms,[547] mostly fungi, many of which are types which produce mushroom-like fruit bodies. Genera involved include *Collybia, Marasmius, Mycena, Clitocybe, Clavaria, Cudonia,* and probably others, including their segregates. Humus may be considered an end product in the decomposition of lignin either from wood or tree and herb litter. In certain areas, such as Finland,[548] the fungi may form a "white rot humus" in which, in association with these fungi, a pale-colored humus is formed in dry grass-herb forests and oak forests which have a thick litter layer. It may be noted that such humus may more accurately be considered litter and fermentation layer materials from which lignin is being at least partially removed by fungi capable of attacking lignin. It may occur in up to 25% of the forest area but only on 0 to 2% of the raw humus.

During a recent symposium in Czechoslovakia,[925] the characteristics of humus and the determination of some effects of humus on plant growth in humus-bearing soils were described in 42 papers.

XIV. DETERIORATION OF COMMODITIES

Practically everything man uses for shelter, nourishment, and communication, especially when these are made of naturally developed materials, is subject to deterioration by fungi. For Pugh,[930] this is merely the result of these types of material being a specialized type of "litter". It has been suggested[1221] that, in the study of some of these commodities and their contaminants, a new concept of association be applied to the fungi. For example, in molding of butter or leather there may be a succession of fungi, one species overlapping another as a particular nutrient is exhausted, as competition becomes more aggressive, or as one species may be inhibited by the by-products of another, which may be staling products. The idea expressed by Garrett,[427] that a fungus succession proceeds on a substrate from a fungus able to use simple products to a fungus able to use complex compounds to, finally, a complete depletion of the substrate with nothing left for the fungus to use, can apply readily to commodities used by man, if the succession is permitted to run its course. However, in most of man's activities, any decaying matter is usually destroyed before the succession is complete; however, this does not necessarily invalidate applicable portions of it. Usually, today, attempts are made to prevent the molding of clothing, foodstuffs, housing materials, paper, insulation, and other substances in use by man. If molding results from a particular type of storage, the condition is usually corrected; if it results from methods of handling between production and market, the methods are modified. It is rare that food poisoning results from the eating of moldy bread or other foodstuffs in this day of precautionary measures, in the form of additives such as preservatives, against infection or deterioration. However, with the dropping of food additives from certain products in response to pleas of certain environmentalists, toxicity as a result of growth of contaminants may again become of importance.

The most elaborate precautions against natural contamination of manufactured commodities are those taken in pharmaceutical industries where unwanted strains of organisms could cause irreparable damage to the final product. In the preparation of foodstuffs for market, the use of mold inhibitors in bread and of relatively aseptic techniques in other preparations tends to reduce the incidence of natural contamination or the development of natural populations in such products. It is no longer a common event for a batch of rye flour to be contaminated with powdered ergot sclerotium which causes the disorder known as St. Anthony's fire or ergotism.

Other products are less fortunate. Cloth from cotton fibers is still subject to molding and mildewing[1050] without superficial chemical physical treatment. The cellulolytic activity of various fungi[953,954,1197,1231] has been studied intensively. In work with the black species of *Aspergillus*,[1232] most strains tested gave no cellulolytic activity, and the strains of *A. niger* mut. *schiemannii* and *A. luchuensis* tested took 28 days to decrease tensile strength of test fabrics by two thirds. This work has been carried over to degradation of woolen fabrics.[1229] At least one of the cellulolytic fungi, *Memnoniella echinata*, has been studied intensively,[1228] and it has been shown[1292] that spores of this fungus and of *Stachybotrys atra*, another cellulolytic fungus, may be produced on the same mycelium.

In tests for mildew-proofing of cotton fabrics,[1230] impregnation with urea formaldehyde and melanine-formaldehyde resins was used, and the fabrics were subjected to attack by *Myrothecium verrucaria* and *A. flavipes* test organisms. Complete resistance to pure culture tests was obtained with 6.4% Resloom®, 5.5% Aerotex®, and 0.5% Rhonite® resins. In the more severe soil burial test, cloth impregnated with Aerotex® gave the best performance. While control cloth lost its tensile strength in less than 7 days, a cloth impregnated with 5.9% Aerotex® M-3 retained all of its tensile strength for the 14-day duration of the experiment.

Cellulases are known to be remarkably stable to pH changes, temperature changes, and chemical inhibitors.[746] They are competitively inhibited by cellobiose and methocel and inactivated by such protein reactants as halogens, heavy metals, and detergents. A type of polymeric leucoanthocyanin found in many plants, notably in the unripe fruit of persimmons, is a powerful but nonspecific inhibitor of cellulase. The susceptibility of cellulases to chemical inhibitors increases with purification. The C_1 component, which allows cellulase to attack insoluble cellulose, is more labile than the hydrolytic C_2 components. Formation of cellulolytic enzymes is repressed by growth on sugars.

Seeds of crop plants are consistent carriers of molds, and when these are stored improperly, the seed can be rendered useless by the activity of many kinds of fungi. Untreated seeds are also carriers of spores or mycelium which can also infect the new crop. Lumber, when stored improperly or when cut from infected trees, is subject to a number of fungus-caused disorders. White pocketrot and other sapwood and heartwood decays develop if the lumber is not properly dried and stored. Various kinds of stain-producing fungi make sapwood for window frames, sills, and other uses unsightly. The natural contaminants of poles, posts, and other wood products which touch the ground, as well as local populations of fungi, can actively destroy the wood in relatively brief periods of time unless it is treated, and sometimes even then. A number of types of fungi can attack wood pulp during storage, causing discoloration and deterioration. In the processing of pulp and paper products, a number of types of fungi can develop which are difficult to control, since some of the ultimate product comes in intimate contact with foods, and control agents usually include control chemicals potentially poisonous to man.

The yeast biota of wheat and flour was studied by Kurtzman and Wickerham[666] and compared with that reported from other cereal grains. A relatively small number of species was represented in both wheat and flour. A similar small number of species was reported on ensiled corn and on stored rice. Many of the isolates on ensiled corn were species of *Candida*, but this was not true of stored rice. *Hansenula anomala* appeared to be a common inhabitant of cereal grain since it was found in the study of wheat as well as in the studies of corn and rice. *Cryptococcus albidus* was found in all six areas on wheat, but not in wheat flour, and was absent from ensiled corn and stored rice, which may be attributed to a possible inability to survive under storage conditions.

Various types of wastes, residues, and compounds are subject to microbial attack either for useful purposes or as nuisance organisms. In the case of feedlot wastes (FLW), which consist principally of cow manure, fungi have been ased in an attempt by Morrison and colleagues[813] to salvage the proteins and other compounds present. Approximately half of the proteins, 250 mg/g, solubized in 0.1 *N* NaOH were recovered by precipitation with sodium or ammonium sulfate. Protein analyses of sieve fractionated manure and direct microscopic counts indicate that 45% of the protein is of microbial origin. The suitability of the protein-extracted fiber residue as a nutrient for *Trichoderma viride* QM 9414 biomass production was enhanced by protein hydrolysis with cellulase. Other treatments — ball-milling, alkali-treatment, and grinding, in order of effectiveness — increased the susceptibility of manure fibers to cellulase. The sugars released by enzymatic hydrolysis can serve as a carbon source for the growth of *Candida utilis*. The theoretical protein yield from extracted and fermented fresh manure was approximately 350 mg/g.

That *Pleurotus ostreatus* can degrade lignocelluloses solid waste as found in cattle feedlot wastes was evidenced by a 10 to 40% decrease in lignin in the fermented residues; cellulose content decreased by 15 to 40% in a study by Kanashiro.[622] However, lignin content remained unchanged in a sawdust-oats medium. The fermented FLW fiber gave a product with 18% mineral ash, 51% readily solubized fraction, and 3% insoluble nitrogen by dry weight. In contrast, *Agaricus bisporus* does not grow well in fibrous FLW mixed with wheat straw, and *Coprinus lagopus* mycelia fermented the mixed fibers to give a product elevated in lignin but diminished in cellulose content.

The use of *Chaetomium cellulolyticum*[186] in the production of single-cell protein from wheat straw has been described, and the ability of this production of cellulase was compared with that of *T. viride*. More single-cell protein was produced in chemically pretreated straw. With the use of a highly purified form of cellulose, it was found that both organisms produced large amounts of cellulase, although *C. cellulolyticum* produced more protein and less cellulase than *T. viride*. The interrelationship between biomass formation, free enzyme production, and substratum utilization seems to be complicated by varying degrees of substrate enzyme adsorption effects, substrate inaccessibility because of the presence of lignin, and by the degree of cellulose crystallinity, and enzyme-complex differences.

Among isolates[301] from raw sewage, by enrichment culture using natural gas, 12 cultures of species of *Geotrichum*, *Phialophora*, and *Acremonium*, among others, were tested for fungal oxidation of gaseous alkanes. *Acremonium* species selected for further study grew vigorously in submerged culture at the expense of either natural gas or ethane. Resting cells of *Acremonium* grown upon natural gas were tested in a Warburg respirometer; these cells oxidized ethane, propane, and *n*-butane in the presence of oxygen. Methane was not oxidized, either alone or in the presence of ethane. Resting cells were also able to oxidize 2, 3, and 4 carbon alcohols, aldehydes, and fatty acids, which are regarded as the corresponding metabolic intermediates in the oxidation of either ethane, propane, or butane.

Various concentrations[1075] of heptachlor dissolved in hexadecane were added to cultures of fungi grown in yeast nitrogen base prepared with synthetic sea water and with deionized water. *Candida maltosa* and *C. lipolytica* showed greatest utilization of hexadecane (20.91%), whether heptachlor was present or not. Compared with low concentration, high concentration of heptachlor appeared to have a slight stimulating effect on utilization of hexadecane by *C. maltosa*, but had no effect with *C. lipoytica*.

Cladosporium resinae, frequently in association with species of *Fusarium*, *Cephalosporium*, and *Penicillium*, was isolated[100] from 18 of 19 commercial jet aircraft and from 8 of 13 fuel storage tanks. Fuel filters showed obvious accumulations of *Cladosporium*-type growth. Samples from aircraft wing tanks indicated that *C. resinae* sporulated within the fuel system. Colonies were associated with joints, crevices, and bottom water deposits within the tanks. Scrapings of polyurethane from the walls contained tightly adhering interwoven fungal hyphae with numerous chlamydospores. A number of laboratory studies yielded data showing that this fungus grew very slowly in jet fuel alone, but more rapidly when water was present. The fungus remained viable but did not grow in the presence of ten times the recommended amount of a boron-based inhibitor. Complete inhibition of the fungus by a deicer compound occurred only when the water content of the culture was 10% or less. *C. resinae* survived in sealed glass jars of jet fuel for up to 10 years.

Ponderosa pine test blocks were studied for weight loss as a result of attack by thermophilic and thermotolerant fungi by Ofosu-Asiedu and Smith.[840] Blocks 0.2 × 1.9 × 4.4 cm were placed on fungal cultures grown for 1 week on Abrams-cellulose or yeast-cellulose medium. Cultures were incubated at 25 to 60°C for 2 to 12 weeks. Maximum wood substance loss occurred during the first 6 weeks. Optimum temperatures were between 40 and 50°C. When blocks of different sizes were used, weight loss was related to volume, not surface area. Decrease in size did not result in an increase in weight loss following fungal attack. Degradation of wood[841] in chips of two species of pine and one of spruce usually stored in piles outdoors in British Columbia resulted from action by thermophilic and thermotolerant fungi. Wood of the pines was degraded more readily than that of the spruce. In stored pulp wood logs and chips, brown- and white-rot fungi, sapwood-stain fungi, thermophilic fungi, and molds were described as deteriorating agents.[1077,1079] The economic significance of these fungi in the manufacture of pulp was discussed.

New shakes of western red cedar (*Thuja plicata*) are free of living fungi, according to Smith and Swann,[1081] and possess a degree of resistance to fungal attack because of the presence of thujaplicin which is progressively leached out causing the loss of fungicidal properties. *Rhinocladiella mansonii* and *Ptialophora hoffmannii* make up 60% of the fungal isolates recovered from leached shakes. The former may attack some of the leachate, while the latter may cause a type of soft rot in the wood. It was suggested that if red cedar shakes are to be protected in an area of high decay incidence, some form of penetrating, fungicidal treatment must be applied before they are placed on a roof.

In a study of fungitoxic potentials of wood preservation, Smith[1078] measured threshold retention for creosote, pentachlorophenol, and chromated copper arsenate using both the respirometer (RTR) and weight loss (WLTR) methods. RTR values were similar to those of WLTR but were obtained after only 4 weeks instead of the normal 12 weeks. The American standard test strain of *Gloeophyllum (Lenzites) trabeum* would seem to be losing its previous tolerance for pentachlorophenol. The RTR method verified the tolerance of *Lentinus lepideus* for creosote and of *Poria monticola* for chromated copper arsenate. The study also proved an oxathiin and an oxathiin captan mixture to be as toxic as pentachlorophenol to some standard test strains of wood destroying brown-rot fungi.

In a study of vinyl acrylic paint films,[1264] it was found that in humidity chambers species of *Aspergillus, Mucor, Alternaria,* and *Cladosporium* were primary colonizers of the film, although *Aureobasidium pullulans* became the dominant organism indicating that its activity was, or needed to be, preceded by that of other fungi. Initial colonization was only delayed by the presence of the commonly used biocides phenyl mercuric acetate (PMA) or PMA with zinc oxide.

Enzyme systems secreted by the fungi colonizing vinyl acrylic paint films were determined[1263] by concentration of culture solutions by ammonium sulfate fractionation and by column chromatography with Sphadex C-100. Cellulases were isolated from *Aspergillus* sp. and *Alternaria* sp.; other individual hydrolyses were not detected from these fungi. It was suggested that cellulases from these fungi probably aid in the initial colonization through the hydrolysis of hydroxyethyl cellulose present within the paintfilm. *Aureobasidium pullulans* excreted significant amounts of cellulase, proteinase, esterase, phosphatase, sucrase, and maltase, and also possessed the ability to hydrolyze extra cellular polysaccharides excreted by *Aspergillus* sp. and *Alternaria* sp. The inability to grow *Aureobasidium pullulans* on paint film itself in a humidity chamber does not seem to be because of its inability to produce extracellular cellulases.

Painted wooden tongue depressors were exposed[1076] to the weather in North Chicago, Illinois and Barcelonales, Puerto Rico to determine the incidence of paint-defacing fungi. To determine populations present on the film, the sticks were scrubbed with disinfectant, immersed in 20% sodium hypochlorite (Chlorox®) and 20% ethanol in deionized water, followed by scrubbing with a sterile cotton swab to eliminate surface fungi, and cultured on a nutrient medium of 0.1% yeast extract and glucose in 0.15% agar. In the early weeks of exposure it was possible to surface sterilize the sticks in this way; in later weeks this sterilization would not be achieved, presumably because fungal hyphae had penetrated the film. The ultimate fungal population at both exposure sites and the populations consisting of *A. pullulans, Alternaria, Cladosporium,* and *Penicillium* on scrubbed sticks were the same as on nonscrubbed controls.

Spoilage of cutting fluids is effected by fungi including species of *Fusarium* and *Cephalosporium.*[982] Using spoiled fluid as an inoculum simulated residual contamination in a sterile fluid. In systems with no sterilizing agent (hexahydrol-3,5,tris-(2-hydroxy-ethyl)-*s*-triazine, or (T) and no contamination, fungal counts were never above 10^4 colony-forming units per milliliter. In systems with contaminated fluid and no T, counts reached 10^6 colony-forming units per milliliter. If 0.058 T was added immediately after inoculation, fungi were controlled; however, 0.2% T did not control previously established populations.

Media tested for the study of fungi in cutting oils[983] included potato dextrose agar, Sabouraud dextrose agar, brain-heart agar with chloramphenicol and cycloheximide, and trypticase soy agar and gentamycin. Before plating, aliquots for sampling were mixed for 10 sec in a micro-Waring blender. The latter medium yielded the best results with field and laboratory samples.

In general, artificially produced material, including many types of plastics, are not satisfactorily biodegradable. Hueck[571] has prepared a review of 30 titles in this area.

XV. SOLID WASTES AND COMPOSTING

Atkinson[56] noted that, while the chief function of fungi is the decomposition of the remains of higher plants, in a type of compost they served as food for those higher plants.

Davey[296,297] found that untreated sawdust applied at the rate of 40 yd^3 per acre depressed the growth of such plants as deciduous and coniferous trees and seedlings, even after the addition of green manures and other supplements. Sawdust amended with ammonia, phosphoric acid, and potassium sulfate was incubated with *Coprinus ephemereus* for 3 weeks, after which the sawdust was converted into finely divided dark-brown material resembling a rich mull-like soil. The nature of this comlost was described in detail. Application of the composted sawdust to a sandy soil produced marked increase in the growth of both coniferous and deciduous seedlings and medium red clover.

Rothwell and Hortenstin[985] compared effects of composting of municipal refuse with various added materials. They found that fungal numbers increased with each added increment of garbage. Garbage compost was second only to cow manure when mixed with soil in CO_2-C evolution.

Dewatered sewage sludge was composted.[1030] Seeded into the process were propagules of viral, bacterial, fungal, and protozoan organisms. The fungus *Candida albicans,* as were the others, was killed out in the first 3 days of the process. There was no mention of what organisms might have been beneficial in the process. *Aspergillus fumigatus* was shown on the cover as one of the "aerobic, thermophilic microorganisms associated with the composting process".

Fungi were present in the air[49] in all stages of a composting plant treating municipal solid wastes and sewage sludge. Air samples yielded numbers of bacterial and fungal colonies (filamentous and yeast-like) which appeared on all plates although the media used were selective for bacteria.

Mixed refuse, paper wastes, and yard and grass wastes were ground and mixed with soil in a study by Mahloch.[740] To this mixture was added a sterile, liquid basal salts medium. These materials were incubated continuously for 45 days. Samples were removed every 4 days starting with the initial day, suspended in a sterile diluent, and dilution plates poured on potato dextrose agar, subculturing on potato dextrose agar and Czapek agar. The numbers of fungi were correlated with physical and chemical parameters of the substrate measured during the incubation period. A negative correlation existed between numbers of fungi and temperature, pH, percentage moisture content, and numbers of bacteria. A positive correlation existed between numbers of fungi and chemical oxygen demand and percentage volatiles. The predominant fungi isolated from the substrates were species of *Trichoderma, Geotrichum, Rhizopus, Penicillium, Aspergillus,* and *Cladosporium.* Gray and Biddlestone[458] summarized information available on solid wastes in a review among other reviews on plant litter decomposition.

A. fumigatus has been found abundantly[803] in a forced-air composting process using sewage sludge and wood chips and in 30-day stationary compost piles in an experimental compost operation at Beltsville, Maryland. In both cases the fungus was found in zones in the pile with temperatures less than 60°C. Compost piles stored out of doors 1 to 4 months after wood chips were removed yielded detectable amounts of *A. fumigatus* in the 0- to 25-cm outer layer. *A. fumigatus* constituted 75% of the viable air spora at the composting site, although at 320 m to 8 km from the site, the fungus constituted only 2% of the viable air spora. Populations in the composting process were reported on the basis of colony forming units (CFU) per gram of dry weight (GDW).

XVI. ENSILAGE

Ensilaged corn fodder[889] fungal populations largely have the appearance of those of other substrates stored under identical conditions, confirming the prominence of ecological factors governing the development of cosmopolitan and nonspecific molds. In the ensilage process, the biotope is characterized by a certain degree of anaerobiosis, a limiting and selective factor. It results in a restricted group of active species.

XVII. DUNG

The fungi which inhabit dung comprise a special group made up of members of several classes of fungi. In most cases the spores of such fungi must be ingested by an animal. During the course of passage through the digestive tract, these spores are exposed to the action of gastric juices which hasten their germinability. Upon passage from the animal they germinate, develop rapidly in the dung, and produce fruiting bodies, on or in which spores are produced. These spores may be liberated in such a way that they may become attached to adjacent or nearby herbage, to be ingested by another grazing animal to continue the cycle. A recent text on saprobic fungi[567] treats them among other groups of saprobic fungi. Termed "passage" fungi by Cooke,[209] these fungi are considered neutral fungi, requiring passage through an animal but causing no harm to the animal nor contributing to its well-being.

To date, attempts which have been made to study the succession of development of these fungi have been weak except for the little attention which has been given to the matter in certain laboratory exercises. The greatest interest has been shown in their morphology,[168] the peculiarity of their fruiting structures,[149,589] and the germinability of their spores in artificial nutrients in the presence of certain growth promoting substances found in manure, artificial media in which other fungi have been grown, or developed synthetically in the chemical laboratory.[910]

Work in England[1049] has used dung of rabbit, horse, sheep, and cow to determine concentrations of unspecified fractions of dung extracts to stimulate fruiting of *Pilobolus* species. Increasing concentrations of dung extracts increases sporangial development in both species although in slightly different ways. For *P. crystallinus*, the descending order of greater effect is rabbit, horse, sheep, and cow; for *P. kleinii*, the order is horse, rabbit, cow, sheep.

Chapter 11

POPULATION GROUPS — WATER

I. FRESH WATER FUNGI

The fungus populations of water have been studied largely for the various types of interesting species of Mastigomycotina to be found therein. Little attempt has been made to obtain complete populations from a single body of water by numerous sampling devices, with the possible exception of Douglas Lake, Michigan.[1088] Sampling for various types of species in other regions of intensive research on aquatic Phycomycetes, such as North Carolina and Massachusetts, has not yielded species lists for definite bodies of water so far as available literature indicates. Chytrid populations of planktonic algae have been summarized in England[172-174] and for North America in a book on fresh water Phycomycetes, now in its second edition.[1089] The fungi which occur in fresh water have been sampled by numerous devices, and usually when cultures developed in which soil populations appeared, they were considered as contaminants.[509] Such fungus spores as may be present in water samples appear to have reached the samples in one of three ways: from spore-bearing populations of the bottom and immediate shores, from spores precipitated from airborne populations, and from runoff water added to the stream from adjacent fields, woodlands, and other surfaces. This type of population has not been studied for clean streams; however, clean portions of several streams,[214,255] reaches of which have been polluted by outfalls from treated and untreated sanitary and industrial wastes, have yielded an interesting population. Of these, some members may be natural to the stream bank and bottom population, and others may be dependent, at least in part, on the pollution load carried by the stream as a source of nutrients. The more heavily polluted portions of a stream appear to contain larger numbers of fewer species than do the adjacent clean portions.

For many fungi, water rather than air is the medium of life, although most of these species require high levels of dissolved oxygen. Many of the Mastigomycotina cannot exist without water, since their nutrition patterns and their method of dispersal are adjusted to this environment. Although some of these fungi are found in soils,[508] they thrive for the most part in bodies of water.[203,604,1024,1089] In some cases it has been demonstrated that elaborate hormonal systems have been built up by such fungi.[940] When they develop parasitism on other aquatic plants, drastic changes in the plankton population may result.[171]

To date, few attempts have been made to develop, so far as the writer is aware, a quantitative sociological study of aquatic fungi in streams, lakes, ponds, or other bodies of water. Many difficulties of technique are involved in such a study. Regional species lists at the ordinal[203,605] or the genus level[604,1024] provide some information about distribution and habitat. Additional information is given in annotated lists and discussions.[1176] One study has attempted to determine something of the distribution of aquatic fungi at various depths at various locations in lakes.[557] It was found that species varied at different depths within the same lake. Suzuki (quoted by Cooke[247,255]) and Willoughby[1257,1258] reported the development of quantitative techniques for flagellated organisms.

It has been noted that in the Mastigomycotina two types of cilia have been produced on the zoospores.[283] These may have various functions in aiding the movement of zoospores through water, but in considering these fungi, their principal use has been phylogenetic, both within the fungi as a group and in attempts to relate fungi with other groups such as algae and protozoans. These types of cilia include: those of the true chytrid line which are simple, posterior, whip-like cilia with a thin lash or tailpiece; those of the hyphochytridiomycete line with a single anterior tinsel-type cilium wherein the cilium appears feather-like; and those of the oomycete line with both types of cilia, an anterior tinsel cilium and a posterior whiplash cilium.

In addition to the Mastigiomycotina, a certain group of Hyphomycetes, some of which have ascomycetous perfect states,[936] occur in water where there is an organic substratum, such as decaying leaves, on which to develop. Such fungi have been of interest chiefly because of their peculiar spores which are thought to be especially adapted to life in an aquatic habitat.[586,589,936]

Based on surveys conducted in southern Georgia and northern Florida,[573] two groups of Oomycetes were found. The first group, consisting of species characterized by centric and subcentric oospores, exhibited a marked seasonal periodicity; the second group, consisting of species forming eccentric oospores, showed no discernible periodicity. This periodicity pattern may be universal based on a fragmentary literature. While temperature may play a part in this periodicity pattern, insufficient data are available in the literature to confirm it. On the basis of literature reports, it appears possible that species with centric and subcentric oospores (regardless of generic affinities) are more northern in distribution, while those with eccentric oospores have a more southern distribution pattern. Insufficient data were available to extend these observations to the tropics.

By incorporating aliquot samples of water from nature in a solid such as cornmeal agar and incubating this under water, Willoughby[1257] could count the number of saprolegniaceous fungi in the surrounding fringe and thus obtain an estimate the number of propagules obtained. Such estimates were probably based on zoospores. All estimates made in developing the technique were made from surface water samples only. At Windermere Lake, England, margin samples gave total Saprolegniales estimates of less than 25 to 3200 per liter, and at mid-lake estimates yielded up to 100 per liter with a mean figure of 11 per liter. At Wraymeres Fish Hatchery, total Saprolegniales estimates ranged from 400 to 4600 per liter. At Windermere Lake margin, there was an indication that there was a correlation between high estimates and heavy rainfall and high lake levels. Using this technique, the results suggested that *Saprolegnia* was the most conspicuous genus, followed by *Achlya*, and *Aphanomyces. Dictyuchus,* and *Leptolegnia* were obtained only occasionally.

In Japan, Suzuki (quoted by Cooke[247,255]) counted numbers of hyphae developing on hemp seeds as a basis of estimating numbers of propagules which were probably zoospores, each hypha arising from a single zoospore. Using the technique described above, Willoughby[1258] found that at the lake in Blelham Tarn there was evidence of high activity in periodically inundated soils at the lake margin but not in bottom muds from the lake itself.

Using Willoughby's technique for enumeration of Saprolegniaceae, Dick[315] studied the occurrence of these fungi in soils at Blelham Tarn in England. He found a series of patterns forming a complex mosaic ranging from 5-cm core samples of soil to samples representing as much as 1 m^2 of soil and finally to the production of a continuum to as much as 100 m in extent.

While on the basis of sampling techniques used, it appeared[313,314] that the distribution of Saprolegniaceae in certain areas resulted in groups of species which could be recovered regularly over a period of many months, the origin of such colonies of fungi in the soil remains obscure. Relative water content, pH levels, and other, as yet undetermined, factors are probably involved in both primary colonization and perpetuation of the colonies, although it is doubtful if any single factor is responsible.

Johnson[603] inoculated several saprolegniaceous fungi with the polysporangiate endoparasite *Rozella achlyae*. Material from *Dictyuchus anomalus* was inoculated in cultures of *Achlya flagellata, A. proliferoides,* and *D. monosporus*. Successful infection obtained in the latter two species was attributed to age of the host and the environment in which the host was cultured at the time of inoculation. Host strain resistance or the existence of biological races of the parasite were among factors considered in explaining unsuccessful inoculations. Johnson[602] found that *A. glomerata* has been parasitzed by a species of *Olpidiopsis*.

Over a period of 20 years, desmid populations in Lake Windermere, England have been observed by Canter and Lund[175] in relation to fungal parasitism. A number of species of chytridiaceous and biflagellate phycomycetous fungi were found. There is evidence that the reported observations were not restricted to a single period or a single location. Parasitism did not alter the seasonal periodicity of the desmid population. There was no evidence that the desmid population must be adversely affected by environmental conditions to become parasitized. The evidence suggests that the parasites commonly infect healthy and growing cells.

On the basis of observations on aquatic fungi in Douglas Lake, Michigan, Paterson[888] postulated three major conditions of host relationship: fungi are parasitic on algal cells, attacking living cells during periods of active growth; fungi attack only senescent cells; and fungi attack only dead algal cells. Those chytridiaceous fungi which attack diatoms apparently do so in upper levels of lake water, completing their life cycle as the diatom cell falls to the bottom of the lake. A common habitat of chytridiaceous fungi is chitinous in nature, probably indicating an important role of these fungi in aquatic habitats.

Scott and O'Bier[1020] have reported isolation of 64 isolates of 13 species of filamentous aquatic fungi from fish and fish eggs in 14 states. These ranged from 14 isolates each of *Saprolegnia* sp. and *S. parasitica* to one isolate each of *S. ferax, Aphanomyces levis, Pythium ultimum,* and *Allomyces anomalus*. Inoculum studies indicated that, of these isolates, six species would grow on molded platy fish, but other species tested could not be induced to grow on fish under the experimental conditions used.

Saprolegniosis of fish in Belgium, India, and Zaire was studied by Nolard-Tintigne[836] in the field and in the laboratory. In the laboratory, guppies and xiphos were used as test organisms. Infection was successful only when zoospores were present in the cultures used, and they were rejected by healthy muscle tissue. All strains of the fungi used proved to be infectious, and no differences were detected between strains of saprobic and parasitic origin. There appeared to be no relation between previous viral or bacterial infection and epidemics of saprolegniosis, nor were there adverse effects of toxins. Histopathological studies were made of diseased fish, confirming infections which were generally systemic and invaded the spinal cord as well as the lumen of blood vessels. It was shown that the fungus grows at a constant rate in the tissues resulting in a relation between the time taken by the fungus to invade the tissues and the time necessary to cause the death of the fish.

Experimental work on the dispersal of spores of aquatic Hyphomycetes, as well as of other types of spores,has been summarized by Ingold.[589] Theoretical considerations concerning convergent evolution in relation to the development of the tetraradiate spore in these fungi have been developed by Ingold.[591]

Other factors in the ecology of fresh water fungi have been summarized by Sparrow.[1090] He considered quantitative studies, occurrence in lotic or stream, and lentic environments, such as temporary pools, ponds, and lakes, from the point of view of open water mass and the plankton and adventitious plankters associated with it, as well as with submerged structures and bottom muds. Among special habitats considered were thermal springs, sphagnum bogs, inland salt lakes and salt pools, and acidotrophic lakes. Among habitat factors considered were turbidity of the water, light, substratum, temperature, oxygen, and pH. Phenology, or seasonal occurrence, evidence for succession, the question of communities, and geographic distribution were also considered.

In a series of papers brought together in *Recent Advances in Aquatic Mycology*, Jones[613] has arranged for a number of specialists to bring their areas of interest up to date. Of the topics discussed, 16 are concerned with subjects which have been considered here. In his contribution, Dick[316] said:

> Our knowledge of the ecology of the aquatic Phycomycetes is incomplete. Many of the problems require solutions. Many of the problems still await definition. Models may summarize our current concepts of the ecology of aquatic fungi, but the balance between an opportunity and an individual, and the genotypic adaptation of a species to particular substrates is hard to gauge. Until there is confidence in the significance of particular aspects of descriptive studies there can be only limited approaches to experimental ecology. For such studies to be worthwhile and successful, it is essential to possess a convenient quantitative method, a knowledge of the life form being sampled, a knowledge of the ecological life cycle, a knowledge of the natural substrates, and a knowledge of the likely origin of both substrate and fungus. Without such information that environment cannot usefully be manipulated.

Sladecekova[1063] in Czechoslovakia has developed concepts concerning the significance of the periphyton occurring on the walls of reservoirs, especially in relation to theoretical and applied limnology. It would be of interest to determine the extent to which periphyton could develop in the absence of any organic enrichment of the reservoir or its influent streams.

Working in Northern Ireland, Park[865,866] found that several techniques used concurrently were necessary for the development of information about total populations of heterotrophic microorganisms in streams studied. He classified the organisms he found in several categories: among the *resident* species there were *indwellers*, which were *permanent residents*, and *immigrants*. Immigrants included *migrants*, whose occurrence was *periodic*, and *versatiles*, whose occurrence was *irregular*. Immigrant organisms usually arrived with already colonized substrata such as plants and plant parts, animals and animal parts, and soil. Among the possible combinations of categories of presence and activity of decomposers in water are

Presence category	Possible activity categories			
	Constant	Periodic	Sporadic	No activity
Residents				
Indwellers	+	+	+	−
Migrants	−	+	+	−
Versatiles	−	−	+	−
Transients	−	−	−	+

From Park, D., Trans Br. Mys. Soc., 58, 291, 1972. With permission from **Cambridge University Press.**

Immigrants arriving in an *inactive* condition either go into an immediate decline or colonize a substrate *in situ*. Those arriving in an *active* condition either colonize a substratum *in situ*, continue their activity which could have been limited to the concurrently colonized substratum, or go into an early cessation of activity resulting in their decline.

II. MARINE FUNGI

There has been an increasing interest demonstrated in the fungi which are able to adapt themselves to a marine habitat. Exposed pilings,[73] special wooden sampling devices, driftwood, and other substrates, including algal hosts of disease-producing fungi, have been found to furnish nutrients to such fungi. Most of these fungi are members of the Ascomycotina and the Deuteromycotina. A fungus parasite of eelgrass (*Zostera*) has decimated the populations of this plant, and as a result of unbalancing of food chains, this destruction has resulted in drastic alterations in other marine populations. Fungus parasites of oyster eggs,[343] crab eggs,[284] and yeast populations of shrimp[906] should be mentioned.

According to Johnson and Sparrow,[608]

> There are species of fungi which occur exclusively in a salt water environment, or so we judge from the literature and from our own observations. These could be considered true marine fungi, and defined as follows: capable of developing to reproductive maturity even though they are exposed at some point in their growth to salinities of 30 o/oo (parts per thousand) or more, either while continuously submersed or intermittently inundated by tidal waters. By this definition — admittedly arbitrary — a fungus in ocean waters is marine irrespective of the duration of time that it has occupied the salt environment or its adaptive magnitude. Therefore, a fungus can be marine whether its tendencies are euryhaline (adapted to a wide range of salinity), or stenohaline (tolerant only to small magnitude changes in salinity), halophilous or halolimnic.

Oceanic and estuarine waters exhibit a striking range of living conditions including temperature from below the freezing point of pure water to 40°C, salinities from less than 5 o/oo to more than 37 o/oo, pressures from 1 to nearly 1000 atm, nutrients from barely detectable amounts to rich solutions, and light from high intensities to complete darkness. The maintenance of the marine biocycle is profoundly influenced by the fact that the ocean water is in constant motion. In addition, the organisms living in the community may play an important role in the development and maintenance of that community.

Johnson and Sparrow[608] described the habitat and its effects on the fungi found there and catalogued those species reported from marine and estuarine sampling programs. Emphasis was placed, at least in part, on several species which cause disease in marine animals.

While the factors listed above are theoretically important in the habitat and living conditions of the fungi, little experimental work has been done beyond some studies on temperature and salinity to determine their effect on the fungi according to Johnson.[606] He summarized information available on habitat and geographical distribution, with emphasis on lignicolous species, nonlignicolous species, and yeasts and yeast-like fungi.

Certain fungi are able to penetrate calcareous substrates, including the tubes of wood-boring animals, shells of barnacles and mollusks, and the surfaces of calcareous incrusted algae.[653] Fruiting structures of the Ascomycetes *Halosphaeria quadricornuta* and *Ramispora salina* and the deuteromycete *Periconia prolifica* form cavities in the calcareous tube linings of teredinid burrows, making them brittle and soft. Representatives of *Cirrenalia* and *Humicola* sporulate on the surface of calcareous linings. *Pharcidia balina,* a fungus also referred to as a lichen, decomposes shells of barnacles and mollusks. *Lulworthia kneipii* parasitizes calcareous algae, probably living in the middle lamellae rather than in the calcareous cell walls.

From three brackish water lakes in Japan, 24 species of marine and terrestrial fungi were isolated.[1160] Spore germination and assimilative phase studies showed that these fungi tolerate conditions equal to 30 to 70% sea water.

Based on the thought that temperatures may play an important role in the distribution of wood inhabiting Ascomycetes and Deuteromycetes in the ocean, Hughes[572] proposed a zonal classification for such marine fungi. The zones he selected were

1. Arctic-Antarctic — limited by the 10°C isothere for the warmest calendar month
2. Temperate — the poleward boundary is set by the 10°C isothere for the warmest calendar month, the boundary toward the equator by the 17°C isocryme for the coldest calendar month
3. Subtropical — between the 17 and the 20°C isocryme for the coldest calendar month
4. Tropical — the 20°C isocrymes for the coldest calendar month north and south of the equator.

Distribution of 14 ascomycete species and 6 deuteromycete species based on these zones was mapped showing that temperature may be as important as salinity or more so in the distribution of marine fungi. It is indicated that these temperature relations may be more accurate indicators of distribution, that there may be physiological races of *Torpedospora radiata*, and that the distribution pattern for *Trichocladium achrosporum* may be temperature independent.

In the study of marine species of marine Mastigomycotina,[411] among ecological and physiological considerations were types of nutrient used, temperature and salinity stress phenomena, and related problems. It was thought that any determination of physiological phenomena in vitro is of little ecological value until these studies were paralleled and confirmed by parallel studies in the natural habitat, the ecological niche of the fungus in the sea. Morphological variability can be related to characteristics induced by salinity, temperature, and other physiological factors of the environment, all of which may effect taxonomic considerations.

Using pollen bait,[410] the lower marine fungi of the North Sea and the Norwegian Sea were investigated in surface waters on four voyages of a laboratory ship. Fungi found in these surface waters varied from low to high number. In a study of the significance of the method, seven replications were made near Helgoland resulting in a standard deviation of only 1.155.

Hughes[575] developed an extensive survey of the fungi of marine habitats including lignicolous, caulicolous, and foliicolous species reported from oceans and estuaries. It was noted that the term thalassiomycetes has been coined for such fungi. For a definition of the marine habitat, Hughes adopted the classification used by Smith in 1950: *marine*, where salinity is 30 o/oo or greater; *marine dominated*, where the salinity variation is moderate, but concentrations part of the time are equal to those of sea water; *typically estuarine*, where tidal and seasonal variation is the most striking characteristic, with salinities as high as sea water and as low as fresh water encountered periodically; *tideless brackish*, where salinities are less than those of sea water and the salinity variation is seasonal and often of long duration; *limnetic* or *fresh water*, where the salinity variation is absent, and salinity is absent at all times; and *special instances* where hypersaline conditions occur as a result of evaporation or where waters have different proportions of ions. Going beyond depths at which studies of many species of marine fungi have been made, Kohlmeyer[652] has found that some wood-decaying fungi are active at 1616- and 2073-m depths off the coast of North America.

Since information concerning the occurence of microorganisms in ocean sediments in areas of potential offshore oil drilling might be important,[205] aerobic heterotrophic bacteria, yeasts, and filamentous fungi were isolated from samples from the ocean floor off eastern North America. Groups of filamentous fungi and yeasts of particular interest were those capable of degrading petroleum and chitin.

III. SEWAGE AND POLLUTED WATER

If sewage is considered a type of dung, as well as a waste product or organic enrichment, on the basis of the amount of fecal matter carried therein, listings of the fungi found in this habitat should be mentioned.[220,257,273] Here the fungi were considered in relation to their ability to exist in the habitat. Coprophilous fungi were thought to be organisms restricted to dung habitats. Fungi and other organisms restricted to the sewage habitat were called *lymabiont;* those which seemed to prefer this habitat but could live elsewhere were *lymaphiles;* those which preferred other habitats but could live in the presence of sewage were *lymaxenes;* those which could not survive in such habitats were *lymaphobes*. Species were assigned to one or another of these categories on the basis of their occurrence in the habitat, the number of colonies isolated from one or another type of sample from sewage treatment plants or sewage polluted waters, and the number of colonies isolated from clean areas in contrast with polluted waters.

After water is used in the household or in industry, it becomes sewage or idustrial waste, now referred to preferentially as "waste water". As such it forms an excellent habitat for fungi, and from the user to the sewage treatment plant, the fungus populations which have been observed are luxuriant. From sampling during the process of treatment[227,228,263,274] or from effluents of a particular process,[90] it is evident that fungi can grow and reproduce, at times vigorously, in this habitat. Occasionally fungi capable of producing disease in animals and man[265] or in plants[219] are found to grow and multiply in this habitat. *Phialophora jeanselmei* (now assigned to the genus *Exophiala*) has been recovered from Lytle Creek and from the Dayon, Ohio sewage treatment plant, as indicated in Figures 1 and 2.

The most abundant organisms in biological sewage treatment are bacteria and fungi. Algae occur in the surface slimes of trickling filters, in slime accumulations on the walls of aerators and other structures, and in the waters or ponds, channels, and other structures built for sewage treatment of one type or another. Protista of a wide variety of types are also found. There may be amoebae, ciliates, and other types of minute

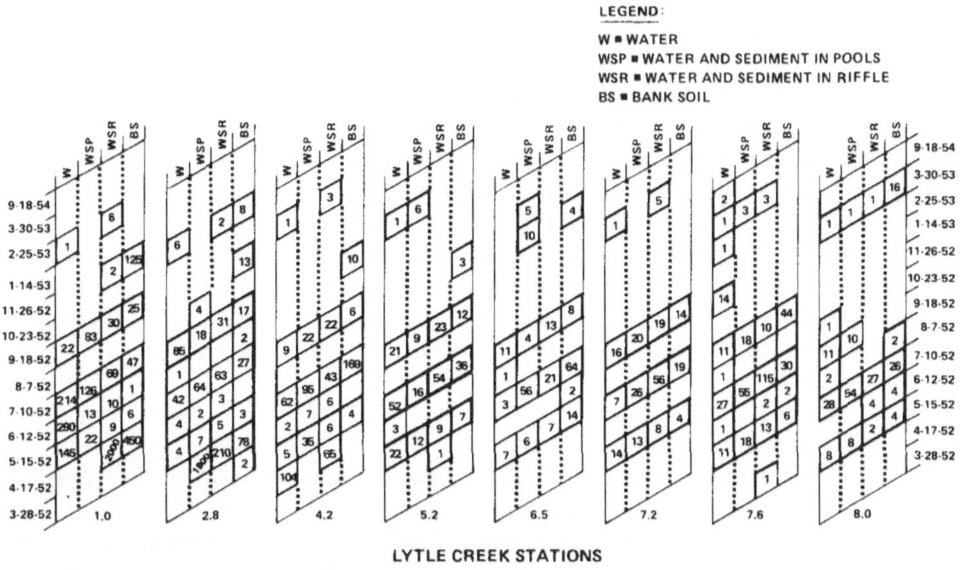

FIGURE 1. Occurrence of *Phialophora jeanselmei (Exophiala)* in Lytle Creek, Clinton County, Ohio. Number in each box refers to total colonies isolated from 5 m*l* of diluted sample.

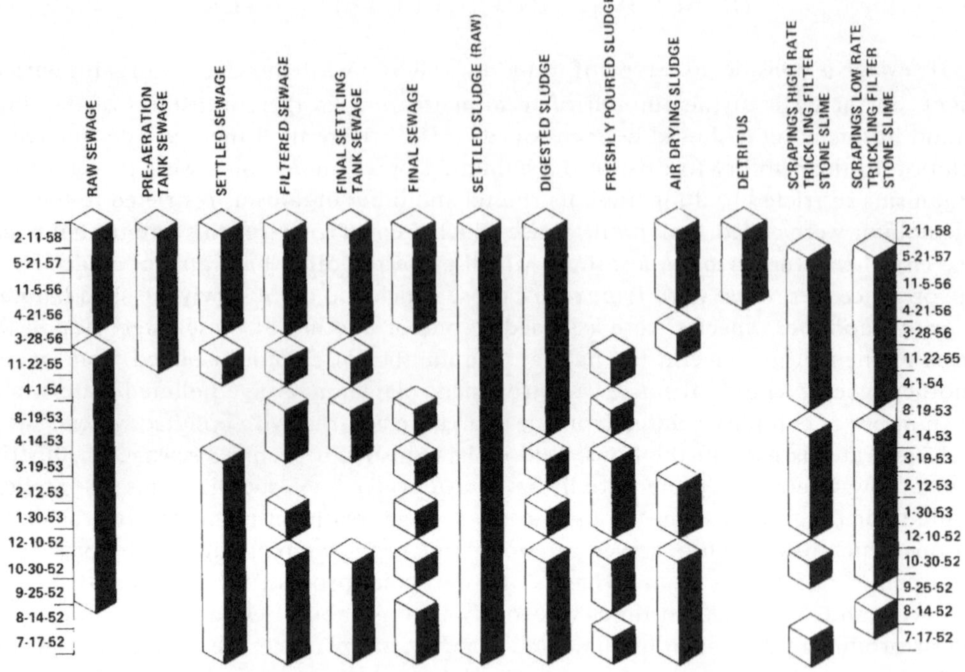

FIGURE 2. Occurrence of *Phialophora jeanselmei (Exophiala)* in the Dayton, Ohio sewage treatment plant. Blocks represent presence, regardless of numbers of colonies isolated.

animals. Higher forms such as nematodes, rotifers, insect larvae, arachnids, worms, and snails, are also found.[228]

Those algae which are found are producers which, under certain conditions, may assume a heterotrophic type of nutrition, especially at night when the DO (dissolved oxygen) is low or absent. At least one genus of colorless algae (*Prototheca*) is encountered,[241,242] and its activities are yeast-like, always heterotrophic.

The animal population, consumers, always ingests food. This may be done by surrounding the food, sweeping it into the gullet, or eating through grazing or browsing. In most cases, the ingestion process affects the bacterial and at least some of the fungal population. Particles of waste matter may also be ingested, and to the extent that these are digested, the consumers may be considered as reducers as well. However, reducer organisms are present which will attack and utilize the waste products of metabolism of all three types of organisms. They are, thus, the ultimate reducers.

The habitat of the sewage treatment plant, whether of the trickling filter or the activated sludge type, or other type of treatment process, such as waste stabilization ponds or sewage lagoons, is ideally suited to the activities of microorganisms. The primary organic feed of these systems is the waste products of the metabolism of higher organisms, especially man, and the waste products of man's multitudinous activities. All these waste products contain organic matter, much of which is easily attacked by a variety of bacteria and fungi.

To date, no deliberate attempt has been made to establish populations of bacteria and fungi in the waste treatment habitat. Casually, to start a trickling filter or an activated sludge tank, inoculum from similar units in other communities has been added to the new unit, or quantities of manure or soil have been introduced in the hope that the proper population will become established more quickly than it could from natural build-up as a result of the continuous flow of sewage and industrial waste. In addition, or separately, certain types of yeast cakes have been used as starters; these were either units of baker's yeast, prepared for a different purpose, or units of special organism preparations containing, hopefully, cells of organisms which would start or continue action of a septic tank or other treatment facility, or enzymatic preparations of such cell activity.

The sewages added to these installations contain a variety of materials in addition to the waste products of man's metabolism. From the bathtub or shower come detergents or soaps, water softeners, and the washings from the surface of the human body, and sometimes those of his pets, From the wash basin comes the dirt from man's hands and face, the residues from his shaving, mouthwashes, tooth scrubbings, and related activities. From the kitchen come residues from the washing of fruits and vegetables, dish washing, cooking, garbage disposals, and, sometimes, washings from dirty hands, kitchen floors, walls, woodwork, wallpaper, and other surfaces. From the basement or utility room come wastes from the laundry, the washing of dirty hands and feet, and a variety of effluvia from a multitude of household chores. If the storm drain is included in the household sewage system, regular supplies of runoff from driveway, exterior surface walls, lawn, flower bed, and other surfaces will be added. From all these sources there is a continual increment of bacterial cells and fungus spores and cells added to the influent to the sewage treatment plant. Many of these additions will find the sewage treatment plant to be a most hospitable place in which to live, since there is a continual supply of food in a liquid environment at a temperature which is not difficult to survive. A few of these invaders of the sewage treatment plant will make it their permanent home; others will not be adaptable to the new habitat or will not be able to meet the competition.

A. Sewage Fungus

In an early review of studies on fungi in polluted water and sewage, Cooke[214] pointed out that several species of bacteria, fungi, and protozoans have been referred to as "sewage fungus". Work on sewage fungus, mostly in England, has been reviewed by Curtis.[289] The ecology, physiology, and nutrition of these organisms — with emphasis on *Sphaerotilus natans*, the bacterium commonly referred to as the sewage fungus — were described, and methods for the control of these organisms were discussed. The occurrences of *Fusarium aquaeductuum, Geotrichum candidum,* and *Leptomitus lacteus* were indicated as near the bottom of the list of those species tabulated as of importance.

B. Nutrition of Sewage Fungi

1. Nutrients

In a consideration of the nutritional requirements of nine fungi commonly isolated from sewage and polluted water,[223] it was indicated that there is a definite group of filamentous fungi which can become acclimated to an aquatic environment. Such fungi, usually considered to be members of the natural populations of soils and litter and usually thought to be inactive under the water-logged conditions of such environments, have become naturally adapted to the specialized environment of the sewage treatment plant through ready availability of nutrients and an apparently more favorable level of competition for that food supply. While such a population appears to have been developed by strain selection from soil-inhabiting ancestors, there is no indication that the majority of species in the group could not return to an existence in the soil at any time.

In England [860], various aspects of the nutrition of which four fungi which commonly inhabited percolating or trickling filters were studied. An apparently undescribed spe-

TABLE 1

Economic Coefficients of Sewage Organisms

Organisms	Maximum growth (mg dry matter)	Glucose consumed (mg)	Economic coefficient of cell synthesis $\frac{\text{growth} \times 100}{\text{glucose consumed}}$
Bacteria			
Zoogloeal bacterium (No. 10)	40	160	25
Zoogloeal bacterium (No. 8)	38	143	27
Fungi			
Geotrichum sp.	85	225	38
Sepedonium sp.	112	206	54
Trichosporon cutaneum	140	230	61
Fusarium aquaeductuum	130	212	61

From *Water Pollution Research*, Report of the Director, England, 1955, 55.

cies of *Sepedonium* required organic sources of nitrogen, as well as thiamin and biotin, for optimum growth. Its pH range was narrow, between 7.0 and 8.5. *Fusarium aquaeductuum* was able to utilize ammonium and nitrate nitrogen and needed no external sources of vitamins. *Geotrichum* sp. was able to utilize ammonium and needed no external supply of vitamins, and *Trichosporon cutaneum* used ammonium but required thiamin. Zn, Ca, Mn, and probably Fe and Cu were required by *Sepedonium* sp., which was the only one of the four fungi inhibited by concentrations of Zn above the optimum in media deficient in other elements. Mn, Fe, or Cu, when added with Zn, counteracted the inhibitory action, while Ca increased it.

Using a standard test system (cotton cloth in liquid medium), fungi known to be capable of using cellulose as a sole carbon source, which were isolated from sewage, were tested for cellulolytic activity.[260] Inocula of fungus mycelia were less effective in aiding the reduction of cellulose (cotton) fabric than inocula of spore suspensions when the growth medium was raw sewage. For other media, spore suspension inocula were less effective. Raw settled sewage appeared to contain organisms capable of utilizing cellulose as a carbon source in the presence of other organisms which may remove toxic substances and/or provide growth promoting substances.

Preliminary experiments on the efficiency of microorganisms in sewage were conducted in England.[33] Typical bacteria and fungi isolated from trickling filters were used. The growth rate and the rate at which glucose was destroyed were determined in still culture in 25 mℓ of medium. The "economic coefficient", the ratio of the dry weight of growth to the corresponding weight of glucose destroyed was reported (Table 1).

2. Nuisances and Problems

Based on water user complaints of unpleasant flavors in British tap waters, tests were reported[34] on presence of growth of fungi in water-conducting systems. Colony counts of 10/100 mℓ were not usually considered significant. Counts of up to 100/100 mℓ or more could indicate growth of fungi in pipes, water, water purification plants, or reservoirs. Tests were made with the membrane filter and a somewhat modified Martin's medium. Actual numbers had little significance since they depend on the number of spores and mycelial fragments washed off the growth in the pipe, while the unpleasant taste in the water depended on the organisms producing it. Identity of the taste-producing organism was not indicated.

A descriptive summary of biological problems encountered in water supplies[734] described a large variety of such problems. In the 301 listed and tabulated reports and papers, only two (0.7%) indicated that fungi contributed to this problem. Occurrence of excessive growth of true fungi in sewage treatment systems can usually be attributed to faulty construction or overloading of the plant with sewages rich in nutrients, resulting in excessive growth of certain fungi.[214,247]

C. Sewage Treatment Systems

1. Trickling Filter Types

To determine the extent, or lack of it, and the techniques used for the study of bare areas and the development of populations on them, with special reference to microorganisms and especially fungi, a survey of the literature was undertaken.[218] Sladecekova[1061] also surveyed the literature with emphasis on periphyton. A technique was then developed[224,263] for the study of fungal populations on trickling filters using glass slides on standard and high-rate trickling filters at the Dayton, Ohio sewage treatment plant, which had been surveyed for fungi in an earlier study.[227] After sterilization, 60 sets of four slides each, mounted on hardware cloth tagged with metal numbers, were placed at angular positions on the surface stone of trickling filters and removed at the end of 24 hr, 48 hr, 5 days, 1 week, and at weekly intervals until all were returned at the end of 57 weeks. On arrival at the laboratory, the slides were air dried, weighed, rewetted, and dismantled. The growth was removed with a rubber policeman and subjected to action in a Waring blender. Portions of it were then plated for fungi and directly counted for algal cells and fragments of algal filaments[228,263] (Figure 3). A diagram of the ecosystem (Figure 4) was developed as well as curves illustrating growth of the organisms found (Figures 5 and 6). A diagrammatic cross section of the slime which developed on the slides and the filter stone was presented (Figure 7).

Work on periphyton in Czechoslovakia prompted the comparison of techniques for obtaining quantitative results from experiments in which growth was compared using fresh weight, air-dry weight, and oven-dry weight.[1060] Based on a statistical analysis, the coefficient of correlation ($r = 0.987$) showed high significance. Thus, a change of air-dry weight by 1 mg involves a change of oven-dry weight of only 1.03 mg. We can assume, therefore, that both values are practically identical, and the use of air-dry weight instead of oven-dry weight determination is satisfactory.

The results of different investigations[519-521] on fungal populations in trickling filters (sewage bacteria beds, percolating filters) in England support the hypothesis that, in bacteria beds having film dominated by fungi, the seasonal fluctuations in the standing crop of fungal mycelium is the result of seasonal changes in temperature and nutrition conditions. This is the outcome of microbial competition and predation under different conditions of nutrition and temperature. Superimposed on these fluctuations are the effects of macroinvertebrate grazing activity. Although the grazers (fly larvae) were not found to be the cause of the seasonal fluctuations in film, it is considered that they play an important role in the economy of the beds by tending to prevent excessive accumulations of film.

The trickling filter[228] is an ecological tool of the sanitary engineer in the treatment and purification of domestic and industrial wastes. Filter bed populations are made up largely of organisms dependent on preformed organic matter. The organisms forming the film develop rapidly at first, but rate of development slows as the film matures. Of the environmental factors observed, only BOD (biochemical oxygen demand) removal is correlated with development of the film. Colony counts of fungi occurring in the film were fitted to a logistic curve which describes the rate of development of the film. Curves developed from the data from both standard rate and high rate filters were roughly parallel.

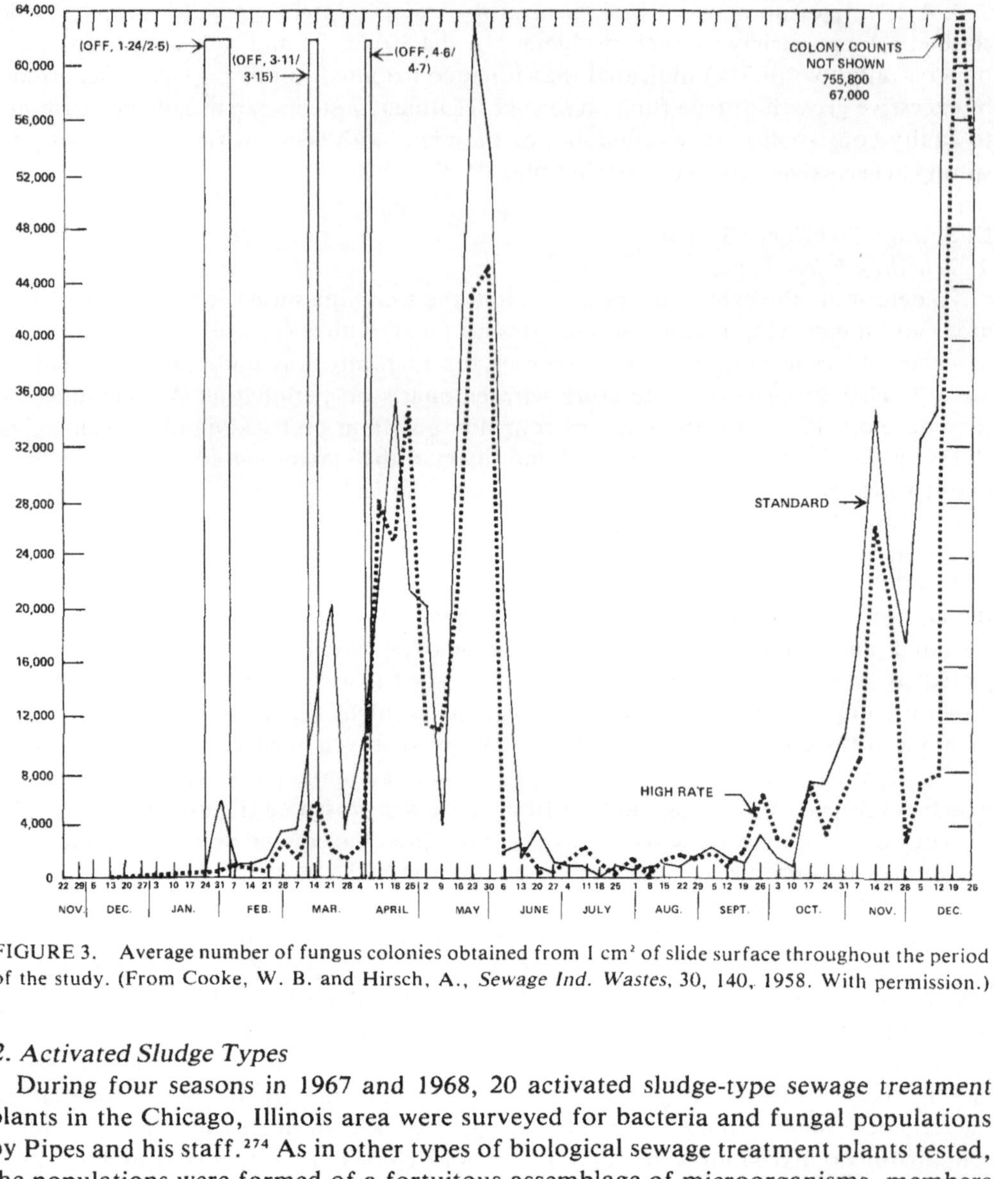

FIGURE 3. Average number of fungus colonies obtained from 1 cm² of slide surface throughout the period of the study. (From Cooke, W. B. and Hirsch, A., *Sewage Ind. Wastes*, 30, 140, 1958. With permission.)

2. Activated Sludge Types

During four seasons in 1967 and 1968, 20 activated sludge-type sewage treatment plants in the Chicago, Illinois area were surveyed for bacteria and fungal populations by Pipes and his staff.[274] As in other types of biological sewage treatment plants tested, the populations were formed of a fortuitous assemblage of microorganisms, members of which were continually dropped out as additional members were recruited. Many of these organisms were active throughout the year. Fewer yeasts than filamentous fungi were found by the plating technique. On a seasonal basis (Table 2), throughout the series of plants tested, larger numbers of fungus colonies were present in the spring, fewest in the summer, with increasing numbers in autumn and winter.

Using monthly samplings between October 1967 and April 1968, the activated sludge sewage treatment plant at Lebanon, Ohio[246,248] was sampled for fungal populations. Among these populations were found 95 species or species groups of filamentous fungi and yeasts. While raw sewage yielded 1430 bacterial cells to 1 fungus cell on the basis of colony counts, effluent from the activated sludge aerator yielded only 41 bacterial cells to 1 fungus cell. Many of the more common fungi present in the Lebanon plant were also recovered from activated sludge plants in the Chicago, Ill. area plants tested,[274] from the Dayton, Ohio trickling filter plant,[224,227] from a waste stabilization

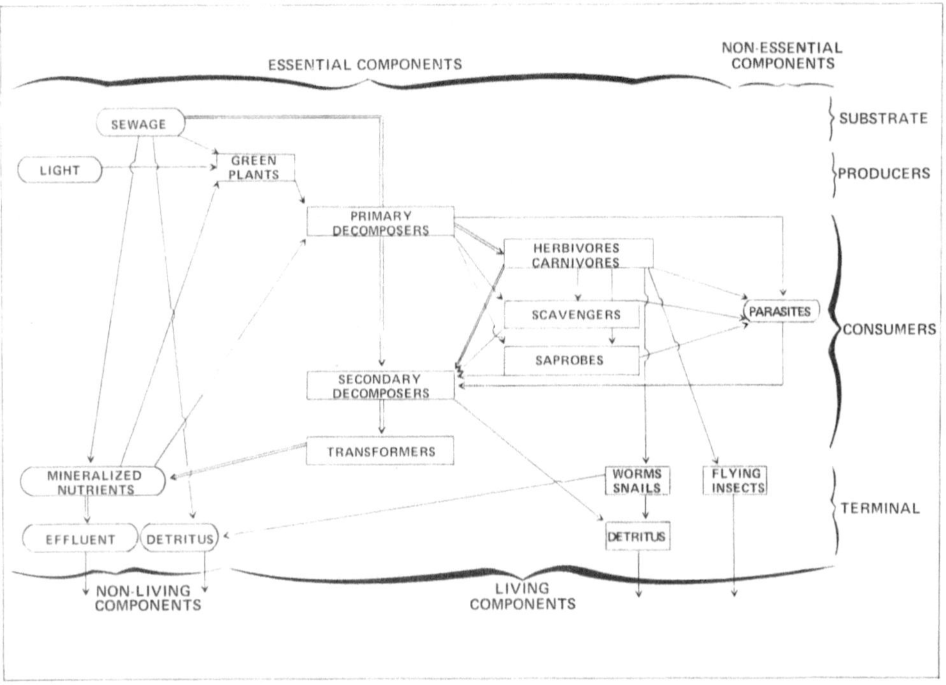

FIGURE 4. Principal steps and components in the trickling filter ecosystem. (From Cooke, W. B., *Ecology*, 40, 289, 1959. With permission.)

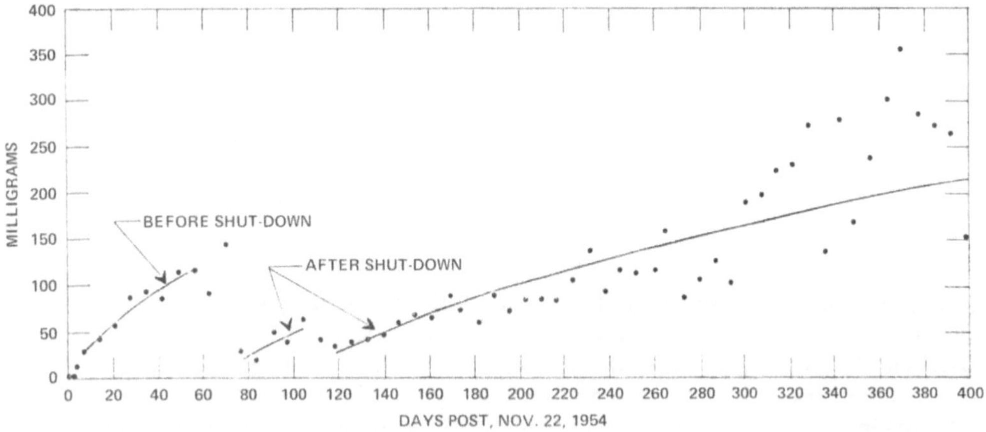

FIGURE 5. Cumulative growth on frames showing influence of shut-downs and rate of development. (From Cooke, W. B., *Ecology*, 40, 285, 1959. With permission.)

pond system near Lebanon[271] (but in a different water shed), and from other sewage treatment plants.[255] Fungi, as reducer organisms, are admirably adapted to a habitat in which their assimilative cells are continually bathed in a nutrient medium such as that offered by sewage in the process of treatment by activated sludge.

The growth of fungi, filamentous and yeast-like, in those structures devoted to the secondary and tertiary treatment of sewage indicates that the liquors being treated contain substances available to these fungi as food and that the fungi are able to use such substances as substrates. As living organisms, these fungi form part of the biolog-

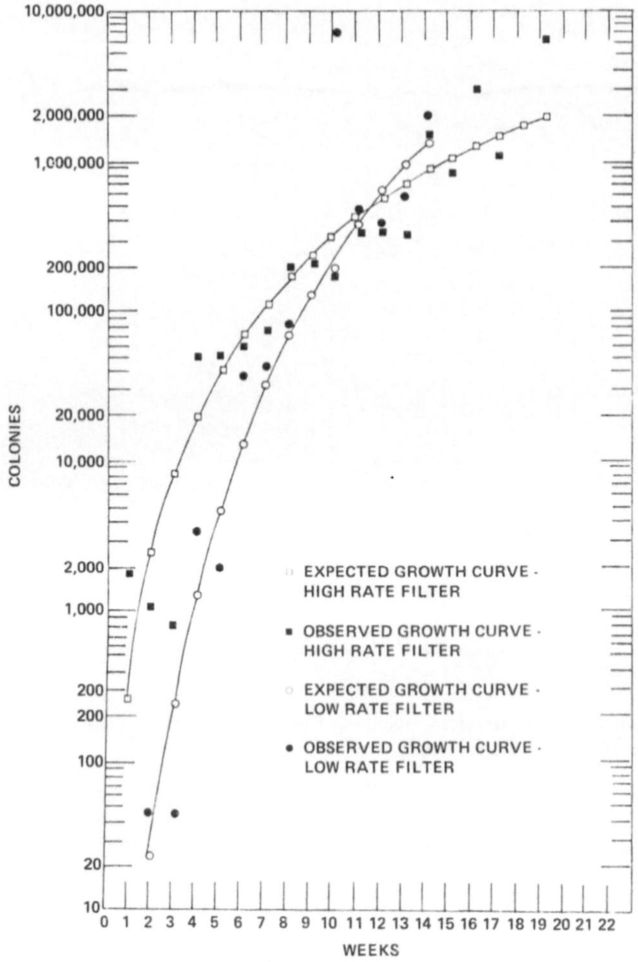

FIGURE 6. Colony counts observed and expected under fitted logistic curve. (From Cooke, W. B., *Ecology*, 40, 282, 1959. With permission.)

ical population the treatment plant was designed to attract and to put in use as the basic unit of the biological treatment system. These fungi, therefore, are performing a useful and necessary function in the biological treatment of sewages being presented to them as substrates on or in which to grow.

In working with activated sludge, Godeau[447] in Rumania prepared an ecosystem diagram (Figure 8) for that biocoenosis. On the basis of certain determinations of enzyme systems regardless of responsible organisms,[1174] it has been suggested that improvements could be made in the activated sludge process.

3. Waste Stabilization Pond Systems

Fungus populations in a small waste stabilization pond system near Lebanon, Ohio[271] were determined using a number of plating and enrichment culture techniques. Throughout the system and in an adjacent stream, 212 species of filamentous fungi and yeasts were recovered. Before the system was put in operation, few filamentous fungi and no yeasts were recovered from samples of bare soil, while after a year of operation, numbers of species increased 5-fold and numbers of colonies counted in-

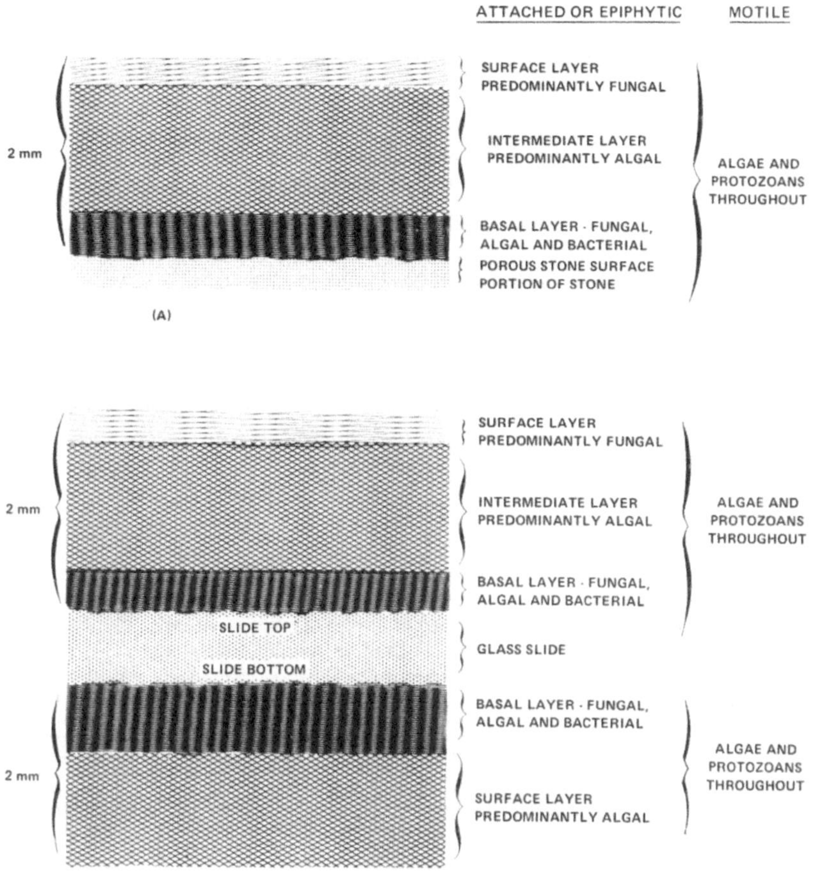

FIGURE 7. (A) Diagrammatic cross-section drawing of growth on filter stone.(B) On slide. (From Cooke, W. B. and Hirsch, A., *Sewage Ind. Wastes*, 30, 147, 1958. With permission.)

creased nearly 200-fold. Additional numbers of yeast cells and filamentous fungi in pond waters and sludges were present but not sampled qualitatively. Except for November, when adverse conditions of undetermined nature affected the fungal populations, the numbers of viable cells indicate a favorable habitat for fungi. It appeared certain that fungi can be counted among those saprobic organisms that play a part in the degradation and, thus, the removal of organic wastes from the waters of waste stabilization ponds.

Species of *Fusarium*[233] and of yeasts[270] (Figure 9) recovered from the Lebanon waste stabilization pond system were considered in separate reports.

Following the observation and study of an oil refinery waste treatment system near Houston, Texas (a treatment system in which the wastes were added to a lagoon or depression and occasionally ploughed into the soil),[331] it was decided that soil microbes can oxidize and decompose a large quantity of petroleum hydrocarbons under a wide range of soil and environmental conditions. Land spreading was thought to be an economical and comparatively fool-proof method of disposal of oily wastes. Natural chemical decomposition of oil in soil may be accelerated by judicious use of lime and fertilizer, artificial drainage, and tillage. Microbiological degradation of oily wastes takes place under a wide range of temperatures, but oxidation is faster in a warm or hot climate than it is in a cold climate.

TABLE 2

Seasonal Distribution of Fungus Groups in Activated Sludge Systems

Season	Raw sewage	Settled sewage	Mixed liquor influent	Mixed liquor effluent	Return sludge	Waste sludge	Final product effluent
	Habitats sampled						
Filamentous Fungi × 10⁶ (Total Colonies Observed)							
Spring	10.3	3.2	141.5	88.4	707.5	123.7	13.8
Summer	6.4	1.2	85.2	169.0	1111.7	731.7	2.4
Autumn	4.5	2.5	87.8	76.0	152.4	750.8	2.3
Winter	0.2	0.3	11.9	22.9	223.7	13.4	0.2
Yeast-Like Fungi × 10⁵ (Total Colonies Observed)							
Spring	0.1	0.03	0.5		3.2	11.2	0.006
Summer	0.07	0.2	10.8		3.3	10.7	0.007
Autumn	0.2	0.4	0.8		1.9	1.5	0.05
Winter	0.2	1.2	1.7		5.7	100.9	0.08

From Cooke, W. B. and Pipes, W. O., *Mycopathol. Mycol. Appl.*, 44, 262, 1970. With permission.

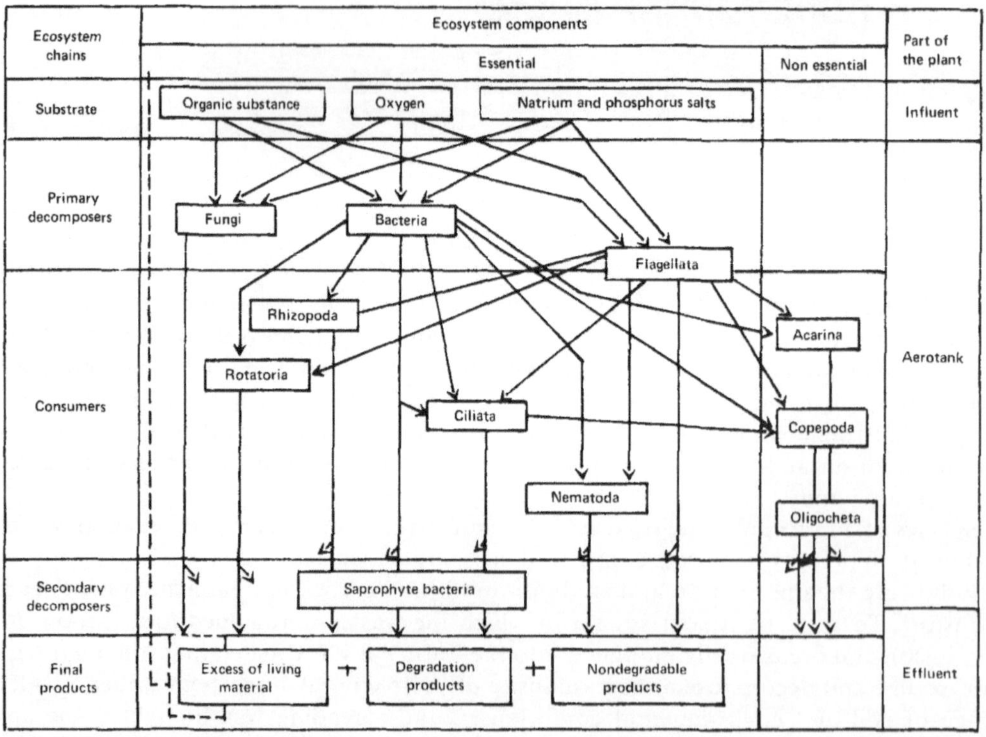

FIGURE 8. Synthetic scheme of the main ecosystem components in aerotanks (activated sludge units). (From Godeau, S., *Hidrobiologia*, 14, 324, 1973. With permission.)

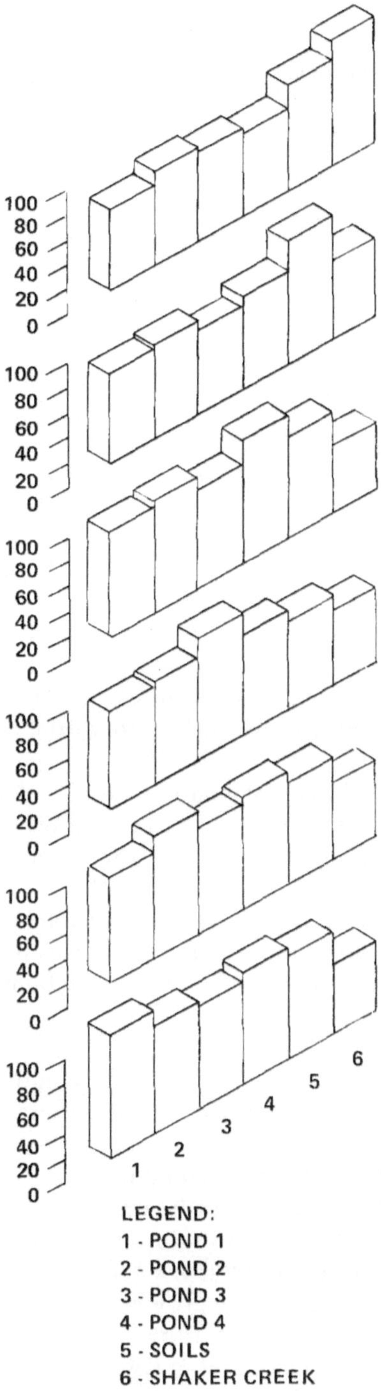

FIGURE 9. Summary of habitat fidelities. Percent of yeast species in one habitat series found in contiguous habitat series. (From *Protoplasma*, 67, 180, 1963. With permission.)

TABLE 3

Growth of Fungi in Motor Oil

Fungus	Ratio of growth in oil to growth in glucose
Fusarium aquaeductuum	
F. oxysporum	0.60
Geotrichum candidum	6.18
Penicillium lilacinum	7.58
P. melinii	0.73
P. ochro-chloron	0.64
Aureobasidium pullulans	0.51
Phialophora jeanselmei (Exophiala)	3.96
Trichoderma viride	0.68

From Cooke, W. B., *Sewage Ind. Wastes*, 29, 1247, 1957. With permission.

The biochemical destruction of oil wastes has been studied[725] in the laboratory using sewage seed. The half-life of the oil samples studied was 1 week, the end products being carbon dioxide, organic acids, and esters. Unstated species of organisms capable of such destruction were considered to be abundant in nature. Their activity was noted in winter but was also evident at summer temperatures. It may be noted that fungi isolated from a stream carrying oil refinery wastes[257] were of the same species as those found in other Ohio streams.

Ahearn and Meyers[5] have reported that filamentous fungi and yeasts are known to degrade crude oil in nature; some species of these have been tested in the laboratory. However, no single organism or combination of organisms is known to be capable of completely degrading crude oil.

In one test it has been noted[223] that certain fungi are capable of using S.A.E. 20 motor oil in place of sugar as a carbon source (Table 3). It was not indicated whether oil would be used in preference to sugar in the same culture medium.

a. Experimental

To determine[272] whether or not species of fungi found commonly in sewage and sewage treatment plants can have a positive effect on the BOD_5 (5-day biochemical oxygen demand) of an artificial sewage in the laboratory, an experiment was devised. Standard BOD test procedures were used. Organic and inorganic nutrients, mold seed, sewage seed, or a mixture of mold and sewage seeds, were added to sterilized distilled water which had been shaken by hand to get into solution as much oxygen as possible, about 8 ppm. Sewage seed was obtained from a trunk sewer serving a residential area. Mold seed was grown in the laboratory, and the growth was washed and processed in a Waring blender. Sewage seed for experiments at pH 7.2 and 9.5 was settled for 24 hr without pH adjustment; that for pH 2.9 and 5.1 was settled for 48 hr after pH adjustment to pH 3.0 about 6 hr after collection of the sample. The synthetic substrate contained 150 ppm each of glucose and glutamic acid. The only oxygen available to the fungi and the organisms in the sewage seed was that in the dilution water to which the nutrients and seeds had been added. The inoculated dilution water was placed in glass-stoppered BOD bottles which were inverted in a water bath for the incubation period. Incubation was effected in a 20°C constant temperature chamber. No oxygen could be introduced by accident to the BOD bottles. Pairs of bottles were removed and tested at intervals of 2, 4, 6, 8, and 10 days.

When compared with earlier results using *Bacillus aerogenes* as a test organism, it appeared that the oxygen depleting capacities of the fungi used equaled or exceeded those of the bacteria at pH 7.2. Fungus activity alone increased as the medium became more acid and decreased as the medium became more basic. Except at pH 9.5, fungi mixed with sewage seed have lower BOD values than sewage seed alone. It was indicated that only with molds could significant reductions in dissolved oxygen be achieved in the first 4 days of incubation at pH 2.9, while the greatest reduction with sewage seed alone occurred at pH 5.1 and 7.2. Molds, in general, were more tolerant of low pH levels. The strains of fungi used were able to compete successfully with other microorganisms for organic materials in solution and for oxygen available only in the dissolved state.

In the study of filamentous waste treatment systems at low pH levels,[145] mixed cultures of filamentous microorganisms were found to proliferate at pH values as low as 3.0 and to remove a volatile substrate. Percent volatile solids were higher at low pH values than at normal levels. A relationship was found to exist between mixed liquor volatile solids and the total organic carbon of the unit. A relationship was found to exist between the chemical oxygen demand and the total organic carbon of the soluble unit contents. Mixed liquor total suspended solids were maintained easily at any desired level up to 7000 to 8000 mg/ℓ.

b. Predacious Fungi

Zoophagus insidians, a rotifer-trapping predacous fungus (Figure 10), was found to be a problem in experimental activated sludge[267] treating nitriles. The sludge was seeded with sanitary sewage. The fungus was controlled with chlorination. The fungus has also been found in activated sludge units fed with sewage in Richmond, California.[914,915] Pipes also found four species of *Arthrobotrys,* a nematode-tapping fungus. *Z. insidians* has also been noted in experimental activated sludge units treating plywood glue wastes at Corvallis, Oregon.

c. Chlorination

In 1936 Smith and Purdy[1080] laid down guidelines for control of sludge floc in the activated sludge process by chlorination. Frequent periodic examination of the floc under the low power of the microscope was necessary to determine the condition of the floc. Bulking may be caused by diffuse structure of the floc, not controllable by chlorination, or by (unidentified) fungus growths or by a combination of both. Chlorination of return sludge at low rates was thought to be a useful method for combatting bulking caused by fungus growth. It was indicated that the proper rate of chlorination lay between 0.7 and 7 ppm of return sludge or between 0.01 and 1.0% of the dry weight of solids, depending on the character of the sludge floc.

No other guidelines for control of fungi or for disinfection of fungi in sewage effluent, sewage treament plant effluent, sludges, or water treatment were given in the literature until recently. Chambers[188] summarized information available on germicidal efficiency of silver, iodine, and quaternary ammonium compounds, without reference to fungi.

Recently Engelbrecht and colleagues[365] found that a yeast (unidentified) was isolated among other organisms surviving chlorination of waste water effluents. This yeast resisted 1.0 mg/ℓ free chlorine for 20 min in contrast with a pure culture of *Escherichia coli* which failed to survive 5 min contact with 0.03 mg/ℓ free chlorine. It may be noted that yeasts do not necessarily serve as test organisms against which to compare all fungi because of the different natures of the cell walls of the two types of organisms.

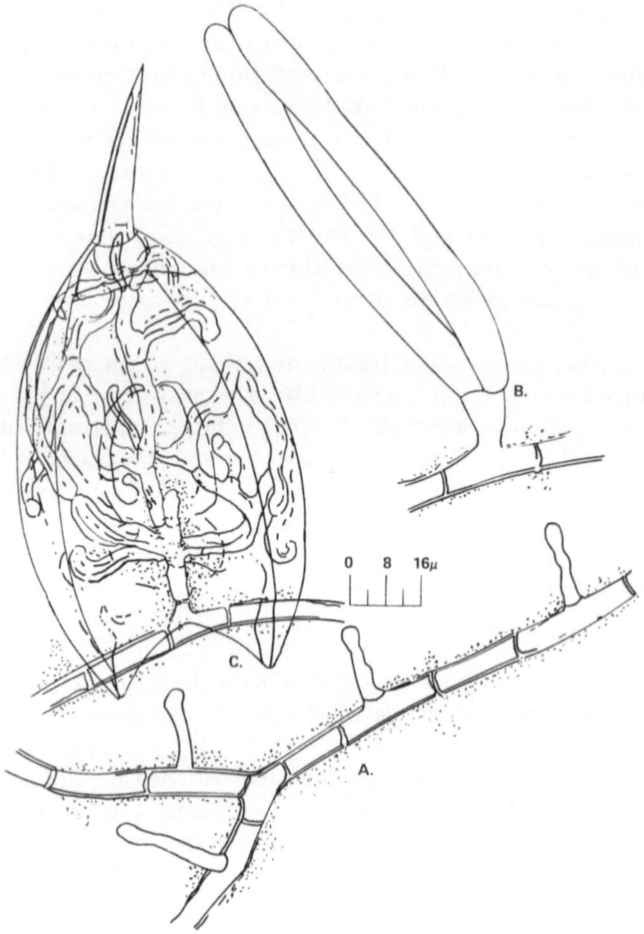

FIGURE 10. *Zoophagus insidians.* (A) Mycelium with mycelial trapping pegs. (B) Mycelial cell with two gemmae. (C) Mycelium with trapping peg which has trapped an individual of the rotifer *Monostyla* and grown into it, consuming the contents of the shell which is about to be discarded. (From A Laboratory Guide to Fungi in Polluted Waters, Sewage and Sewage Treatment Systems, No. 999-WP-1, U.S. Public Health Service Publications, Cincinnati, 1963, 75. With permission.)

D. Polluted Streams

Until the late 19th century, a popular method of sewage disposal, if not the only method, was direct flow of sewers into nearby streams or other bodies of water. As populations increased and pressure to relieve previously clean streams from such unsightly loadings grew, albeit slowly, efforts were made by communities and community groups to clean up the streams by concentrating the sewage through a system of trunk sewers and directing its flow to sewage treatment plants. At first, these were small unit cesspools or septic tanks. The primary-type sewage treatment plant was followed by variously designed secondary-type sewage treatment plants, such as those using trickling filters or activated sludge processes. Presently, the tertiary-type sewage treatment plant has come into favor. With each design improvement, cost to the community has increased, and the service has been extended to all the facets of suburbia. If population increase is not met by increased sewage treatment plant size or efficiency, the quality of treatment, as demonstrated by the effluent, decreases. Whatever level of treatment is adopted by the community, the effluent from the treatment system is still directed to a receiving stream. The organisms present, or those attracted by the organic enrichment, act as those in a trickling filter or an activated sludge tank and serve to aid in the purification of the water.

1. Lytle Creek, Ohio

From April 1952 through March 1953, Lytle Creek, Clinton County, Ohio,[230] was sampled at monthly intervals to determine the nature of fungal populations in the stream which received the effluent of the primary-type sewage treatment plant serving Wilmington, Ohio. Prior to the construction of the plant, the stream had an intermittent flow pattern. In 1951, several sets of samples[509] yielded few aquatic fungi, but in 1954, after construction of an activated sludge-type sewage treatment plant, larger numbers of such fungi appeared at all stations. The 1952 to 1953 survey sampled soil fungi only. The fungi found in the water and streambed samples of Lytle Creek habitats show a definite response to stream conditions resulting from pollution produced by direct dumping into the stream of raw sewage or of sewage which has passed through a primary-type sewage treatment plant (Figure 11). The creek received raw sewage above the treatment plant, and this enrichment of the stream bed resulted in an increase in the population of fungi below this accidental outfall. At the outfall of the treatment plant, the populations of soil fungi were depressed as a result of septic conditions. They recovered downstream, and in the lower end of the creek, in the lower clean water zone, they surpassed the populations of the clean water zone above the treatment plant. All habitats sampled in the stream and its bed were affected in a similar manner by the polluting materials. However, since the more common fungi found in the creek were self-sufficient in all respects (except that they are heterotrophic), it may be assumed that the fungi — at least, the members of the permanent population — were contributing to the purification of the stream and its water as well as to the food supply of other organisms present in the habitat.

2. Other Streams

On the basis of studies at the Lebanon, Ohio sewage treatment plant[248] and the stream receiving its effluent, Turtle Creek, it was concluded that — as in the case of Lytle Creek, Ohio,[230] the Bear River in the Cache Valley, Idaho and Utah,[239] the Cache la Poudre River, Colorado,[243] and elsewhere[257] — a group of fungi made up of many

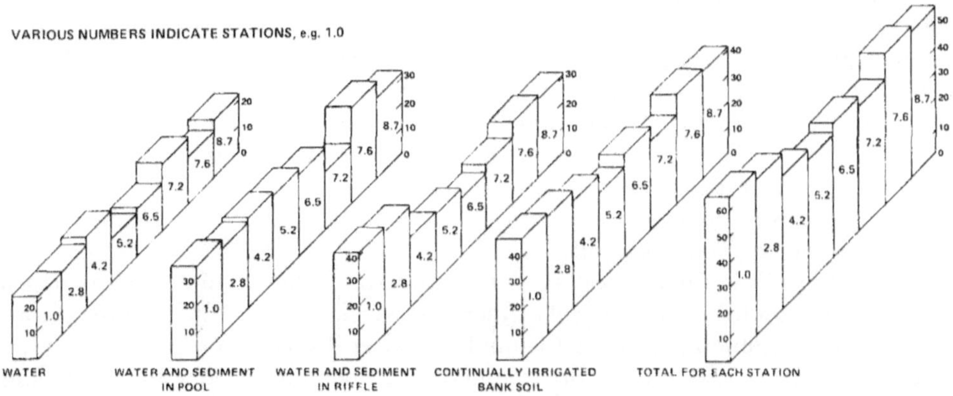

FIGURE 11. Lytle Creek. Percent of total (127) species in each sampling site. (From Cooke, W. B., *Ecology*, 42, 13, 1961. With permission.)

TABLE 4

Number of Fungal Colonies Recovered from Habitat Types in the Bear River Basin by Plate Count Methods (Autumn Survey)

Code	Habitat type	Number of samples	Total species	Estimated averge number of colonies per volume of sample
	Cache Valley			
A	Above sources of pollution	3	44	38,000/g
O	Obvious sources of pollution	10	44	270,000/mℓ
B	Below sources of pollution	2	35	220,000/g
V	Valley streams after absorbing pollution	12	77	209,000/g
S	Soils (mostly agricultural, irrigated or not, sometimes with polluted water)	5	50	402,000/g
	Logan Valley			
AP	Animal pollution (pasture or grazing)	3	37	200,000/g
RB	River bank (no domestic livestock)	3	19	34,000/g
PT	Path and toilet area (dry)	3	23	18,000/g
S	Spring water (Rick's Spring)	1	3	8/2L
U	Unpolluted (apparently) by man or livestock	4	32	250,000/g

From Cooke, W. B., *Proc. Utah Acad. Sci. Arts Lett.*, 44, 309, 1967. With permission.

of the same species is adaptable to the conditions resulting from organic enrichment of streams resulting from sewage or industrial waste additives to the streams. These organisms contribute to the removal of this polluting organic matter. At least monthly sampling from any group of stations is needed to describe the populations more accurately and more meaningfully. Hawkes[518] has discussed aspects of fungus growth in a consideration of biological aspects of stream pollution.

Among streams receiving various polluting effluents which have been sampled is the Cache la Poudre River, Colorado.[243] Samples of water, bottom sediments, bank soil, and upland soil (above high water mark) from nine stations (two times from five stations and one time from four stations) were sampled on agar plates, with baits, and in liquid culture. Sample locations were primarily in lowland stations below a sewage treatment plant, and especially a sugar beet mill, but also included relatively clean stations on Poudre Lake at 10,750 ft in Rocky Mountain National Park. A total of 128 species was recovered. While human, animal, and plant pathogenic fungi were present, species found were of widespread occurrence, and no serious involvements traceable to the river and its populations were noted. However, if streams such as this continue to receive organic enrichment, it is possible that increasing loads of potentially pathogenic fungi may be found.

The Bear River[239] in the Cache Valley in southeastern Idaho and northern Utah was sampled in relation to beet sugar mill effluents as well as those of the sewage treatment plant of a small community and effluents from several small food processing plants. In the Cache Valley (two to three times) and on the Logan River (one time), 46 sample sets were tested (Table 4), and 109 species, including filamentous fungi and yeasts, were recovered using three basic procedures and five types of media. A fidelity chart (Figure 12) shows the percentage of species in one habitat type found in each of the other habitat types represented in the surveys on the Bear River and Logan River valleys. In the Bear River valley, stream samples taken above obvious sources of pollution (A), at obvious sources of pollution (O), and just below those sources (B) show more restrictive conditions because of lack of nutrients or excess amounts of rather specialized nutrients, than do streams after pollution has been absorbed (U) and agricultural soils whether these are irrigated with stream water or not (S). As the absorbed pollutants in the stream water become more dilute, they may become more available to a larger number of organisms. The agricultural soils may include species not adaptable to conditions in the streams. About 40% of the species were isolated from only one habitat type, the remainder from two to six habitat types. This may be a reflection of the low number of samples from any one station studied, it may be a result of the seasonal nature of the principal sources of organic pollution, or it may be a direct response on the part of an increasing number of species to more favorable nutrient conditions in a modified environment.

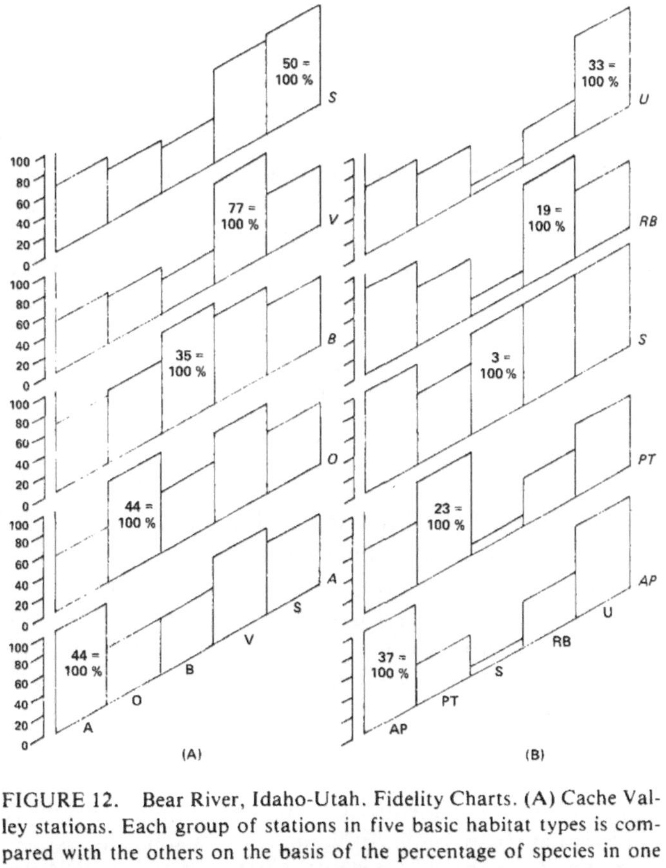

FIGURE 12. Bear River, Idaho-Utah. Fidelity Charts. (A) Cache Valley stations. Each group of stations in five basic habitat types is compared with the others on the basis of the percentage of species in one habitat type occurring in adjacent habitat types. Letter code: A—above pollution sources, O—obvious pollution sources, B—below pollution sources, V—valley stations, S—soils. (B) Logan River samples. Letter code: AP—animal pollution, PT—soils in pathway and around pit toilet doors, S—spring water, RB—river bank soil, U—apparently unpolluted areas. (From Cooke, W. B., *Proc. Utah Acad. Sci. Arts Lett.*, 41, 308, 1967. With permission.)

Hynes[585] has surveyed the biology of polluted waters with special reference to streams and conditions in England. To introduce the subject, he surveyed the history of water pollution. Natural waters and natural qualitities are considered as a prelude to background information. Additional background factors include ecological factors as they relate to river floras and faunas. A variety of human influences on natural waters is discussed. The chemistry of various types of effluents is discussed in relation to physical and chemical effects on rivers. Some biological effects of effluents, such as those of poisons, simple deoxygenation, suspended solids, and organic matter, are discussed. The effect of heat, salts, and other types of pollutants on lakes is also considered.

A direct correlation between the increase of pollutional organic matter and the development of aquatic Hyphomycetes was noted during a "starch campaign" (waste release from the manufacture of starch products) in Czechoslovakia.[1062] These fungi colonized artifical substrates in addition to standard wood exposure materials, porcelain, and plastics. Of the 16 spore types observed, the more frequent species in the streamlet studied were *Anguillospora longissima, Clavariopsis aquatica, Piricularia aquatica,* and *Lunulospora curvula.*

3. Acid Mine-Drainage

A study initiated in the spring of 1964[238,256] was based on samples of water, water-creek bank interface soil, and upland soil (from above the high-water mark) collected along streams carrying acid mine-drainage in Ohio and West Virginia. Relatively large numbers of fungus colonies were recovered using dilution plate and enrichment culture techniques. Many of these species have also been found in streams draining areas which produce no acid pollution and in streams receiving sewage effluents. Some of the species may be present by accident and inactive; others, although present by accident, may have found the habitat conducive to active growth processes. Active species may be beneficial in removing organic pollutants from the water and, at the same time, also be involved in the production of organically stabilized colloidal iron. Apparently, true fungi are not directly involved in transformations leading to the type of pollution referred to as "acid mine-drainage" or to reduction in acidity of the habitats in which they can survive and grow.

In McCreary County, Kentucky[1211] a study was conducted on Cane Branch of Beaver Creek draining an area which had been strip mined between 1955 and 1959. Helton Branch was used as a control. More numerous and more diversified fungal biota was found on Cane Branch, finally including representatives of 42 genera of filamentous fungi and yeasts. Of these, 21 were isolated only from samples collected on Cane Branch. A species of *Rhodotorula* was found consistently on Cane Branch, never on Helton Branch. Seasonal differences in fungal populations appeared to be relatively insignificant.

E. Pulp and Paper Mill Wastes

A number of species of fungi have been isolated from materials collected in pulp and paper mills in Sweden (from paper mill air, white water, slimes),[777] Nova Scotia,[131] New York,[1203] and elsewhere. It was determined[132,133] that some of the fungi involved in slimes had relatively broad temperature and nutrition requirements and could be stimulated by at least two unidentified chemicals in the white water.

As a means of disposal of spent sulphite liquor[231,235] from a pulp and paper mill at Niagra, Wis., the waste water was sprayed on a 40-acre field in northern Michigan. After 3 years, samples of the soil were obtained. Samples were taken at 0, 2, and 6 in and 5, 10, 15, and 20 ft in areas receiving the waste and in an adjacent area not receiving waste water. It was found that the area was essentially an abandoned river valley which had been filled with glacial debris and the 20 ft sampled was largely a bed of coarse sand. Wastes percolating through this sand were essentially free of BOD-producing substances by time they left the area in springs which feed the Menominee River. Much smaller numbers of fungal spores appeared in lower levels sampled and in the control area. However, larger numbers (Figure 13) were present in treated soils. Amounts of organic matter (Figure 14) were only slightly larger below the 5-ft level in the sprayed area, but Folin-Dennis reacting materials (Figure 15) increased rapidly below the 5-ft level. Compared with untreated soil, the numbers of fungal colonies were greater at all levels below 1 ft. The isolations indicate that, as long as nutrients are available, fungal cells are recoverable from all levels down to 20 ft in a coarse, sandy soil.

F. Effect on or of Chemicals

Hexadecanol has been used as a producer of monomolecular film on water to retard evaporation. In carboys in a 20°C constant temperature chamber, using raw Ohio River water, hexadecanol pellets were found to have developed encrusting layers of fungal growth. It was suggested[723] that, as a result of the decomposibility of this material in nature, this should be considered in the economics of its use. There appeared to be little build-up of extractable organic material in the substrate as a result of hexadecanol decomposition.

Ludzack and Ettinger[724] have presented a list of chemical compounds which, at the time of its compilation, were not known to be biodegradable and, therefore, could

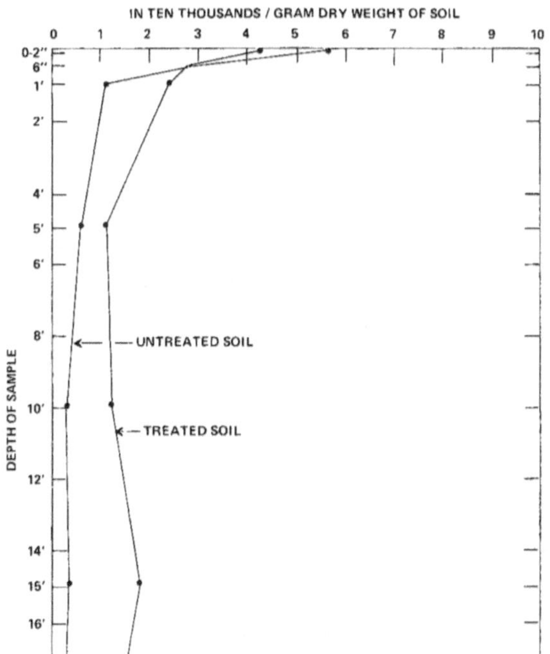

FIGURE 13. Colonies of fungi in treated and untreated soils. (From Cooke, W. B., *Eng. Bull. Purdue Univ. Eng. Ext. Ser.*, 45, 43, 1961. With permission.)

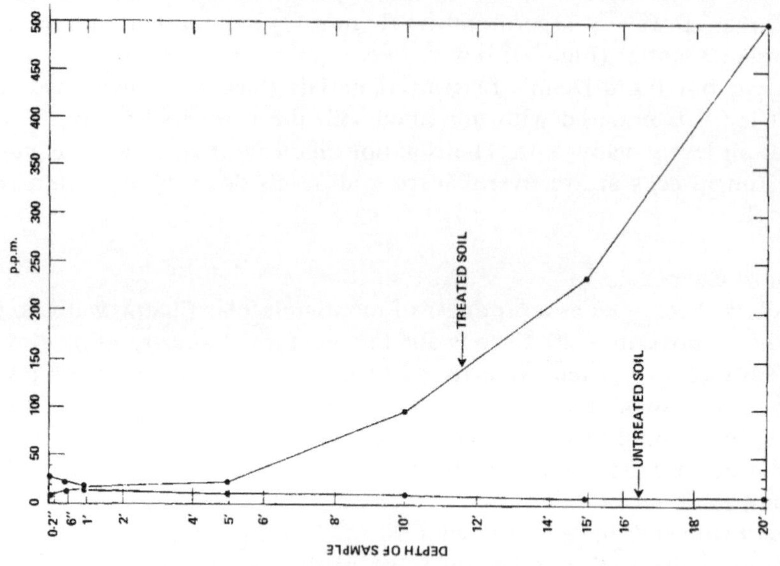

FIGURE 15. Folin-Denis reagent reacting materials in treated and untreated soils. (From Cooke, W. B., *Eng. Bull. Purdue Univ. Eng. Ext. Ser.*, 45, 40, 1961. With permission.)

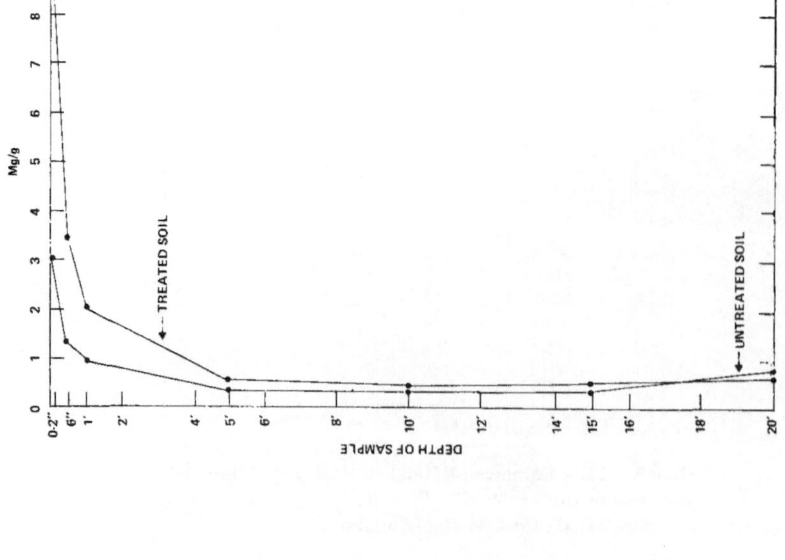

FIGURE 14. Organic matter in treated and untreated soils. (From Cooke, W. B., *Eng. Bull. Purdue Univ. Eng. Ext. Ser.*, 45, 39, 1961. With permission.)

present problems to the waste treatment plant operator. Possible strains of fungi could be developed to degrade some of these compounds.

Using 0, 1, 3, and 9 mg/ℓ zinc sulfate, Williams and Mount[1251] studied the effects of zinc on populations of small streams. As concentrations of zinc increased, populations of phytoplankton decreased, but those of fungi and slime-forming bacteria increased. The only fungus specifically noted was *Alternaria* sp. It was assumed that these organisms were using the dead plankton organisms as food.

G. Use of Sludge on Soils

Anaerobically digested sewage sludge[245] was spread on a field in northwestern Pennsylvania, and aerobically digested sludge was spread on a field in southeastern Pennsylvania. Soil samples were taken before and after sludge addition and at weekly intervals for 5 months thereafter, and were plated for fungi. From these samples, 95 species or species groups of fungi were recovered of which several were known to represent species known to produce plant disease. The extent to which populations of such fungi could build up in soils to which sewages or sludges are added is unknown.

In a greenhouse experiment[250] using two soils (Clermont silt loam and Otokee fine sand), two grasses (*Dactylis glomerata* and *Phalaris arundinacea*) were planted in sufficient separate pots so that aerobically digested sludge could be added to one set and anaerobically digested sludge could be added to the other set, both at the rate of ¼ in per week or ¼ in every 3 weeks. At the end of 5 and 9 months, soil samples were removed and plated for fungi of which 56 species were isolated. Of these, 13 species were present in the original Clermont silt loam, 17 in the Otokee fine sand, 18 in the aerobically digested sludge, and 12 in the anaerobically digested sludge. The build-up of fungal populations indicated that, in addition to a rhizosphere effect which may have been present, the fungi were actively reducing the organic matter added to the soil in the digested sewage sludges.

H. Laboratory Guides

Work on fungi in waste water, waste water treatment systems, and polluted waters has been summarized by Cooke,[234,247,249] and a laboratory guide and systematic list of fungi found in such habitats has been prepared.[257]

Chapter 12

POPULATION GROUPS — AIR

While no fungi are known to grow and reproduce in the air, the spore load of certain areas, especially in relation to allergies and epiphytotics, is usually considered a "population". Most fungi with "dry" spores are transported by air, and when spores of a certain species are released to the air in large enough numbers by explosive forces from asci and basidia, from conidiophores, or by simple release from pressure of numbers or by air currents, they can result in serious infestations of material, crop plants, allergies of man, or even more serious diseases of man and animals. Splash dispersal can also release spores, especially those which are produced in slime, into the air during periods of precipitation.

Surveys of fungus spore populations have been carried out in a number of places over the surface of the earth. These usually followed an intensive pollen sampling program and sometimes were conducted simultaneously with the pollen sampling. They were usually made in connection with studies of allerogen populations. Among the common genera found by the gravity technique was *Cladosporium,* and its presence in the air over London has been emphasized. In Israel, in addition to a series of annual gravity plate surveys for allerogenic species,[642] the spore content of orange groves and packing sheds has been sampled with particular concern for species of importance in the pathology of the Shamouti orange.[95] Several surveys of the mold spore populations of the air in New Zealand have been prepared separately or in conjunction with studies of pollen incidences. Two such studies[349,694] may be cited in addition to the one on media cited under techniques. These are of interest not only for the species lists presented, which are similar to those of other areas, but also for the differences in results obtained by the methods used. In Canada, one report[1034] listed a Mastigomycete spore as having been found on gravity-collecting plates exposed in Toronto.

Canadian workers have been intensively interested in the incidence of fungus spores in the air over seemingly unlikely areas.[855,856] Most of this sampling has been carried out on routine missions by Royal Canadian Air Force planes or in special planes obtained for the study. Plates have been exposed over the Atlantic Ocean and over large areas of the Canadian Arctic regions in an effort to learn something of the distribution of fungus spores in air masses. Fungi of most importance in such studies included rusts, smuts, and other fungi capable of causing damage to crops vital to Canadian and U.S. farmers. An early report on fungi of the upper air in the U.S. was published from work carried out in Tennessee.[1275]

The relation of large numbers of spores in the air to the incidence of epiphytotics of cereal grasses has received attention in Minnesota[1096] and Kansas.[854] The techniques of sampling rust and smut spore populations in areas where wheat is a major crop are important to the forecasting of disease control measures.

Fungal populations of the air are of great concern to man as possible sources of allergies resulting in asthma or vasomotor rhinitis. One of the more important centers for research in this type of study in the U.S. has been the University of Texas.[814] English workers[765,766] have studied this problem from the point of view of the quantitative aspects of allergy to house dust in which the spores in house dusts of various types have been analyzed in relation to sensitization risks from the inhalation of fungus spores. In the air of houses infested with the dry rot fungus *Serpula lacrimans,* spores can be found in large numbers[478] during the time of their active production, especially in the cellar and lower floors. This is important not only in the spread of the fungus itself but also as a possible source of allerogenic sensitization.

In most publications dealing with allerogenic fungi and fungi in the air, the authors are content with presenting lists of species found, together with statements concerning their abundance or rarity. Such work relies on texts based on general monographic studies, regional biotas, or Saccardoan compilations. One departure has been made in Argentina where a manual[1288] (published in Spanish) is devoted to the listing and description of the fungi associated with allerogenic studies.

The fungus spora in the air over Kansas was sampled by Pady[852] using a Pady-Bittis sampler in which 256 silicone-treated glass slides were exposed. An exposure of 30 min was made daily during the summer and once or twice a week in the winter. Results were reported on a cubic foot of air basis. Fungus spores were present throughout the year but were distributed seasonally, with peaks in July and August and low concentrations in winter. Summer numbers ranged from 30 to 700 per ft^3, with occasional peaks to over 1100 per ft^3. In the winter numbers dropped to 5 to 20 per ft^3.

Using a slit sampler and a General Electric electrostatic air sampler placed on the top of a building in Manhattan, Kansas[858] from September 1956 to September 1958, spores were obtained on silicone covered slides and colonies on nutrient plates. More than 168 species of fungi were identified. The maximum number of spores recovered was 2877 per ft^3, the number growing as colonies on the agar 567 per ft^3. The commonest colonies were of *Cladosporium* (44.5%). Other groups included nonsporulating (17.6%), *Alternaria* (12.6%), yeasts (8.4%), *Penicillium* (6.3%), and *Aspergillus* (5.4%). The most numerous spores were of *Cladosporium* (40.9%), basidiospores (24.3%), yeasts (7.3%), smuts (5.9%), two-celled hyaline spores (4.4%), and *Cercospora* (1.0%).

Pady[853] developed a continuous quantitative slit-type spore sampler that gave 24-hr-long deposits on a 1 × 3 in. slide. The sampler is watertight, compact, and has performed satisfactorily under all weather conditions. At a flow rate of 0.4 ft^3/min, the efficiency was reported to be at least 50%.

The Kramer-Collins spore sampler[660] is a continuous type spore sampler. Fungus spores 3 μm or larger are collected on glass slides, in bands, one band per hour. The sampler may be run continuously or intermittently through the hour, on a household current or car battery. At an air flow of 0.8 ft^3/min, efficiency is 100% at wind speeds of approximately 2 mph, declining with increasing wind velocity. The sampler is adaptable to a variety of field and laboratory studies of air spora.

The air spora of New York state was sampled for fungi using Durham samplers placed on top of buildings at 35 locations so that they were 50 mi apart. They were in use during 6 months each in 1953 and 1954.[843] The total numbers of fungus spores for the seasons usually closely approached, and often surpassed, the total number of pollen grains. *Cladosporium* usually gave the highest counts, often giving more than half the total count. Other high numbers were given by *Alternaria, Fusarium,* and *Curvularia.* Because of the difficulty of identification of other types of fungus spores, they were lumped in a miscellaneous category.

To determine background levels of air spora in six major cities in the U.S., Adams and co-workers[2] used Anderson samplers in each of three different seasons in each city. It was found that high levels of bacteria and fungi are present in normal air. These numbers vary with time, season, and location. Fungi represent the major component of this biota, often overgrowing colonies of other organisms. With short collecting periods, the lower stages of the Anderson sampler gave more meaningful results. With prolonged sampling periods, up to 2 hr, inhibitors must be used in the medium for meaningful results.

According to de Vries,[1187] *Aspergillus fumigatus* is frequently isolated from air indoors and out of doors. Thermophilic fungi and bacteria were better isolated on media incubated at 37°C, especially when plates incubated at 25°C were overgrown by rapidly growing fungi.

Kramer and Pady[659] found that if rose bengal is incorporated in the isolation medium and if the medium is exposed to direct sunlight, the rose bengal is inhibitory to fungal growth. If the plates are protected from direct sunlight, the photodynamic action of sunlight on the rose bengal will be prevented.

Pady and co-workers[857] exposed silicone-coated glass slides in a continuous type sampler from May to July in 3 separate years. Considerable variation in air spora was found from day to day and from hour to hour with diurnal periodicity found in all 3 years.

Spore germinability was determined[661] using 25 24-hr samplers and 10 48-hr samples of air. Overall average germinations were obtained for the following spore groups: *Alternaria* (80%), *Cercospora* (53%), fusoid ascospores (43%), *Cladosporium* (45%), rust urediospores (32%), and basidiospores (6%). It is suggested that the possible day to day variation in spore germination is related to the age of the spores.

Gregory and Montieth[479] have edited the papers presented in a symposium on airborne microbes, including a number of papers based on sampling methods, the physics of airborne particles and their deposition, members of populations, and activities of populations once successfully deposited.

Chapter 13

POPULATION GROUPS — LIVING PLANTS

I. INTRODUCTION

The literature of plant pathology[524,565,1110] is full of data and information concerning plant disease fungi, both obligate and facultative. A somewhat different approach to plant disease has been presented from Switzerland in both German[413] and English[414] editions. Here, the ability of the plant to accept or withstand infection is considered important.

A few examples will suffice to indicate the breadth of activity required in the study of fungi in relation to the diseases they cause in higher crop and wild plants. In monographic treatments of various groups of pathogenic fungi, much information can be drawn concerning life cycles, host plants, geographic distribution, and cross-inoculation studies. Such monographic treatments as those dealing with rusts,[52,288] smuts,[382,383] and the Plasmodiophorales[630] give comprehensive information concerning the species they treat. Certain generic monographs present different types of approach to the various problems presented by fungus genera to the plant pathologist. A study of the genus *Pythium*[788] presents, in addition to morhological considerations, the host range and geographic distribution of the species in this widespread and important genus. The genus *Synchytrium* has presented problems in method of treatment. One worker based his classification on the reaction of the host tissues and the galls which were produced. More recent workers adopt a more "natural" method[629] of approach. In studying the diseases caused by members of the genus *Septoria* on grasses,[1094] the possibility of one species causing a disease on many related or unrelated species of grasses was explored extensively through field collections and greenhouse cross-inoculation studies. In studying the genus *Mycosphaerella*,[53] it was found that a large number of species had been proposed unnecessarily on the basis of the idea that one species of fungus is confined to one species of host plant. An example of the type of study carried out in relation to the effect of one fungus species on its host plant was the study undertaken[31] on the disastrous effects of the *Endothia parasitica* infection of *Castanea dentata*, the American chestnut.

Root rots and root-disease fungi have been studied intensively. Reviews of the various aspects of this work have appeared and need only be mentioned here. The relation of soil fungi to root infection was discussed in an early review.[157] Root rots of cereals[1036,1037] form one of the more important groups of this general type of disease, but root rots of noncereal crops[99] and of deciduous fruit trees[275] are also important to these groups of crop plants. Most of the current knowledge of these types of disease-producing fungi has been summarized in recent texts from England.[427,433] Here it is thought that the more primitive types of root-disease fungi are those which invade juvenile or senescent root tissues that have not yet gained or have lost the ability to resist such attack. As the pathogenicity of fungi evolves, it progresses through those root-disease fungi which attack only vascular tissues producing wilts, to those fungi which produce hypertrophy of root tissue without killing the host, and finally to those fungi which live in harmony with their host, causing minimum distortion of host tissues but apparently unable to live without the host plant, as in certain types of mycorrhizae.

Specialized studies dealing with various problems of plant pathology, such as the cytology[961,962] and physiology [146,195] of host-parasite relations, and physiologic specialization of parasitic fungi[950,951] have been adequately reviewed. General host-parasite relations have been discussed by Ainsworth.[7] Because of their importance in relation to loss of crop by disease incidence, the rusts[609] and the smuts[195,561] have been studied intensively in relation to physiologic specialization, hybridization, and genetics. In attempting to control certain diseases by breeding disease resistance into a host, it is also important to learn how rapidly a pathogen can mutate to a disease-producing organism on hosts resistant to earlier strains. Vegetable crops, too, may be subject to breeding for disease resistance.[1200] Progress in various types of breeding experiments for disease resistance in wheat, oats, barley, and flax[58] and in breeding for resistance to *Verticillium* and *Fusarium* wilts[1032] has been reviewed.

II. ENVIRONMENT

Environment is very important in the development of epiphytotics caused by fungi. A crop may not be economically introduced to a new area if the climatic conditions are conducive to the development of certain types of disease-producing fungi. Various phases of the relation of the environment to cereal smuts have been reviewed.[383,1134] Maximum, minimum, and optimum temperatures in which many disease-producing fungi are known to grow have been summarized in a book published in Tokyo.[1145] The relaton of weather to the fungus diseases of plants has been critically reviewed in England.[392,393]

Microclimatic data, including the geography of spore dispersal over leaf surfaces, temperature, and relative humidities on leaf surfaces, are important types of information for use in plant pathology and for use in epidemiology. Techniques are available[1190] for synthesizing such information from limited tests, since a complete survey of a plant or a population would be a lengthy and time-consuming process.

III. FOREST PATHOLOGY

In the field of forest pathology, the diseases of forest trees and the fungi associated with staining or deterioration of forest products have been treated extensively in several texts and many separate reports and papers. Three texts published in the U.S.,[86,122,566] one in England,[182] and one in Czechoslovakia[990] cover the field extensively. The occurence and type of heart rot in living trees has been reviewed.[1189] In studying the fungi which cause decay in forest trees, the ability of the observer to identify the causal fungus is important. Keys to the identification of cultures of fungi obtained by culture from heart and sap rot specimens[832-835] have been published. In the further removal of dead wood from the forest, other wood decay fungi such as molds are operative. The studies of a worker in France[747] help to show the quantitative and qualitative occurrence of various filamentous fungi in decaying woods. The content of nitrogen in sound and decayed conifer wood has been related to the loss of weight during decay.[543] The decay caused by *Coriolus versicolor* is speeded up in the presence of nitrogen. It was found in Maine[543] that, as wood decays, its potential heating value decreases.

Many fungi cause stain reactions in wood used for building, crating, and other purposes, resulting in unsightliness. Some of the species which cause such stain reaction have been discussed[299] as a result of studies on cultures of species from the southeastern U.S. Work with these fungi has been speeded up by the realization that a serious disease of the American elm (*Ulmus americana*) is caused by one of these fungi (*Ceratocystis ulmi*).[483,583] Control of the growth of fungi in wood products during seasoning is an important factor in developing a commercial product. Problems of the control of such fungi during seasoning processes have received attention.[1184] Extensive researches on the decay produced by one of the dry rot fungi have been carried out in Germany.[371]

Based on studies in one of the older forest plantations in Michigan, one with the longest continuous records of disease patterns in plantations in the U.S., Baxter[87] surveyed a pure stand of Norway spruce (*Picea alba*) and a mixed stand of Norway spruce and locust (*Robinia pseudoacacia*). Physiological disorders were associated with cultural practices. In the mixed planting, natural splitting of trunks of spruce provided infection courts fo *Hirschioporus abietinus*, resulting in decay and, indirectly, in windfall. *Phellinus pini*, while not observed in the pure spruce stand, was associated with dead branch stubs in the mixed stand. The observations resulted in several suggestions for improvement of management practices.

Gilbertson and colleagues have surveyed wood decay fungi of various groups of woody plants in Arizona and the Southwest, expecially conifers,[438] ponderosa pine,[439] junipers,[443] and mesquite.[442] An advantage of this work is the use by the authors of the latest acceptable nomenclature for groups of difficultly identified species of fungi and the thorough searches made for species attacking the concerned hosts.

Ehrlich[351] reviewed the use of certain terms in describing various phases of fungus activity, especially in relation to forest pathology. Parasites have been described as either dependent upon or growing on living hosts. Necrophytes grow on dead tissue. They include two types of organisms: perthophytes grow on dead tissues of living plants whether or not the organism was responsible for the death of the tissue, and saprophytes grow on other types of dead tissue. Chromatophytes add a stain to the tissues in which they grow. Pathogenicity, the ability to produce disease, is demonstrated by pathogens; saprogenicity, the ability to produce decay, is demonstrated by saprobes; and chromogenicity, the ability to produce pigments, is demonstrated by chromagens. The action of a parasite enabling an abiotic agent to adversely affect a host was termed thryptogenicity and was produced by a thryptogen.

In addition to adding significant amounts of nitrogen to the soil, *Alnus rubra* may be useful in forest management by aiding in the control of *Poria weirii*, an important root pathogen of Douglas fir.[1136]

Studies of root exudates[691] of *A. rubra* and *Pseudotsuga menziesii* in a search for chemicals which might be inhibitory to *Poria weirii*, on the one hand, and not inhibitory on the other, yielded ferulic acid and syringic acid from the former, *p*-hydroxybenzoic acid from the latter, and vanillin and *p*-coumaric acids from both. Research on the hydrolytic mechanisms of both pathogen and hosts is necessary for an understanding of the possible role of these substances in nature. In a later study,[1152] it was found that in combination all hydrolysates from alder roots inhibited both tested strains of *P. weirii*, while in combination the two hydrolysates from Douglas fir inhibited one strain and stimulated the other.

From data obtained in culture, a tentative and incomplete theoretical model for the direct and indirect effects of red alder (*Alnus oregona*) on *P. weirii* and other conifers has been developed by Trappe.[1151] Alder produces *Poria*-inhibiting compounds that could reduce the pathogenic longevity in buried inoculum. These compounds, together with the relatively high nitrate nitrogen levels under alder, result in a selective increase of populations of organisms that actively compete with, inhibit, or parasitize *P. weirii*. As a *Poria*-resistant species, moreover, alder provides no satisfactory food base for maintenance of the pathogenic viability. The net theoretical potential in these phenomena is a decrease in viable *Poria* in the soil; if alder prevails on a soil long enough, *Poria* might well be eradicated from the site. This has implications for forest management in suggesting ways to control forest succession.

Using 1946 and 1972 aerial photographs of *P. weirii* root-rot pockets[828] in the high

Cascades of Oregon, progress of the disease was measured in ten infection centers. Annual extensions over the 26-yr period averaged 24 cm in radius. Relative growth rate of isolates of the fungus on malt agar were not correlated with the field activity.

Wood cubes containing *P. weirii* were placed in soil in which urea had been mixed or broadcast. The *Poria* failed to survive, and treatment of harvested areas with urea was suggested as a possible control measure. *P. weirii* survived better[826,827] in alder blocks inoculated and placed in soil treated with urea when urea levels were low. *P. weirii* numbers were greatest after 4 weeks, after which it was gradually replaced with a *Trichoderma*.

Canker diseases,[546] from several causes including debarking and fungus attacks, of aspen in the western U.S. increase with stem age, requiring more efficient management. Stands may be limited to 80 to 100 years in age by the use of clear cutting, prescribed burning, and managed wildfires, from which regenerated stands will be produced.

Peterson and Smith[902] have edited a handbook on forest nursery diseases in which 22 persons have contributed data on 33 such diseases. In general, the scope includes disease diagnosis, disease control, and methods of use and safe handling of pesticides. In particular, discussion of diseases usually includes an introduction, distribution, hosts, symptoms, life cycle, and control methods. Vaartaja[1172] has surveyed methods of chemical treatment of soils in seed beds to control nursery diseases, especially of forest trees.

In the breeding of forest trees for disease resistance,[105] especially in relation to such diseases as white pine blister rust, chestnut blight, Dutch elm disease, stem rusts of hard pines, and *Dothiostroma* needle blight, it is recognized that coexistence of host and pathogen are facts of nursery and plantation forest life.

IV. FUNGI OF PULP, PAPER, AND RAYON INDUSTRIES

In the development of the pulp, paper, and rayon industries, fungi have found a different medium for growth. In many cases, pulping processes are carried on seasonally, and the raw materials are stored until the manufacturing plant can use the excess supplies. During the storage process, if conditions are right, fungi causing stains in the pulp can develop relatively rapidly thereby increasing costs, especially for bleaching problems. Many of these fungi can cause deterioration of the cellulose; others cannot. In some cases, the stains are caused by a select group of soil fungi whose spores contaminate the air of the plant; in other cases, the usual fungi causing blueing and other stain reactions in finished lumber can cause the troubles. Control of humidities, moisture, and temperature of storage bins has been recommended for protection against such loss-producing fungi. Much research on this subject has been carried out in Sweden[777] and Italy,[448] as well as in the U.S.[86,122,566]

V. CROP PATHOLOGY

A. Synecology of Plant Disease

As a result of information gathered on plant disease fungi in a number of areas in the world, from the nothern temperate regions to desert and tropical regions, an idea has been developed[120] in which an attempt was made to apply synecological techniques to the study of plant disease fungi. The author's examples were mostly obligate parasites and, rather than relate these to specific hosts and host crops, he related them to general areas over which the weather is relatively uniform. For instance, he found that powdery mildews are more prevalent in areas approaching desert conditions than in tropical areas, while rusts present a reversal of this pattern.

B. Populations of Plant Pathogens

In a review based on the populations of plant pathogenic fungi,[1223] such fungi were considered under obligate parasites and nonobligate facultative parasites. The fungus enters the plant through wounds, natural openings, or by penetration of continuous plant surfaces. Within the plant, the fungus develops spatial relations with the host and adapts to the nutrition it finds there. The successful fungus synchronizes its life cycle with that of the host while growing in the host. Outside the host this is less pronounced, but it is necessary for survival, for establishing contact with the host, and for successful infection. There are parasites with single hosts and pathogens capable of attacking more than one or many hosts. In the case of heteroecism, the pathogen requires at least two different hosts on which to complete its life cycle.

C. Rhizomorph Studies

The growth of *Armillaria mellea* rhizomorphs[428] through soil from small pieces of infected wood used as inoculum to potato tubers used as potential host was characterized by progressive decline in weekly growth increments. The initial growth rate was related to the size of the inoculum food base; the subsequent decline was attributed in part to the depletion of nutrient reserves through fungal respiration and growth and partly to the competition for food supplies between the apex and subordinate branch apices. Vigor of infection increased with size of inoculum and with decreasing distance between inoculum and tuber.

A test of Bliss's hypothesis[429] that *Trichoderma viride* is not seriously affected by carbon disulfide fumigation of the soil and can compete favorably with or infect *A. mellea* rhizomorphs in fumigated soil was carried out. When small woody inocula of *A. mellea* had been incubated for 3 weeks in soil that had been previously fumigated with carbon disulfide at the rate of 828 ppm and then kept for 3 weeks before use, 30% of the inoculum segment ends failed to produce rhizomorphs when placed in tubes of fresh soil for a viability test. The indirect effect of soil fumigation does not wholly account for the total loss of viability when inocula of *A. mellea* are directly fumigated in soil. It seems that carbon disulfide must directly damage at least the peripheral mycelium of *A. mellea* in the woody host tissues, and this facilitates invasion of *T. viride* and/or other fungi. With this additional postulate, Bliss's hypothesis seems to account well for all the observed facts.

Additional work with *T. viride* and *A. mellea*[430] showed that, with a pure culture of *T. viride*, incubation of inoculum segments of *A. mellea* for 23 days in autoclave sterilized soil resulted in a 95% kill. Dilution of the pure culture to one fourth and one eighth strength with unsterilized soil caused a progressive retardation in this lethal effect. On the other hand, incubation of inocula of *A. mellea* in a pure culture of *Penicillium wortmannii* for the same period resulted in no loss of viability, but rather in a slight increase in rate of rhizomorph growth from the inoculum.

D. Exobasidium Studies

The genus *Exobasidium* includes parasites of a number of species of Ericaceae. Sundström[1121] studied pure cultures of 72 strains belonging to nine species representing 16 host-races, parasitizing the same host species and causing similar disease symptoms. Useful taxonomic characters were found in the ability of strains to use various carbon sources including the splitting of arbutin. Two host-races were able to use hydroquinone. Morphological characters were generally constant. Some strains regularly grew in a mycelial state, others in a yeast-like state. Host-races with mycelial growth were less stable; sulfydryl compounds or appropriate chelators could transform the mycelial to the yeast-like state. Inoculation on *Vaccinium vitis-idaeae* using a uninucleate monospore culture of *E. vaccinii "vitis-idaea"* showed that that host-race is homothallic, since infections led to hymenium formation. Inoculations with this host-race on hosts of other host-races or with other host-races on *V. vitis-idaea* gave no infection. Inoculations made on mature host plants gave better results than those made on sterile seedlings, while inoculations made on tissue cultures of hosts gave no infection. Serological studies showed that the strains of the same host-race have identical antigens, while there are large antigenic differences between most host-races.

E. Rhizoctonia Studies

In an extensive survey[885] based on 14 papers presented in a symposium on *Rhizoctonia solani*, Kühn summarized the known information concerning morphology and pathogenicity of this important fungus. Surveys of the imperfect state, with four obligate and nine probable synonyms, and of the perfect state, *Thanetophorus cucumeris* (Frank) Donk, with 15 synonyms in *Hypochnus, Corticium, Botryobasidium, Pellicularia,* and *Ceratobasidium*, were explored in detail using light and electron microscopy. The mechanisms of variation were also described. The fungus persists in the soil as a saprobe, and the physiology and metabolism of this stage were summarized. The colonization and growth of the fungus in the soil were described. *R. solani* is an important producer of disease in a large number of crop plants. The mechanisms and physiology of host penetration by the fungus were discussed. The epidemiology of these diseases were considered in detail. The control of the fungus and special methods for the study of that control were described.

F. The Rust Cycle

Petersen[899] has given a thorough review of the morphology, taxonomy, phylogeny, and biology of the rust fungi.

G. Coffee Berry Disease

Field inoculations were conducted in Kenya by Mulinge[819] to determine the stage in development of coffee berries when they were most susceptible to infection. Young berries in the pinhead stage, abour 4 weeks after flowering, and mature berries (14 weeks old) took longest to be infected, while 10-week old berries which were expanding most rapidly were infected more quickly at both medium (5200 ft) and high (6000 ft) altitude sites. Additional experiments showed that berries inoculated after flowering on March 1 during development of a "long-rains" crop developed lesions in 2 to 4 weeks, while those infected after flowering near the end of October during development of a "short-rains" crop developed more slowly.

H. Field Grass Diseases

In Norway Årsvoll[51] found pathogenic fungi in 75% of the 2576 fields of timothy or timothy-meadow fescue grassland meadows. Temperature-growth studies made with isolates of 33 species of these fungi show that most of them grow well at temperatures of 0°C, several even at −6°C, and thus should be considered as psychrophiles or psychrotolerant species, well adapted to the soil surface temperatures that are usual during a Norwegian winter. The major overwintering grass pathogens in Norway, with the percent of fields from which they were isolated, are *Fusarium nivale* (59%), *Typhula incarnata* (17%), *T. tschikariensis* (30%), and *Sclerotinia boreale* (10%). The remaining 29 species included several weak pathogens, or secondary invaders, and may be able to cause significant damage to cultivated grasses. The predisposition of plants to attack during the winter by lowering of vitality may explain the pathogenic ability of these poorly specialized psychrotolerant fungi.

I. Sooty Molds

Sooty molds, which occur more commonly in subtropical and tropical regions, appear to be leaf parasites to the casual observer. Species similar to those in the southeastern and the southwestern U.S. have been studied in Australia,[400] and it has been shown that, in at least some circumstances, they are more closely associated with insect parasites such as aphids, which produce honeydew, than with the leaf tissues on which they are found. It has also been suggestd that they may be associated with leaf exudates.[577]

J. Fungi on Seeds

The presence of fungi in seed has been of concern to plant pathologists, since various types of disease may be carried in this way and different types of fungal infection may be initiated through seed. Types, methods, and reasons for seed treatment have been reviewed.[689] The types of fungi found in association with seeds have been monographed in Canada, starting with a paper on *Stemphylium*.[484] The significance of seed-borne fungi was of importance to the president of the British Mycological Society as indicated by his 1949 presidential address.[824] A number of species of *Fusarium* have been found in close association with cereal seed in Canada.[452]

VI. CONTROL

A. Flood Fallowing

Flood fallowing of land in an attempt to control Panama disease of bananas caused by *Fusarium oxysporum* var. *cubense* has received much attention from workers in Central America.[1115,1119] To understand the problem in the field, the colonization of banana roots by the pathogen and other soil fungi has been studied. The effect of soil moisture on the pathogen and related species of *Fusarium* has been examined in various laboratory tests. Techniques of flood fallowing[1119] and the effect of various factors in the altered environment, incuding that of oxygen supply,[1117] have been described. While flood fallowing has achieved a certain amount of control, a resident inoculum appears to remain in the soil or in dead and decaying plant parts, so that infection of a new crop may take place. Breeding of a root-disease resistant strain of bananas has practically solved the problem.[1118]

B. Usual and Unusual Techniques

Control of plant disease has been discussed for many diseases in the textbooks of plant pathology and in the literature of the control of each disease. Various control methods, which are departures from usual methods, have been described.[1108,1109] Some workers have approached plant disease control from the point of view of chemical control agents. The chemotherapy of plant disease has been summarized.[1116] Various chemical elements are required in very small amounts by most living organisms, but some can be used in plant disease control when applied in large or toxic amounts. The nature of fungicidal action by copper and sulphur has been considered extensively.[771] The use of fungicides for control of disease in forest and shade trees and for control of fungal activity in forest products has been discussed.[507] The definitive test on the action of fungicides was written in Connecticut[564] and was updated in chapters in a comprehensive survey of many fields in plant pathology.[565]

C. Host Relations

Plant hosts themselves play an important role in the development of plant disease.[85] The most obvious gross effect is the change from mixed plant communities to uniculture in forest, field, and garden. The breeding to improve the host plant, from the point of view of the crop it may produce, may result in the activation of as yet unknown strains of the pathogen. When strains of the pathogen are activated by host plants other than those proper to the pathogen, the hosts are called "bridging hosts". "Our subject is ecology, and in this context the mechanisms underlying that fact (new strain of plant creating its own pathogen) is, perhaps, less important than the fact that all that is needed to get a different population of parasites, not only of different species but also of types within a species, is to change the plant population."[85] For another approach, Carl[179] has reviewed the area of plant disease control by crop rotation.

D. Sclerotium Control

The drying of sclerotia for short periods of time,[1067-1069] followed by their introduction to moist soil, causes nutrient leaching, which does not prevent their germination but promotes the growth of other soil microorganisms which attack the sclerotia and use them as nutrients. This is suggested as a biological control mechanism for diseases caused by sclerotium-producing organisms.

E. Certain Chemical Treatments

In work in Norway,[739] isolates of *Pyrenophora* species which cause leaf blotch of oats were subjected to the action of organic mercurial fungicides. Strains were inhibited on malt agar plates by 0.45 ppm mercury as methylmercurydicyanimide and by 25 to 50 ppm mercury as methoxyethylmercury acetate. Isolates of *P. avenae* varied in tolerance to methylmercurydicyanimide; sensitive isolates on mercury were completely suppressed by 0.45 ppm mercury, while three isolates grew at concentrations suppressed by 3.0 ppm mercury. It should be noted that, in some regions because of environmental degradation possibilities, the use of mercurials and certain other chemicals is no longer tolerated.

In New Zealand,[407] captan, the fungicide most commonly used for dressing maize seed, was not effective in controlling smut. Seed dressings containing captan with benomyl, carboxin, or the experimental fungicide BAS3192F gave the greatest degree of control. Soil applications of fungicide were less effective than the combination seed dressings. The use of an appropriate seed dressing for all seed produced in the Gisborne area was recommended.

F. Leaf Protectants

Tests were made by Lukens[726] to determine the effectiveness of various films as leaf protectants against fungus penetration. Copper sulfate applied in emulsions of acrylate vinyl, menetieae, and *Xanthomonas* gum were as toxic to spores of *Alternaria solani* as either a fixed copper or Bordeaux mixture. Of the copper preparations, only the cupric acrylate was not toxic to the tomato leaves. Cycloheximide in acrylic film resisted leaching and protected tomato leaves against infection at nonphytotoxic dosages. The resins alone allowed fungal growth but gave some protection against foliage infection. The copper acrylate complex represents a new copper fungicide of high tenacity. The protective action of cycloheximide in acrylic films demonstrates a means of controlling release of fungicides from foliar residues.

G. Nonpathogenic Infection

There are situations in which nonpathogenic fungi, or pathogens which infect other hosts, can penetrate tissues of plants not susceptible to their attack, according to Matta.[763] Morphophysiological alterations are found in plants infected by nonpathogenic fungi. An effect of previous infection by nonpathogenic fungi is a type of protection from disease produced by other fungi. Immunity may be induced by simultaneous or sequential infection by nonpathogens.

H. Tolerance

"Tolerance may be defined as that capacity of a cultivar resulting in less yield or quality loss relative to disease severity or pathogen development when compared with other cultivars or crops," according to Schafer.[1011] Thus, in accepting this definition of tolerance, the plant pathologist must admit that the fungus causing the disease is potentially always present and that the host must be developed to the extent that an acceptable or better than acceptable crop can be produced in spite of it. Among examples of tolerance cited are those to diseases caused by species of *Septoria,* by viruses, and by nematodes. Explanations are given of the phenomenon, and various types of qualitative measurements are discussed including equivalence of disease, comparison of differences, the characters measured, the separation of tolerance from reduced disease development, and techniques for evaluation. Problems in using tolerance, breeding methods for tolerance, and the future of the concept were developed in a discussion by Schafer.

VII. SOIL

Garrett[425] has surveyed the etiology, environment, and microbial ecology of root disease fungi. His two more recent texts[427,433] consider the fungi causing root disease more extensively. Menzies[779] has reviewed the successful survival of fungal plant pathogens of many types in soil.

VIII. AIR

The effect of air currents on the transportation of spores of airborne fungi, especially those which cause plant disease, is important in the etiology of plant disease. Study of the sporal populations of the air[1096] yields much information for plant pathologists concerning the origin and distribution of various types of plant disease. In confined areas an abundantly sporulating fungus can fill the air with its spores. In England[765,766] the spore content of the air in buildings infested with the dry-rot fungus *Serpula lacrimans (Merulius lachrymans)* has been found to be high, a potential source of allerogenic reactions in the persons associated wth such habitats.

With detailed knowledge[842] of the life cycle of certain pathogens, and especially the means of their aerial dissemination, it can be concluded that reduction or elimination of the primary inoculum will afford almost complete control of those plant diseases occurring in the semiarid environment in California.

A. Dispersal and Spread

"A pathogen spreads where it goes and infects," Van der Plank[1179] has said. It may go along a stem or across a continent. A lesion anywhere means that the pathogen has spread. It is unrealistic to confine the meaning of spread to large distances. Dispersal is the movement of the propagule from the lesion to healthy susceptible tissue or, for systemic disease, from an infected to a healthy, susceptible plant. The restriction of the use of the term is convenient because dispersal to healthy, susceptible tissue is the only dispersal relevant to spread.

IX. MARKET PRODUCE

Kamat and colleagues[621] have developed a method of describing diseases of market produce. The fungi which develop on fruits and vegetables from the time of harvest, during transport, and in storage and their presence in the marketplace and their effects are considered here. Emphasis was placed on produce in the markets of India. The terminology adopted had been developed earlier in France. *Homotopic* diseases are those in which both the fungus and the host have the same distribution; *amphitopic* diseases are those in which both fungus and host have a wide distribution; *oligotopic* diseases are those in which the fungus and the host occur in few localities; and *meiotopic* diseases are those in which the fungus has a narrower distribution than the host, the host having a wider distribution than the fungus. In this classification, physiologic specialization is not yet considered, and much information is needed to perfect the classification.

X. OCCURRENCE, QUARANTINE, AND LEGAL ASPECTS

The occurrence of plant disease in the U.S. is subject to fluctuations according to climatic conditions. soil variation, crop rotation, and other factors. These factors and their effects on disease fluctuations have received much attention.[1113] Plant quarantine is designed to prevent the introduction of plant disease into a region or a country, in this case, the U.S. The methods of operation and their successes and failures have been described.[773] Plant diseases which have not yet been introduced into the U.S. and problems associated with the prevention of their introduction have been adequately covered.[584]

The fact that plant disease control can become the subject of legislation has been pointed up[408] in an extensive review of the laws man has erected to aid in the control of black-stem rust of wheat by removal of barberry, the control of white pine blister rust by removal of susceptible species of *Ribes* (currants and gooseberries), and the control of apple rust by the removal of juniper or red cedar (*Juniperus virginiana*) trees. Methods and results of forecasting the occurrence and incidence of plant diseases have been reviewed by Miller and O'Brien.[799]

XI. RECENT TEXTS

Among recent tests on plant pathology may be mentioned that of Wheeler[1225] in which the approach is largely physiological and genetic. The major subject areas consider the mechanisms of pathogenicity, the response of plants to pathogens, disease resistance mechanisms, and the genetics of pathogenesis. Three definitions are of interest. Biotrophs are "organisms which, regardless of the ease with which they can be cultured, in nature obtain their food from living tissues on which they complete their life cycles." Rusts, smuts, and mildews typify this group. Hemibiotrophs are "organisms which attack living tissues in the same way as biotrophs but continue to develop and sporulate after the tissue is dead." This group is typified by the leaf spotting fungi. Perthotrophs are "organisms which kill host tissue in advance of penetration and then live saprophytically." This group is typified by *Sclerotium rolfsii* which utilizes sclerotia in the course of initial attack. Wheeler noted that this classification can be subdivided and may be useful where formal terms are required. Informally, parasites can be simply termed strong or weak.

In his text *Principles of Plant Infection,* Van der Plank[1180] considered topics such as inoculum potential, the effect of other variables, epidemics, the spread of plant disease, and the genetics of the relation between host and pathogen. In considering the term inoculum potential, Van der Plank noted that in 1932, Horsfall's original concept carried the idea of mass action, the greater the amount of organism present and the greater its virulence, the more severe the disease. Thus inoculum potential carried the idea of the quantity and the quality of the infectious material ordinarily called inoculum. He gave a full discussion of the later modifications of this concept. For Van der Plank, "Flor's enunciation of his gene-for-gene hypothesis (in 1965) was one of the really important events in the history of plant pathology." The hypothesis was proposed for the relation between resistance in flax and virulence in the rust fungus that attacks it. It has since been extended, and more than 17 host-pathogen combinations have been explained by use of it. The hypothesis may be stated as: For each gene conditioning avirulence or virulence in the pathogen, there is a corresponding gene conditioning resistance or susceptibility in the host plant. Analysis of the terms epidemic and endemic in relation to the diseases described in these terms in plants and animals is discussed, as are the various facets of vertical and horizontal resistance in plants to disease.

Robinson,[974] in his book *Plant Pathosystems,* has defined the pathosystem as "any subsystem of an ecosystem which involves parasitism. A pathosystem may be natural (wild pathosystem, disease triangle), or artificial (crop pathosystem, disease square). The term may be used at any disease level which should be stated. The conceptual, geographical, or other boundaries of a pathosystem may be defined as convenient." The application of the pathosystem concept is principally at the crop level. A glossary of 306 terms is presented in which terms are described or redescribed in view of the pathosystem concepts used or introduced in this volume.

Chapter 14

PATHOGENS OF FUNGI AND FUNGAL PATHOGENS OF LOWER ORGANISMS

I. VIRAL INFECTIONS OF FUNGI

A survey of viruses infecting fungi has been made by Hollings and Stone.[559] While viruses have been reported in other Basidiomycetes, the more important kinds are those associated with the cultivated mushroom. Methods of virus transmission and the epidemiology of the diseases produced are described. The physiology of the virus infection and the detection, purification, and identification techniques are given. Practical control measures are discussed. Viruses are known to cause infection in certain species of Deuteromycotina, Ascomycotina, and Phycomycetes. Certain other disorders of fungi are thought to be related to viral infection.

Work with viral infections of fungi has been limited but appears to be increasing in importance. Sinden[1044] discussed viral infections of cultivated mushrooms and concluded:

Is it possible that the intense irritation of the host to the presence of the virions in relatively low titer as compared to that of viruses of other fungi indicates that adaptation of the virus to the host has not progressed very far? If this is the case is it possible that the mushroom is not the natural host for these viruses? Perhaps other hosts for these viruses should be sought as harbourers from which transmission to the cultivated mushroom sometimes occurs causing the severe epidemics experienced in the industry since their first appearance in 1948.

Statolon is a fermentation product of *Penicillium stoloniferum*[650] whose antiviral activity has been shown to occur through the production of interferon in cells. The active moiety was found to reside in a virus, the first mycovirus found in the Fungi Imperfecti. The fungal viruses and their dr-RNAs as inducers of interferon protect against lethal doses of infectious mammalian viruses.

A virus (or virus-like particles) causes an infectious, degenerative disease of the cultivated mushroom *Agaricus bisporus* transmitted through anastomoses or spores. Several different virus-like particles (VLPs) have been reported by other workers. Saksena[1004] has found three morphologically distinct sizes: spherical particles of about 26.7 nm and 34 nm and bacilliform particles approximately 20 × 51 nm, associated with diseased mushrooms. Apparently these are the more common forms elsewhere and possibly the most damaging to crop production. Techniques for large-scale production yield highly purified virus in quantity, and this is necesary for work leading to characterization of these viruses and for production of specific antisera for diagnostic purposes.

An in vivo replication system from *Penicillium stoloniferum* strain NRRL 5267 has been developed to test various fungal products[309] and other antimetabolites against *P. stoloniferum* fast-moving virus (PsV-1) replication. Cycloheximide at 1, 5, and 10 mg/mℓ blocked PsV-1 replication 40, 75, and 90%, respectively, compared to untreated controls. At these levels, fungal growth was demonstrated by biomass and dry weight and was unaltered, showing that cycloheximide is a rather selective inhibitor of PsV-1 replication. Mycophenolic acid, an antiviral metabolite produced by virus-free strains of *P. stoloniferum* and *P. brevicompactum*, at 300 mg/mℓ blocked PsV-1 replication by 25% but also reduced growth by 30%. Of the fungal metabolites tested, only patulin and gliotoxin inhibited PsV-1 replication.

Viruses isolated from yeast cells[4] are morphologically and biochemically similar to most of the viruses extracted from fungus mycelia. They are isodiametric particles 33 to 40 nm in diameter containing ds-RNA genomes. A correlation between ds-RNA species with a molecular weight of 1.4×10^6 daltons and the ability to produce killer toxin has been found in killer strains of *Saccharomyces cerevisiae*. The possibility is suggested that foreign viruses can be transmitted to and replicated within yeast cells.

II. FUNGI PARASITIC ON OTHER FUNGI

Durrell[346] has shown that *Trichoderma viride*, whether or not it exudes an antibiotic, attacks other fungi parasitically, especially Phycomycetes, directly penetrating the host mycelium and digesting its contents. Madelin[737] discussed the known cases of mycoparasitism in all classes of fungi and lichens. The host-parasite relationship was also considered.

For Cooke,[209] there are three types of relationships among fungus species thought by others to represent mycoparasitism. These include direct competition for nutrients, saprotrophism, necrotrophism (the ability of one fungus to utilize dead mycelium of another fungus), and biotrophism (the direct invasion of the mycelium of one fungus by another).

III. FUNGI PARASITIC ON LIVING ANIMALS — INVERTEBRATES

Insects and spiders[741] are subject to fungus diseases of various types. The Entomophthorales includes several fungi which are parasitic on numerous species of insects whose bodies eventually become mere masses of fungus tissue.[1105] Members of the Laboulbeniales[94,1026] are entirely parasitic on insects. Members of this order are usally attached to one or another chitinous part of the outer skeleton. Members of the Cordycipitaceae sclerotize insect bodies and pupae, from which elongate stalks producing stromata with clusters of perithecia are produced. It has been suggested[376] that, because of their susceptibility to certain diseases, insects causing diseases of plants may be contolled by the introduction of fungus epizootics of these insects. Insects and spiders are also prey to a number of hyphomycetous fungi.[741] Except as the host insect is an incidental member of the biota of one or another location, it is difficult to place such fungi among others in given population.

The nature of the role of soil microorganisms[436] and soil-inhabiting, macrophagous invertebrates in the decomposition of plant residues is closely interrelated. Under the combined influence of both groups of organisms, the decomposition of organic residues is more rapid than under the influnce of microorganisms alone.

An unannotated bibliography based on 300 references based on papers published since the mid-19th century dealing with insects that feed on fungi[389] has been prepared. The indications are that insects may be primary dispersal agents for both pathogenic and saprobic fungi.

Batra[80] has described the ectosymbiosis between beetles and the ambrosia fungi with which they are associated as a natural relationship in which one organism cannot live without the other. The beetle larvae exclusively feed on specific fungi and take shelter in wood weakened by the fungi. The insect in turn shelters the fungus during its quiescence, transmits it into a suitable plant host, and fertilizes it with its excrement.

Insects are unique among animals in having developed mutalistic relations with fungi.[84] This may have come about because so many insects and fungi share the same microenvironments. In addition, most insects are equipped to carry living spores of fungi either in their gut, in folds between their joints that contain waxy secretions, or among their bristles. Such relationships are of two types: (1) In the gardens of wood wasps, ambrosia beetles, termites, ants, and probably those of midge galls, the fungus extracts nourishment from a substrate, and the insect feeds either on the fungus, the substrate predigested by the fungus, or both. The fungus is prevented from producing sexual fruiting structures in some cases but is supplied by the insect partner with a suitable habitat and a means of dispersal. (2) In the colonies of *Septobasidium* and scale insects, the situation is reversed. The insect feeds on the substratum and nourishes the fungus, and the fungus provides shelter for the castrated or normal insect partner.

In developing a study of the relation between tissues of host plants and ambrosia beetles and their associated microorganisms, Barras and Perry[77] produced a 244-title annotated bibliography. Their own work has been on ambrosia beetles and their fungi, especially with transport and inoculation of beetle galleries in the bark and wood of host trees. A species of *Sporothrix*, probably the imperfect state of a *Ceratocystis*, and a Basidiomycete have been isolated from the mycangium (the fungal transport pocket of the beetle), and the recovery of spores from release sites on the beetle has been described. In the mycangium, the *Sporothrix* may grow as a yeast, since spores in the sympodulae in culture are similar to the yeast-like cells in the mycangium.[75, 76, 498]

Fungi, including yeasts, are common in the honey stomachs [81] and provisions of various kinds of wild bees. Associated with the velura of Apoidea are at least 124 species of fungi fo which 49 are associated with alkali bees. These fungi may be parasitic, commensals, or mutualistic. Yeasts, singly or in association with bacteria, are pioneer colonizers during a microbial succession in larval cells of many subterranean bees. They are followed by species of the genera *Aspergillus, Penicillium, Emericellopsis, Sartorya, Pseudoarachniotus, Gymnoascus, Carpenteles,* and *Fusarium. A. flavus* and *Saccharomyces* spp. are pathogens in many species of bees, and fungi are the main cause of decline in alkali bees.

In an intensive study of the fruit bodies of 27 polypore species between June 1972 and July 1973,[838] 3443 beetle individuals (imagines only) were collected. They belonged to 153 species of 28 families and were caught on sporophores of 20 polypore species. Among these the Staphylineae were represented by 63 species and 1890 individuals; of these, 54 species were found almost exclusively on soft-flesh polypores such as *Laetiporus sulphureus* and *Polyporus squamosus.*

The classification of the fungus beetle fauna in relation to the condition classes of the fungal sporophores is based on that of Höfler and Scheerpeltz.[555,1012] The young-fresh up to the mature-fresh condition is considered to be the *Gyrophaena* phase, the oligo- to mesosaprobic condition as the *Atheta* phase. The number of individuals of the fungus beetles diminishes with the height of sporophores above the ground.

In addition, the food of 180 beetles was studied. Microscope slide mounts of 100 showed that at least one part of the digestive tract contained so many parts of hyphae or spores or both that the beetles concerned were considered to be fungus eaters. In the microscopic study of the food of a fungus gnat larva, large quantities of spores were found which suggests that some of the beetles found called fungus eaters had picked up parts of hyphae and spores by consuming their prey which had eaten the spores. Thus, it is suggested that beetles may be involved in fungal dissemination.

The life history of the forked fungus beetle *Bolitotherus cornutus* (Panzer), an insect which feeds on *Ganoderma applanatum,* was studied by Liles[695] from material observed and collected in Cheboygen County, Michigan. Eggs were deposited singly on the upper and lower surfaces of the fungus fruit body at the rate of one or two eggs a day for a total of 8 to 12 eggs. Eggs were laid in mid-June by overwintered mature adults. Overwintering larvae matured in mid-summer and started ovipositing in late August. The larvae bored into the host fungus, and pupae were formed in an enlarged chamber in the larva tunnel. Larvae reared in the laboratory on pulverized *G. applanatum* went through four instar stages before pupation; in the field, apparently five instar stages were passed through. Feeding and mating of mature beetles wee largely nocturnal.

In their book on fungus beetles and beetle fungi, Scheerpeltz and Höfler[1012] recognized a relationship between beetles and fungi, and this was categorized by Höfler[555] as (1) mycetobiont or fungus restricted, (2) mycetophile or with a strong, friendly or weak relationship with fungi, and (3) mycetoxene or fungus visitors. In addition, Scheerpeltz and Höfler recognized five stages in the development and decomposition of a fungus fruit body with which fungi were associated: young-fresh; mature-fresh; oligosaprobe, weakly decomposed; mesosaprobe, decomposed; and holosaprobe, strongly or completely decomposed.

Some studies have been carried out on a collection of 36 species of fungi cultivated by leaf-cutting ants.[539] In addition to natural media, they grow well on a complete synthetic medium. Tests for antibacterial activity were all negative. Four isolates, each cultivated by a different species of ant, produced fruit bodies of a *Lepiota*-like fungus; 18 species produced bromatia like those of *Attamyces bromatificus* Kreisel.

Madelin[736] has reviewed the subject of fungus parasitism of insects in which he considered the host, the parasite, ectoparasitism and endoparasitism, and the life cycles of the parasites. Far too little is known about the relationships involved.

Feeding habit and habitat preferences are discussed[682] for 14 species of North American microlepidoptera known to occur in sporophores of wood-rotting fungi. Members of the Occophoridae, including *Omophila* v- *flava,* are thought to be general scavengers using fungi only occasionally, while the Tineidae include obligate fungus feeders, occurring mostly on Polyporaceae. Among the Tineidae some species, such as *Merophaga cryptophoria*, appear to be host specific, while others, including *Nemapogon defectellus*, are ecologically restricted but utilize a number of host fungi. Certain fungus species such as *Phellinus gilvus* and *P. munzii* are frequently inhabited by Tineidae, while *Coriolus versicolor* and certain others serve only as incidental hosts.

In a test of the relationship between four species of amoebae and 35 species of fungi,[523] the fact that 19 species of yeasts were eaten to various degrees by amoebae indicated that yeasts are a possible food source for soil amoebae. While conidia of 16 fungal species were ingested, only those of *Paecilomyces elegans* and *Polyscytalum fecundissium* were actually digested; only "wet" spores could be ingested; ingestion did not markedly affect digestion of spores of *Mucor hiemalis;* and soil amoebae may serve as a dispersal agent for fungal spores. Species of Trichomycetes inhabit the hind gut of worms and insect larvae. These have been surveyed by Lichtwardt.[692,693]

Predatory and parasitic species of Acarina (mites)[627] may inhibit strong development of bacteria and fungi in soils. Uropodina feed upon bacterial slime, yeast cells, conidia, and fungus spores. Species of trombidiform mites feed upon fungal hyphae and spores or certain organic substances being liquified by bacteria and fungi. If oribatid species appear in increasing numbers in compost and soil, it is indicated that there is a change from anaerobic to aerobic processes; in such cases it is indicated that the putrefactions attributed to Acaridiae are diminishing.

In Germany, Kühnelt[665] has noted that there was a para-development of shell-bearing amoebae, rotifers, and Collembola in areas of the litter dominated by the mycelium of fungi. It was suggested that the low pH and tannin content of such areas prevented the normal development of these animals except for certain species of Collembola which were abundant in some areas rich in mycelium.

Bärlocher and Kendrick[62] used a crustacean, the sand flea (*Gammarus*), to test food leaves infected with fungi found in streams in decaying litter. Of the ten species of fungi used, a species of *Humicola* most improved the flavor of ash leaves, while a species of *Anguillospora* most improved the flavor of maple and oak leaves. *Tetracladium* had the least effect on all three kinds of leaves. Sterilized leaves were incubated in sterilized stream water with 10 mg $(NH_4)_2SO_4$ and 2.5 mg KH_2PO_4 per liter of water. Ten species of fungi were used as agents of decomposition. In the case of five species, there was no significant difference from the controls after 2 weeks, four species yielded significant differences at the 5% level, and one species yielded a significant difference at the 1% level, both in respect to percent loss of weight of the leaves and percent of protein content. Fungi are thought to be very useful components of the stream biota in removal of added organic matter under simulated natural conditions.

In Poland, Pieczynska[911,912] has noted that aquatic animals use Aufwuchs-type growth occurring in streams as food. Harman[506] in New York state noted that fresh water mollusks used slime incrustations on plant stems and elsewhere in bodies of water as food. There was no indication in either case as to the nature of the attraction except as a food source. Since fungus mycelium probably forms a portion of the slime involved, it might be of interest to determine whether any differential flavor could be involved.

Slime develops on the surface of stones making up the menstrum of the trickling filter. This has been studied especially at Dayton, Ohio as to species content of some segment of the biota. While none of the slides in the continuous sampling project developed by Cooke[224] were grazed by *Physa*, it was observed by Ingram and colleagues[594] that in certain seasons as much as 7 tons of snails were removed from the trickling filter field (20 ½ acre filters, 7.5 ft deep) each day. Other trickling filters in southwestern Ohio reported snail problems which are probably not restricted to that area. There was no indication that the snails were selective in their selection of food resources, but no attempt was made to determine whether one type of organism was preferred over another.

The control of the principal fungal, bacterial, and viral pathogens and weed molds on a modern mushroom farm[1043] depends largely on the manipulation of environmental factors. All factors from the arrival of component materials and their handling to disposal of the spent substrate are involved. Once the technology of setting up the mushroom beds, air movement, air conditioning, insulation, and the conditions most favorable to the physical and chemical environment of the mushroom are established, the control of the environment for the control of pathogens and weed molds is not considered to be difficult.

IV. PREDACIOUS FUNGI

In a series of reviews, Duddington[335,336,338-340] summarized information concerning predacious fungi. Species, mostly of Hyphomycetes but including some Zygomycetes and one species of Ascomycetes, attack amoebae, rotifers, and nematodes in soils and nematodes in water. Some are internally parasitic, while others capture their prey alive by various trapping devices. They are common in most soils, but until recently little has been known of their activities under natural conditions. Attempts have been reported[337] to use them for the biological control of plant pathogenic fungi. In Russia, Soprunov[1086] has published a book dealing with many facets of the predacious fungi, and this has been made available in the cited English edition.

The genus *Zoophagus* entraps rotifers on apparent secretions on spicule-like branches of the mycelium (Chapter 11, Figure 7). The fungus grows into the animal, consuming it and discarding the shell. The occurrence and control of this activity in sewage treatment processes has been described.[267,914,915]

Powell[924] has summarized information available on the interaction between fungi and nematodes in wilt diseases, root rots, seedling diseases, and in fungus-nematode root complexes. Some bases for fungus-nematode complexes in plant disease are discussed. Some consideration was given to etiological problems.

Experimental studies[749] with the cosmopolitan fungivorous nematode *Aphelenchus avenae* yielded no root damage in sterilized soils. The use of sterile nematodes in cultures of phytopathogenic fungi, including *Rhizoctonia solani, Thielaviopsis basicola, Pyrenochaeta* sp., *Verticillium alboatrum, Periconia* sp., *Sclerotium* sp., *Armillariella mellea,* and several *Fusarium* species, indicated that these fungi may be the preferred food of this nematode. Pythiaceous fungi were apparently toxic to this nematode. Poor populations developed on some saprobic fungi tested. Occurrence of this nematode in the rhizosphere and diseased tissue in certain plants may have been because of the presence of fungus hyphae.

Domsett and Reid[329,330] have studied the myceliophagus nematode *Aphelenchus avenae* Bestian and the predacious hyphomycete *Dactylaria candida* (Nees) Saccardo by phase contrast microscopy, scanning electron microscopy with whole mounts, and transmission electron microscopy with ultrathin resin sections. Adhesive trapping knobs were produced profusely on the mycelium of this fungus. Nonconstricting rings were rare and were nonfunctional. The knobs were effective trapping devices. Methods of penetration of trapped nematodes were observed as well as the phenomenon of "post-infection" breakout when all but the shell of the nematode had been consumed.

In farmyard soil in Ontario, Barron and Dierkes[78] have found isolates of the nematophagous fungus *Nematoctonus* which, on potato dextrose agar, produced sporophores assignable to the agaric genus *Hohenbuehelia,* a segregate of the Friesian genus *Pleurotus.* The gilled mushroom-like sporocarps discharged basidiospores in the presence of nematodes. These basidiospores germinated to form short aerial stalks with terminal adhesive knobs. Basidiospores adhered to the cuticle of nematode bodies where were initiated monokaryotic infection mycelium. Multiple basidiospore infections on single hosts resulted in dikaryotization within the host body. Crossing single basidiospore cultures demonstrated bipolar sexuality in this species of *Hohenbuehelia.* It is suggested that, before additional species of *Nematoctonus* are described, the imperfect states of all 26 known species of *Hohenbuehelia* should be carefully obtained and studied.

Cooke[209] considered two types of predacious fungi: those which are facultatively antagonistic symbionts of protozoans, rotifers, and nematodes; and those which are obligately antagonistic symbionts of these animals. Both groups include Zygomycetes which attack protozoans and rotifers, and Hyphomycetes which attack nematodes.

Chapter 15

PARASITES OF VERTEBRATES AND MAN

I. FISH

Fungi that cause disease in fish have been known for a long time. Taxonomic studies including such species[203,1024] have been published. These fungi usually form a white fuzzy mass on the fish which may be parasitized, or the fungus may be a secondary invader in diseased tissues or in wounds. A recent study from Germany[955] indicated that fish may be subject to systemic diseases caused by fungi. Material thought to be similar to *Alternaria, Cladosporium,* and two other filamentous fungi has been obtained from tissues of fish in fresh and salt waters in northern Europe. Reference may also be made to the sections on fresh and marine waters.

II. BIRDS

That fungi cause important disease problems in a number of kinds of avian hosts and may be associated with environments commonly inhabited by birds is indicated by an annotated bibliography of 700 titles.[198a]

The "transmissible enteritis" disease of Guinea fowl chicks in France,[810] which appeared in chicks less than 12 weeks old, was characterized by a yellowish diarrhea, a general dehydration, a swelling of the caecums, and a large mortality. Based on the mycological analysis of food, it was pointed out that this fungus (probably *Geotrichum candidum*) does not produce a classical mycotoxicosis. Having been isolated from the intestines of dead animals, it is presumed that it plays an important role in the disease even though experimental infections have been unsuccessful. There is no transmission by egg, and it is suggested that the origin of contamination could have been dung, litter, or drinking water.

In the study of the control of psittacosis in parrots,[626] 655 birds of several species were imported from Latin America, Australia, Malaya, and Ghana. Of the 45 birds which died and whose necropsies showed evidence of possible fatal aspergillosis, 13 deaths were found to be caused by *Aspergillus fumigatus* infection, 16 by *A. oryzae,* and 3 by both fungi. This could be attributed to methods of capture and handling en route to the laboratory, by the administration of chlortetracycline used in the control of psittacosis, or in the administration of cortisone acetate used in an attempt to activate existent latent psittacosis infections. Possible causal relationship of these factors was discussed without satisfactory conclusions.

III. OTHER ANIMALS

The diseases of animals caused by fungi are best known in relation to those caused in man. Veterinary mycology originated as an offshoot of medical mycology and is rapidly developing in response to the recognition of fungi as producers of important diseases in animals. An early text on animal pathogens[780] emphasized the importance of bacteria and viruses but devoted only 18 of its 850 pages to fungus diseases. Löliger[720] has more recently published a text on diseases of fur-bearing animals in Germany in which proportionately more space is devoted to fungus pathogens and in which greater attention is paid to tissue effects and other aspects of mycotic and other infections.

TABLE 1

Taxonomic Distribution of Species of Fungi Pathogenic for Vertebrates Compared with that of the Plant Pathogenic Fungi

	Vertebrates			Plants	
	Pathogenic			Pathogenic	
Fungi	Species	Percent	Poisonous Species	Species	Percent
Phycomycetes	26	12.7	0	91	7.1
Ascomycetes	28	13.6	4	289	28.3
Basidiomycetes	0	0	23	274	21.3
Deuteromycetes	151	73.1	13	634	49.3
Totals	205			1288	

From Ainsworth, G. C., in *The Fungi, An Advanced Treatise,* Vol. III, Ainsworth, G. C. and Sussman, A. S., Eds., Acadenic Press, New York, 1968, 212. With permission.

Among diseases of marine animals, Alderman[24] has discussed diseases of marine fish including one caused by *Ichthyophorus hoferi* on herring, mackerel, and possibly other species. Johnson and Sparrow[608] have devoted considerable space to fungal diseases of marine animals.

IV. MAN

During and since World War II, the literature of the science of medical mycology has developed from a ponderous prewar text[324] in which many hundreds of separate fungus cultures were considered to be etiologic agents of different fungus diseases to more simplified postwar texts[207,364] in which many strains and multiple species were reduced to synonymy thereby decreasing the number of earlier recognized disease-causing fungus species to a relatively small number. Handbooks[207,364,1040] and reviews discussed the broad phases of the subject, while various more restricted treatments such as that of Silverman and colleagues[1035] indicate the approaches possible when studying the various phases of particular diseases.

Among the earlier, if not the earliest, manuals for medical mycology students was one prepared by Castellani.[183] The diseases considered were largely superficial mycoses, and each disease, together with the causal fungus, was discussed at length.

Ainsworth[8] summarized information available on fungal pathogens of vertebrates, including man. Under endogenous mycoses he discussed candidiasis; under exogenous mycoses, he discussed coccidioidomycosis, adiospiromycosis, histoplasmosis, other systemic mycoses, and ringworm. Among host-pathogen interactions were considered environmental effects, effects of the host on the pathogen, dimorphism, and effects of the pathogens on the host. The taxonomic distribution of the more important pathogenic fungi is summarized in Table 1.

A. Natural Occurrence

Most of the fungi causing important fungus diseases of man and animals have been isolated from soils. This indicates to some workers[16,363] that the soil is a natural reservoir of disease agents. In some cases, the air one breaths is definitely the source of infection, especially when it contains large numbers of the spores of the etiologic agent.

In other instances, the airborne nature of the infectious agent is still only suspected. Study of such fungi and of the contaminants which may be associated with them has resulted in the publication in Argentina of a unique manual.[1288] In some cases, the body population contains cells of organisms capable of causing disease, and when the balance between the different members of the population is upset by therapeutic agents or other means, the cells present are reported to be able to produce the disease.

Whether or not these fungi play an important role as members of soil fungus populations is as yet undetermined. They are found only with special techniques, or when located with the usual soil sampling techniques, their importance is minimized, or their presence is unrecognized, since many of them take longer to fruit, some may be considered sterile mycelium, some may be unknown or confused with other species, and some are of only sporadic occurrence. Using routine sampling techniques for soil fungi, species like *Histoplasma capsulatum* are not found, while soil fungi occurring with them are found in abundance.[126,127]

According to Emmons et al.,[364] (first edition)

> The fungi that cause systemic mycoses are normal and more or less permanent members of the mycoflora of the soil, compost, or organic debris. They survive the competition of other fungi, bacteria and the microfauna as long as nutritional, thermal, and other as yet unknown factors are favorable. They cause disease in man when a susceptible person is exposed to them, but we know of no evidence to indicate that saprobic survival in soil is dependent upon periodic infections, with parasitic passage in man or animals. If the substrate is adequate in other ways, the reservoir does not require reseeding from a living host. These fungi are vigorous and self sufficient saprobes as long as environmental conditions are favorable. They are parasites by accident.

Ajello[17] summarized information about the natural occurrence of species of fungi belonging to the genera *Epidermophyton, Microsporum,* and *Trichophyton.* Gaps in knowledge of these geophilic, zoophilic, and anthropophilic fungi are suggested. The search for potential disease-producing fungi in nature has been extended to the Galapagos Islands, Ecuador[22] where, from 39 soil samples from ten islands, seven species of fungi in five genera were isolated, including *Arthroderma quadrifidum, Chrysosporium indicum, C. keratinophilum, C. tropicum, Ctenomyces serratus, Gymnoascus reessii,* and a species of *Malbranchea.* The collecting sites were located in dry areas, the soil being essentially made up of disintegrated lava which was low in organic matter.

In a search for pathogenic fungi on Oahu, Hawaii,[646] sand samples from beaches of low and high swimmer density and soils associated with bird droppings were processed. In the 361 samples, more pathogens or pathogenic fungi were found in areas where swimmer density was very high, while few were isolated from low swimmer density areas. The more important species listed were *Cladosporium werneckii, Epidermophyton floccosum, Geotrichum candidum, Microsporum canis, M. gypseum, Trichophyton mentagrophytes, T. rubrum,* and *Trichosporon cutaneum.* In the case of soils associated with bird droppings, 61 isolates of *Cryptococcus neoformans* were recovered from 170 samples.

Müller[818] took samples of beach sands periodically through the bathing beach season in Europe and checked for the presence of dermatophytes. *Epidermophyton* was present in sands of beaches on the Baltic Sea from June through September and at Estoril near Lisbon and on the Adriatic Sea until October. It was not present in sands covered by tides. There was no apparently natural or long established deposit of spores as was the case with *Trichophyton mentagrophytes* and *M. gypseum.*

B. Aspergillosis

A chapter in the text of the Raper and Fennell monograph on the genus *Aspergillus,* contributed by Austwick,[59] described the known cases of aspergillosis produced by different species of *Aspergillus*. Many of these were apparently the result of the accident of the presence of the fungus which is usually considered to be opportunistic. This does not reduce the importance of the disorder produced in the patient.

C. Candidiosis

Winner and Hurley[1262] have summarized available information on *Candida albicans*, the etiological agent of the widespread disease candidiasis, also called thrush or moniliasis, which until recently has been rarely isolated from nature.

D. Coccidioidomycosis

Ajello[18] edited the papers presented in a symposium based on the fungus *Coccidioides immitis* and the disease it produces, coccidioidomycosis.

Recent work by Baker[64] has shown that fungi are as avid travelers as man. In a survey based on swabbings taken from shoes of persons arriving at the Honolulu airport, a number of fungi have been reported as arriving in the Hawaiian islands from other parts of the world, increasing the biota or adding to existing populations.

Candida albicans and *Cryptococcus neoformans* cells were exposed for 8 weeks[738] by storage in sterile sea water to determine the effect on their viability and pathogenicity. Both viability and infectivity were unaffected as indicated by mouse passage tests.

Growth of *Aspergillus fumigatus* was tested[1188] to determine whether strains isolated from human patients could be distinguished from saprobic strains from nature. No characteristic was found to correlate with one group only.

E. Histoplasmosis

The fungus *Histoplasma capsulatum* has been studied intensively in relation to its natural environment and to the disease histoplasmosis[1256] which it causes in man and animals. The disease has been thoroughly reviewed,[1035,1256] and Cooke[244] has published a bibliography covering most reports on it published between 1906 and 1970. In the present survey, it might be well to mention some of the types of study carried out on this fungus in relation to its environment. It is only within the last 30 years that we have learned[362] how to obtain this fungus from its natural environment. The technique used at first has been modified and expanded so that more recently a revision[680] has been described. That the technique is not too effective is shown in that, even after shaking or beating the soil sample suspected of carrying spores, the spores and carrying mycelium are so tightly adherent to soil particles that only 10% recovery is effected. To be sure that the strain of the fungus being sought is pathogenic, it is routine to inject suspensions of the sample into four or more mice,[680] with the added precaution of using antibiotics, and to permit the fungus to incubate within the mouse for as long as 4 weeks. Following the incubation period, the mouse is sacrificed and portions of its liver and spleen are streaked on Sabouraud agar or Emmons' American Sabouraud agar, usually made with neopeptone. After such slants have been incubated under room conditions for 2 or more weeks, the fungus is considered to have been present in the sampled substratum in a pathogenic form when it develops on the agar. In addition to the mouse technique, the membrane filter has been used,[451] but with this method the spore is collapsed, is not viable, pathogenicity cannot be proved, and the spore could have been that of a species of *Sepedonium*. Spores of this fungus have been studied in reference to size, since large spores may not be effecively inhaled. In the strains studied, most of the spores were small, although large ones were present, and there were both smooth and tuberculate spores. Various soil studies have been carried on with relation to this fungus. In one type, soil samples were surveyed[680] for the presence of the fungus; again, an attempt was made to correlate the more common soil types[1289] in which the fungus has been found with its potential distribution. In areas of its known occurrence, large numbers of different kinds of soil fungi have been found with it in piles of bat droppings[126] and in bird droppings under starling roosting areas.[127]

In a study of the survival of spores or cells in water,[264,967,1256] it was shown that *Histoplasma capsulatum* yielded high rates of infectivity in mice after storage in tap water and in both sterile and raw Ohio River water, at 5°C and at room temperature for 500 days and for 300 days at 37°C. Infectivity was retained as shown by mouse passage.

In *Emmonsiella capsulata,* the perfect state[667] of *H. capsulatum,* a search for mating types among isolates recovered from nature and from patients with histoplasmosis from 13 states in which the organism is endemic[668] was instituted. Of 933 soil isolates tested, the ratio of (+) to (−) strains was 1:1. Of 184 clinical strains the ratio was 7 (−) to 1 (+). The mycelia of the (−) type converted to the unicellular form in vitro more readily than that of the (+) type.

F. Zygomycosis

Diseases produced by species of Zygomycetes[19] have been summarized, the survey precipitated by a case of disease produced by *Saksaena vasiformis.* Problems of identification were based on the ability of the fungus to sporulate on primary isolation media used in medical mycology. Ten species of Zygomycetes, classed in two orders and six genera, were considered to be well authenticated as producers of disease in humans.

Rhizopus oryzae has been found to produce a mucormycosis during use of Elastoplast® dressings and bandages in a hospital in Minnesota.[643] Since the dressing is applied outside bandages, its sterility has not been considered important. In concerned hospitals its use has been withdrawn.

G. Sites of Infection

Concern has been shown for sources of possible infection as a result of handling and wearing of contact lenses.[900] in regular sampling programs, most organisms found are the same as those found in the upper respiratory tract and on the fingers. . Cultures from contact lens storage fluid were sterile. Organisms transmitted to the lens as a result of handling the lens did not survive more than 24 hr. The only fungi cited were "yeast", *Candida* sp., and *Trichosporon* sp.

A study based on eye cosmetics (mascara) sought potential pathogens in these substances.[6] Over 200 samples of mascara of many brands and using both water emulsion and oil-based anhydrous products were tested. In only 1.5% of unused mascaras, but in 60% of used mascaras, bacteria and fungi were found. These organisms were more commonly found in the applicator brushes. The organisms found were more commonly associated with the skin of the face and the fingers of the user, but some potential pathogens were isolated. *Candida parapsilosis* was isolated from mascara. *Aureobasidium pullulans, Cladosporium* sp., *Candida lipolytica,* and *Cryptococcus albidus* were occasionally isolated from mascara, and *Pityrosporum* sp. was isolated from skin but not from mascara.

Cultures obtained by culturing swabbings[1178] from nostrils and from the conjunctival sac from people without evidence of mycotic disease showed that a high percentage of these people were contaminated by various fungi including pathogenic species. Samples tested included 338 nose specimens from Belgium, India, and Zaire and 349 eye specimens from Belgium, India, Tunisia, and Zaire. All samplings were made in hospitals in Kinshasa, Zaire; Madras, India; Brussels, Belgium; and Nabeul, Tunisia. Results compared favorably with Hirst spore trap samples from air, especially in relation to *Alternaria* and *Cladosporium*.

Samples of sputa[623] from 29 patients in a Montreal hospital known to have yeasts were surveyed for yeast populations. In the 32 samples, colony counts varied from 300 cells to over 1 million cells per milliliter of the emulsified sample. *Candida albicans* was found in 23 samples as the only yeast present, and *C. tropicalis* was the only yeast in four specimens.

C. albicans has been shown to be the most frequent fungus in high concentrations in all areas sampled[202] in a study based on 86 specimens including 23 propharyngeal, 26 jejunal, 20 ileal, and 17 fecal samples from 27 normal adults. Both the frequency and concentration of *C. albicans* increased from the propharynx 30% to 54% in the jejunum, 55% in the ileum, and 65% in the fecal specimens. The stability of the mycobiota pattern in the small intestine was demonstrated in five subjects who were resampled 5 to 9 months after the initial sampling; quantitatively and qualitatively the fungal pattern was unchanged.

In Tokyo, Japan, Okudaira et al.[844] isolated fungi from lungs of 81.1% (129 of 159) autopsy cases. From 113 of these, 918 strains of filamentous fungi dominated by *Aspergillus* and *Penicillium* were isolated, and from 58 cases, yeasts dominated by *Candida* were isolated. Of these fungi, few were pathogenic, and none caused opportunistic infections in the autopsied individuals.

H. Dimorphism

One of the more interesting phenomena associated with fungi pathogenic for man and higher animals, especially those fungi causing systemic mycoses, is their ability to develop as yeasts in the body or in vitro at body temperatures in the presence of special media or cultural conditions and their ability to develop as filamentous fungi at room temperature on standard fungus culture media.[1013] *Histoplasma capsulatum, Blastomyces dermatitidis,* and *Coccidioides immitis* are among the more important fungi able to develop in this way. This phenomenon has been described above.

I. Epidemiology

A survey is given[625] of the epidemiology of the principal systemic mycoses of man and animals, including histoplasmosis, blastomycosis, coccidioidomycosis, and cryptococcosis, with special reference to animals. Distribution of the diseases and the fungi causing the diseases are described.

J. Immunology

In the *Manual of Clinical Immunology,* Kaufman[631] has contributed a chapter on serodiagnosis of fungal diseases. Sections are devoted to (1) detection of hypersensitivity, (2) collection, preservation, and shipment of specimens, (3) detection of circulating antibodies and antigens, and (4) in vitro and in vivo identification of fungi by fluorescence antibody techniques. Diseases considered are aspergillosis, blastomycosis, candidiasis, coccidioidomycosis, cryptococcosis, histoplasmosis, paracoccidioidomycosis, and sporotrichosis; for each disease, clinical indications, the test, reagents, and interpretation are considered. The practical application of fluorescent antibody procedures in medical mycology was described by Kaplan.[624]

K. Allerogens

The air spora includes spores of a number of fungi which cause allergies. Regular sampling of the air spora (described above) has yielded a population of spores in the air which is fairly constant on a worldwide basis as noted, e. g., in the paper[1178] based on eye and nose populations in Belgium, India, Tunisia, and Zaire.

The immediate retention of particles[334] in various parts of the human respiratory system can be calculated by the application of the physical laws governing sedimentation, impaction, and diffusion. This has been demonstrated experimentally by the use of aerosols including noninfectious spherical particles.

It has been pointed out[1092] that many industrial, agricultural, and municipal processing facilities possess the potential to produce hazardous aerosols containing pathogenic microorganisms. The definition of the potential hazards associated with these various activities has resulted from investigations of agricultural practices, textile mills, abbatoir and rending plants, and sewage treatment plants.

V. PREVENTIVE MEASURES FOR FUNGUS DISEASES

There are three principal reasons for considering measures for the prevention of fungus growth. All are based on man's health, economy, and the interruptive nature of cerain fungus activities in these areas of man's existence. The literature of plant pathology[524,564,565] and of human and animal pathology[207,364] is full of methods for controlling diseases caused by fungi. That control of human diseases, especially of the types referred to as "deep mycoses", has progressed less rapidly than control of plant diseases is based on the lack of availability of adequate amounts of test materials, as few people are willing to undergo the elaborate and extensive tests used on thousands of plants annually in plant disease control experimentation. Utilization of copper and sulphur compounds[564,686] or organic compounds[564,565] and of heavy metals in fungus nutrition and the combatting of fungus disease are reviewed and discussed in the literature. Legalistic aspects of plant disease control have been reviewed,[408] although little information is yet available in relation to discontinuance of some of these materials as a result of environmental protection activities.

A more recent method of soil treatment to prevent disease organisms from attacking crop plants is sterilzation. Much of the work on this subject has been reviewed by Garrett.[427,433] This type of treatment is applied on an increasing scale, not only in the laboratory and greenhouse but also in the field. It is interesting that the ubiquitous *Trichoderma viride* is one of the few fungi to survive some type of sterilization treatment, that under certain conditions it can be used to control other fungi, while under other conditions it needs control, and that following such treatments the soils can be "reseeded" with fungi harmless to crops but which can remove crop debris from the soil without infecting the living crop. Vaughan and colleagues[1183] have studied the relation between germinating seeds and developing roots on the reestablishment of microorganisms in soils which have been fumigated.

With the realization of environmental damage resulting from the wide usage of various types of chemicals containing organic complexes using heavy metals and certain types of chlorinated hydrocarbons, legal restrictions were imposed on their use and the indiscriminate and careless use of other chemicals as pesticides of various types. Chapters in most plant pathology texts have yet to be updated to include the revisions necessary, but agricultural experiment stations and cooperative extension services in most states have published guidelines for various segments of the agriculture industry. Some of the Bulletins of the Ohio Cooperative Extension Service are introduced with the following statement:[1253] "An amendment to the Federal Insecticide, Fungicide, and Rodenticide Act was passed by Congress and signed by the President on October 21, 1972. This regulation requires pesticides to be classified for 'general' or 'restricted' use by 1976. It requires anyone who applies a 'restricted' pesticide to be certified. The new regulation makes it a criminal offense to misuse a pesticide. You should remain aware of these changes and be alert to any specific requirements which affect your operation." Some of the areas covered by bulletins in Ohio include commercial fruit spray,[1253] bedding plants,[1138] commercial floral crops,[923] woody ornamentals,[922] commercial vegetables,[802] home vegetable gardens,[800] turfgrass,[753] backyard fruits,[599] soil fumigation,[372] and ornamentals.[801]

A. Food Sanitation

Without specific application but demonstrating the presence of viable fungal cells in milk and milk products, Cooke and Brazis[259] showed that yeasts and filamentous fungi are present in all types of raw milk and in processed milk products tested.

Kuehn and Gunderson[663, 664] have shown that fungi are present, sometimes in large numbers, in frozen fruit and meat pastries, demonstrating the necessity for immediately cooking these foods upon removal from the freezer or destroying them if the electric current has been accidentally disconnected for inordinate lengths of time.

In their book *Keeping Food Safe,* Bradley and Swedberg[124] have given a wide range of procedures to be used in preparing foods in the home, whether these are home processed or commercially processed, to prevent spoilage or the development of poisons or toxins.

Chapter 16

SYMBIOSIS

I. INTRODUCTION

There are varying degrees of social relationship among organisms. These have been described above for the fungi and are described in detail by Cooke.[209] Some of these relationships have been indicated in earlier paragraphs. For the fungi, two important aspects of symbiosis are evident, mycorrhizae and lichenization. Buchner[147] has pointed out a number of instances of insect-fungus symbiotic relationships. Similar symbioses or symbiotic-like phenomena have been reported by Barras[75,76,498] and by Cooke.[226]

A different type of symbiotic relationship is demonstrated by the achlorophyllous vascular plants which are usually thought of as saprophytic in habit. Furman and Trappe[409] have brought together information on representatives of 50 genera of these plants in seven families of Monocotyledonae and Dicotyledonae. These plants parasitize mycorrhizal fungi as shown by radiological techniques in several cases. The association has evolved to a high degree and is very successful.

Ardini[47] has reviewed the various factors affecting the germination of orchid seeds. In the process of symbiotic germination, the presence of a fungus is required. In most asymbiotic germination trials, fungus extract is required. A synthetic medium has been developed, but this is usually supplemented by or includes the required fungus extract.

II. MYCORRHIZAE

According to Harley:[503]

> Mycorrhizal infection is defined as an association of fungus and host in which destructive disintegration of the host does not occur and which is a prevalent and usual condition of the host plant in natural habitats, is very common and widespread. Indeed, in this respect, roots provide an important habitat for certain special ecological groups of fungi. Mycorrhizal association takes many physiological and morphological forms. Any given kind or form can be analysed in respect of the effect of each partner upon the major and minor nutritional supplies of the other and in regard of the morphogenic effect of one upon the other.

The types of mycorrhizae[500,501] include ectotrophic mycorrhizae and endotrophic mycorrhizae, the latter type being formed by fungi with both septate and aseptate mycelium. The physiology of mycorrhizal fungi is considered separately for Basidiomycetes with septate mycelium and for Phycomycetes with aseptate mycelium. Susceptibility and resistance of the host to mycorrhizal fungi was discussed. The physiological interaction of the host and the fungus varies according to whether one is dealing with ectotrophic mycorrhizae, endotrophic mycorrhizae of orchids, mycorrhizae of the Ericales, or vescicular-arbuscular mycorrhizae. Harley's review is necessarily a summary; a more detailed account is given in his book[502] and in those edited by Marks and Koslowski[750] and Sanders et al.[1005]

Basidiomycetous and ascomycetous macrofungi were studied[489] in 25 stands of Norway spruce on acid soils and on soils with nearly neutral pH ranges. Among the 264 species identified, three groups were recognized: lignicoles, those growing on wood; humicoles, those growing on humus; and mycorrhizal. For the humicolous and mycorrhizal species, two groups were recognized: acidophiles, those growing on soils with pH 3.4 to 4.0, and neutrophiles, those on soils with pH 4.8 to over 6.0. If the data for the 25 stands were ordered according to the percentage of infected stems, it is evident that with increasing disease incidence the number of neutrophile fungi increases. If the data were ordered according to the soil pH values, the acidophilic species are aggregated in the acid soils and are almost totally absent from soils with pH above 5.0. The neutrophile species are absent from all acid soils with pH 4.6 or higher. In acid soils, mycorrhizal species account for more than 50% of the species present and are mostly acidophile species. On soils of decomposed limestone, no differences in the percentage proportion of the mycorrhizal fungi could be distinguished for mature spruce stands following hardwoods, in which there was 10% heart rot incidence, and mature spruce stands developing from oldfield afforestations in which there was 45 to 50% heart rot incidence. There was no clear connection between the species composition of mycorrhizal fungi and the large differences in heart rot incidence. Presumably the acidophilic mycorrhizal fungi have a protective function to deter heart rot fungi. However, the role of neutrophile mycorrhizal fungi on weakly acid and neutral sites in relation to heart rot attacks is less clear.

Starting a forest nursery on an agricultural soil yielded poor results with seedlings; these results were not corrected by direct fertilization.[1155] When soil from a nursery in which ectomycorrhizae were present was added to that of the unproductive nursery, the seedlings improved in growth and started to develop mycorrhizae by the end of their third year.

Lengthy lists of fungi known to produce mycorrhizae and suspected of being mycorrhizal were presented[1147] for the U.S., particularly the Northwest. A number of Ascomycetes, mostly members of the Tuberales,[1149] are known to produce mycorrhizae. It has been noted that only a small amount of work has been done in experimentation with these fungi.

Cenococcum graniforme is an ectomycorrhiza-forming fungus on ectomycotrophic plants.[1148] It has an extensive host range, and its geographic range includes Northern and Southern Hemisphere temperate and arctic-alpine habitats. A technique for isolating the fungus in pure culture has been developed by Trappe.[1150] On the basis of anatomical characters it has been suggested[1149] that *C. graniforme* is the imperfect state of the tuberaceous *Elaphomyces anthracinus*.

A miniature forest made up of species of *Dryas, Loiseleuria,* and *Salix* was recognized[375] in the Swiss National Park alpine zone. These shrubs formed ectomycorrhizae with a number of fungi also found in lower forest zones as mycorrhiza formers. Many of these fungi responded to the alpine environment with a form of nanism expressed in smaller size of fruit bodies. This may have been a result of their association with the dwarf plants.

Working in the Cordillera Pelada of Chile, Singer and Moser[1048] found ectotrophic mycorrhiza forming fungi only in association with *Nothofagus.* Where these or similar fungi occurred above 500 m but without *Nothofagus,* it was assumed that the original forest had been altered by man's activities. Ectotroph-free forests below 500 m were entirely different from those above 500 m. These were true rain forests with fungus populations related more to those of other similar forests than to those of the higher elevations.

Fogel[390] collected sporocarps of hypogaeous fungi in a 40- to 60-year-old stand of *Pseudotsuga menziesii* in western Oregon between March 1972 and March 1975. There were 24 species of fungi represented. Species which each accounted for 5% or more of the total sporocarps were *Tuber murinum,Hysterangium crassum, H. separabile, Hymenogaster parksii,* and *Truncocolumella citrina* var. *citrina.* Sporocarp moisture content ranged from 17.4 to 88.6% as determined for several species and was probably controlled by age and soil. Hypogaeous sporocarps had substantially higher macronutrient contents of N, K, and P, plus the micronutrients of Fe and Al, than did the epigaeous sporocarps of *Fomitopsis pinicola*. Estimates of yearly productivity ranged from 11,052 to 16,753 sporocarps per ha^{-1} representing 2.3 to 5.4 kg dry weight per ha^{-1}. The productivity curve was bimodal as a result of temperature and moisture effects, with peaks in May to June and in October. Sporocarp numbers increased exponentially with distance from the nearest live Douglas fir stem to a peak at 160 to 200 cm, beyond which numbers dropped sharply. The optimum sporocarp zone was slightly less than the average midpoint between tree stems (205 cm).

Stunted apple trees in soils high in arsenic in eastern Washington were found to be poor in mycorrhizal roots,[1154] although good mycorrhizal production could be obtained in such soils resulting in healthy trees.

Ectendotrophic mycorrhizae have been formed in pure culture[1285,1286] with *Arbutus menziesii* and *Arctostaphylos uva-ursi*. It was suggested that arbutoid mycorrhizae may be produced by the same fungi that occur with forest trees. While the mycorrhizae vary in size and shape depending on the host, the fungus mantles and the attached mycelium and rhizomorphs are the same.

Seedlings of *Festuca* sp.[551] were grown in pots, and the culture was inoculated with fungal spores carrying ^{14}C. *Endogone* sp. formed endomycorrhizae with the grass. It is not known to grow saprobically in the soil. ^{14}C recovered from *Endogone* spores is an indication that the radioactive carbon was translocated to those spores.

Using *Glomus mosseae* spores, Ho and Trappe[552] have shown that the fungus possesses a nitrate reductase enzyme system and, thus, might be able to assume increased symbiotic effectiveness in terms of nitrogen assimilation and translocation to the host. This system occurs in other ectomycorrhiza-producing fungi as shown by their reduction or utilization of nitrate ions. Results of recent experiments indicate that two species of widespread vescicular-arbuscular mycorrhizal forming fungi, *Glomus macrocarpus* and *G. mosseae*, are able to reduce nitrate.

Fourwing saltbrush (*Atriplex canescens*) is a useful shrub[25] which can be planted on coal mine spoil reclamation projects in areas receiving 250 mm (10 in.) precipitation or less. Inoculation of saltbrush seedlings with *G. mosseae* improves transplanting success Vescicular-arbuscular mycorrhizae have been observed in association with at least 14 arid area shrubs. Williams and Aldon[1254] have described the endomycorrhizal association of a number of arid-zone shrubs.

Marks and Koslowski[750] have edited the results of a symposium devoted to problems related to ectotrophic mycorrhizae. Topics covered included structure, classification, distribution, growth, mineral nutrition, carbohydrate physiology, hormonal relationships, rhizospheres of mycorrhizae, mycorrhizae and feeder root disease, and application in forest practice. Sanders and colleagues[1005] have edited a companion volume on endomycorrhizae in which a similar series of topics is considered at length. Harley[504] opened the symposium, concluding his statement as follows:

> The outstanding problems in the study of mycorrhiza are common to mycotrophy in general and to the whole subject of symbiosis between ectotroph and heterotroph. Mycorrhizasts have mainly concentrated up to now on describing the effects of infection on the growth and mineral nutrition of the host plant. The problems of carbon nutrition of symbiotic systems are equally important and deserve increased attention. The importance of symbiosis in ecosystems has been much underestimated and the extent and effects of diversion and short-circuiting of carbon and nutrients throughout symbiotic systems requires further evaluation. The problem of the effects of each symbiont on the selective permeability of the membranes of the other and of the movement of nutrients within the symbiotic system is a central and essential problem for the future.

III. LICHENIZATION

In Holland,[931] Switzerland,[1143] and the U.S.,[28] the role of the fungus in the formation of a lichen has been intensively studied. The fungus components of lichens require complete vitamin supplements when grown in pure culture outside the lichen thallus. Furthermore, it is difficult to synthesize a lichen from its known fungal and algal components when those are grown in pure laboratory culture. The relationships of lichens with their environment and other organisms have been described by Smith[1066] and Hale[491] among others. The biological and economic significance of lichens has received attention from Llano[713] and others.

Chapter 17

OTHER POPULATION GROUPS

I. TEMPERATURE RELATIONS

Most fungi are mesophilic, developing or capable of developing their life cycles in the milder ranges of temperature. Even those species pathogenic for man and warm-blooded animals develop normally and regularly in the usual temperatures of the surrounding environment. However, there are two groups of species, each including unrelated species, which have become adapted to cold or to warm environmental temperatures.

A. Thermophilic Fungi

Cooney and Emerson[276] defined thermophilic fungi as those fungi which grow at 15 to 60°C, while thermotolerant fungi grow at lower temperatures as well. Emerson[358] has surveyed these fungi with emphasis on such physiological factors as temperature range, nutrition, and carbon dissimilation, oxygen relations, and the basis of thermophily. Among other topics considered were natural distribution and occurrence in composts, self-heating organic materials, nesting materials, dung, diseased animals, and soils. In England, Apinis[40] studied these fungi in the surface layers of field and natural grassland soils. These fungi have both activities of economic use, such as composting, and retting of fiber- and rubber-producing plants, and detrimental activities, such as overheating in hay resulting in serious loss by fire. Tubaki and colleagues[1161] found a number of thermophilic and thermotolerant species of fungi in aquatic sediments. These, with their cardinal temperatures, included the following:

	Temperature (°C)		
Fungi	Minimum	Maximum	Optimum
Thermophilic			
Mucor pusillus	20	45	30—42
Humicola grisea var. *thermoides*	20	45	30—42
Malbranchea pulchella var. *sulfurea*	25	45	37—42
Talaromyces emersonii	30	55	42—45
Thermoascus aurantiacus	30	55	42—45
T. lanuginosa	30	55	42—45
Thermotolerant			
Talaromyces trachyspermus	15	37	30
Hamigera avellana	15	42	32—37
Neosartorya fischeri var. *glabra*	15	45	25—42
N. quadricinata	15	45	25—42
Rhizopus sp.	15	45	25—42
Emericella nidulans	15	45	25—42

B. Psychrophilic Fungi

Frozen cherry and blueberry pastries were sampled for fungal populations;[663] species recovered on pour plates at 0 and 5°C were considered psychrophilic, and those recovered at 10 and 20°C were considered mesophilic. *Aureobasidium pullulans* was dominant in the crust at all temperatures. *Penicillium thomii* was the dominant mesophilic fungus in filling, while *A. pullulans* was the dominant psychrophilic fungus in the filling. Blueberry filling had 440 psychrophilic fungi per gram of sample at 0°C, blueberry crust had 65 per gram, while cherry pastries had 77 per gram. Additional work[664] yielded similar populations. Blueberry pastries had from 347 to 1586 psychrophilic fungi per gram, 90% of which were *A. pullulans;* cherry pastries had 70 to 110 psychrophiles per gram, and apple pastries had only 10 to 92 per gram. The psychrophilic population of chicken pies was 50% species of *Penicillium* and 46% *Chrysosporium pannorum.*

In leaf litter in Finland,[547] psychrophilic fungi were isolated. The mycelium in the laboratory grew at levels below 25°C and down to below 0°C. The temperature of isolation was 5°C. Those fungi growing under the snow had formed white patches under the litter, and the leaves in the litter had higher acidity, higher respiration, and higher decomposition rates, and contained five times as much readily soluble nitrogen

Deverall[312] reviewed work published on both arctic and antarctic polar fungi and alpine fungi, plant pathogenic fungi growing at low temperatures, and fungi on refrigerated foods.

Seven species of corticolous corticioid fungi that fruit under snowbanks in southern Arizona were discussed by Gilbertson.[440] Basidiocarps of *Coniophora corrugis* and *Corticium lepidum* apparently fruit only in snow. *Athelia decipiens, A. epiphylla, Leptosporomyces galzinii, Confertobasidium olivaceo-album,* and *Byssomerulius hirtellus* produce actively sporulating basidiocarps in snow, but also fruit readily during the late summer and autumn rainy season.

II. FUNGI ASSOCIATED WITH VARIOUS SURFACES

A. Rhizosphere, Rhizoplane, and Root Surface Fungi

The rhizosphere is that area of the soil adjacent to the root of the plant. When many plants of the same crop are planted in a field, it may be difficult to differentiate between the roots of the individual plant. If roots in an association involving a number of different plants are to be studied, it would be difficult to determine to which plant any one root belonged. The rhizoplane is the surface of the root adjacent to the soil, and the root surface could include one or more layers of cells of the root itself. Fungi are present in all of these areas, encouraged in part by exudates from actively growing roots, in part by dead cells sloughed off of the surface, and in part by decomposition products not consumed by bacteria and fungi present in an actively growing association of microorganisms. In the early study of the rhizosphere, the soil in the immediate vicinity of the roots was sampled grossly. This is still done. As knowledge of the habitat and techniques for its study became refined, this early technique was discarded by some for techniques to obtain more information about those fungi more intimately associated with the roots. The organisms studied here are distinct from those forming mycorrhizae which have their own "rhizospheres".

It has been noted[901] in Canada that plant age and soil type influence the nature of the fungal biota associated with plant roots, which themselves had an effect on those populations. In England in a coastal area,[929] fungi associated with the roots of plants in sand dunes and in marsh areas were compared. Many more isolates per sample were obtained from the sand dunes than from the salt marshes. However, the differences were more quantitative than qualitative. The results of studies[882] on fungi colonizing the roots of barley, cabbage, and dwarf beans during the first 10 days of growth indicate that initial root colonization may be by any of a wide range of soil fungi, but that this mixed population rapidly gives way to a stable and typical root surface mycobiota dominated by species of *Fusarium, Cylindrocarpon radicicola,* and species of *Gliocladium* and *Penicillium.*

In the study of rhizosphere and nonrhizosphere populations of fungi[883] in association with mature dwarf bean plants, three methods were used: soil dilution platings, soil platings, and soil washing followed by plating of washed soil particles. Compared with other methods, the soil washing technique allowed a more accurate assessment of fungi active in the rhizosphere.

The soil washing technique[884] showed small but consistent differences between rhizosphere and nonrhizosphere fungal populations. *Fusarium* spp., *Mucor* spp., *Cylindrocarpon radicicola*, certain sterile forms, *Varicosporium elatum*, and *Humicola grisea* were more common in rhizosphere soils, while *Trichoderma viride* was more common in nonrhizosphere soil and did not appear in rhizosphere soil. Slight qualitative changes were demonstrated with increasing plant age. No corresponding changes were observed in the mycobiota of nonrhizosphere soil.

Fungal populations in the rhizosphere of eight varieties of peanut (*Arachis hypogaea*) were compared[937] by the dilution plate technique. Four varieties had more fungi than the others, and one variety showed very little rhizosphere effect. Age of the plant had a very significant influence on the number of fungi present in the rhizosphere. High rhizosphere/soil ratios were obtained when the plants were 30 days old, and these gradually decreased until the plants were 3 months old when a slight increase occurred attributed to activity of fungi around dead or senescent roots. No correlation could be established with root nodules.

The rhizosphere of roots of *Rhizophora mangle*, the red mangrove, was studied in Hecia Swamp, Oahu, Hawaii[686] using a variety of techniques. Macrofungi were studied by direct observation. There was a pronounced rhizosphere effect with greater numbers of fungi in the rhizosphere than in the surrounding saline soil. Of the fungi recovered, 87.7% were Fungi Imperfecti and the remainder Ascomycetes and Phycomycetes, but no Phycomycetes were recovered from the rhizoplane or from macerated root tissue. Of the Fungi Imperfecti, 31.5% were common to the rhizoplane and rhizosphere, and 23.7% were common to the rhizoplane and macerated root tissue.

Parkinson[878] has summarized types of research carried out in developing concepts of the relation of fungi to the biology of the rhizosphere. The continuing development of this type of study will involve microbiological, physiological, and biochemical studies which must necessarily be coordinated.

B. Phyllosphere Fungi

A pioneering study on the fungus populations of the surfaces of growing parts of living plants was carried out in Philadelphia[632] on the fungus populations of formed, dormant, and opening buds of 12 species of horticultural deciduous trees. More species were associated with opening buds, fewer with dormant buds. The occurrence of fungi on these surfaces was regarded as fortuitous, even though a continuity of biota existed between the aerial surfaces of these hardwood trees and the soils in which they grew.

Ruinen[989] applied the term phyllosphere to the microenvironment of the leaf surface and first used it in a study of the fungi found in such habitats in the humid tropics. The populations on the leaf surface form a continuous layer varying in thickness and composition of species which were found on both surfaces of the leaf. A gradual change in species composition occurred during physiological aging of the leaf. As an environment, the phyllosphere is subject to influences from the whole plant as well as the atmosphere. Climatic factors have direct as well as indirect influences. Free water is essential both as precipitation and as dew.

It has been pointed out that an important group of fungi in the phyllosphere is the sooty molds.[577] While they do grow in association with honeydew-producing insects on leaf surfaces, they are also associated with glandular trichomes producing exudates from the leaf. As many as eight species may grow together in harmony, forming a complex community. The original sooty mold was *Fumago vagans,* a combination of *Cladosporium* and *Aureobasidium.* The sooty mold environment includes honeydew, leachates, glandular secretions, and metabolic by-products of other organisms living on the same substratum, including sometimes nitrogen-fixing bacteria, results of the dung of some of the inhabitants of the niche, the diluting effect of deposited water, and the concentrating effect of drying from frequent exposure to direct sunlight.

Preece and Dickinson[926] and Dickinson and Preece[317] have edited volumes resulting from presentations at two symposia based on studies on phyllosphere fungi.

C. Caulosphere Fungi

The term caulosphere[424] was introduced for the habitat of fungi growing on bark and in the surface cells of bark. Representatives of 40 genera of Phycomycetes, Ascomycetes, and Deuteromycetes were isolated from aspen (*Populus tremuloides*) bark in Minnesota. Such populations could be the source of populations of fungi in wood pulp. There may be differences in populations found on the surface and within the cells of the bark, as well as those found at different heights on the tree. Garner's aspen study was concerned only with species readily plated out from bark surfaces. A number of species of macromycetes also occur on bark without causing damage to the host plant. The nature of their relationship with their host tree requires further study.

D. Gemmisphere, Spermatosphere, Carposphere, and Geocarposphere Fungi

Under a variety of names, possibly including more than those given here, habitats of the fungi associated with seeds, seed coats, seed grains, seed germs, caryopses, and fruits have been studied. Information about seed-borne fungi has yet to be synthesized in symposia. However, Malone and Muskett[744] have produced a convenient manual of seedborne fungi in which techniques were described for recovery of filamentous fungi from seeds in a quantitative and meaningful way. The text deals with 77 species of fungi which were described in detail and are illustrated with photographs of the cultures and photomicrographs of conidia and critical structures.

That most seeds are accompanied by molds of one type or another has been known to students of agriculture and ecology for a long time. It is almost impossible to run a seed germination test without getting a weak to abundant crop of a variety of fungi. In a review, Christensen[194] has drawn attention to the nature of the deterioration of stored grains of cereals. Dykstra[350] described efforts to produce seed free from disease-producing organisms.

In New Zealand[1209] turnip seeds were used to measure changes in inoculum potential of spermosphere fungi following soil fungicide applications and invasion of treated soil by these fungi in the field. In untreated soil, *Fusarium* spp. and *Pythium* spp. were the dominant spermatosphere fungi. Treatments used for the control of damping-off fungi gave fair to good results, but high spermosphere inoculum potentials of *Penicillium, Trichoderma,* and *Actinomyces* spp. which followed some of the treatments were always associated with low inoculum potentials of damping-off fungi. Seasonal fluctuation,[1210] i.e., fluctuation of wet and dry periods, gave marked variation in the inoculum potential of spermatosphere fungi. *Fusarium* and *Pythium* species dominated spermatospheres in soils which had been at field capacity for more than a few days. Drying of soil in the summer decreased the inoculum potential of *Pythium* spp. considerably and *Fusarium* spp. slightly and increased the inoculum potential of *Mucor, Aspergillus, Penicillium,* and *Trichoderma* spp.

In barley stored in concrete silos in England,[820] in thermophilic fungi from both grain husks and dehusked grain, mycelium was confined to the husk tissue at the beginning of storage. However, after prolonged storage the tissue of the grain became invaded, especially when self-heating took place. The population of thermophilous fungi was reduced to a certain extent by formic acid treatment. Various techniques have been described for isolating fungi from moldy grain.[821]

Joffe[611] noted the presence of aflatoxin in ground nuts (peanuts) from crops produced during the previous 5 years. He isolated *Aspergillus favus* from such kernels and soils in which they were produced. Because these fruits are produced in the soil, the habitat was referred to as the geocarposphere.

In considering the fungi causing biodegradation of stored barley grains,[42] Apinis proposed a series of terms to describe the temperature and humidity relations of the fungi he found:

1. Mesoxerophilic — relative humidity maxima below 80%
2. Mesophilic — relative humidity 80 to 90%
3. Hygrophilic — on moist barley grains, relative humidity near 100%
4. Mesohygrophilic — intermediate between mesophilic and hygrophilic
5. Thermophilic — no growth below 18°C, maxima between 55 and 60°C
6. Psychrotolerant thermophiles — active between 18°C and below and up to 48°C
7. Microthermophilic — grow well between 25 and 35°C with maxima up to 42 to 45°C
8. Transitional thermophiles — maxima at 40 to 41°C, all grow at 37 to 38°C, optima as in mesophiles
9. Mesophiles — maximum temperature for growth below 37°C
10. Psychromesophiles — mesophiles in field will grow at temperatures as low as 0°C

In the storage of *Vicia faba* var. *minor* (bean) seeds at 20 to 22°C in unventilated atmospheres,[178] tests were made for mold populations above 10^5 spores per gram at varying relative humidities (RH). Storage would be maintained for 280 days at 65 to 75% RH, for more than 240 days at 85% RH and for 2 to 4 weeks at 97% RH using dry seeds, and using moist seeds would be less than that.

That fungi produce toxins capable of causing sickness or death in man has been known for some time, but not until the toxins produced in peanuts after harvest and during storage and transport, noted early after ingestion of peanut butter, was any systematic study of them attempted.[1006] The term aflatoxin was coined on the basis of the toxin produced by *Aspergillus flavus* (or a fungus assigned to that species).

In a postharvest analysis of the fungi associated with peanut samples in France,[809] it was found that fungi infect the surface of the pods; most of these grow through the shell, but only a few reach the kernel. The most common were *Aspergillus flavus* and *A. niger*. Following treatment in oil, the originally very contaminated kernels were processed through a series of heat treatments which progressively eliminated all contamination. Following production of feed pellets, a recontamination occurred, and a large development of filamentous fungi occurred as a result of subsequent storage conditions. The cakes fed to cattle were usually heavily contaminated. A detoxification of cakes by ammoniation permitted the destruction of aflatoxin and also had a fungicidal effect against many molds, including *Aspergillus flavus.*

E. Dermatosphere Fungi

The fungi which are adapted to growth on the skin of man and animals or on the hair of man and animals, including horn, hooves, and other keratin-containing materials, and which in these habitats cause superficial mycoses may be termed fungi of the dermatosphere. Other keratinolytic fungi may be included here such as species of *Onygena* which grow on hooves, bone, owl-pellets, and other keratin-containing substances. Of these habitats, the more important were treated with the medically important fungi. The members of the dermatophytes are usually mentioned only in lists and monographic studies as members of a population of an unusual habitat.

Chapter 18

ENERGY STORAGE AND RELEASE

I. PRODUCTION: BIOMASS

In Estonia, Kalamees and Kollom[620] found that a fungus producing rather large fruit bodies, *Lactarius pyrogalus*, reached only 1.9 kg/ha, and the production of a more delicate species, *Mycena pura*, was 62 g/ha. These data are based on production of fruit bodies. This was in contrast with production of *Oxalis acetosella*, a dominant plant in a forest-grass layer which was 0.37 T/ha, of which the aboveground biomass was 0.09 T/ha, and the total annual biomass of *Cirriphyllum piliferae*, a dominant of the moss layer, was 45.5 kg/ha. It was warned that in spite of this relatively low value for biomass, the fungi are not unimportant members of the ecosystem since they are important for a variety of transformations.

II. ACTUAL PERIODS OF FRUITING

A. Opportunity of Observation

Observation of growth and development of fungi in the field is a matter of personal convenience, governed by one's personal schedule and climatic conditions. Experiments can be set up to accomodate any schedule. Places of observation can be arranged to fit any arrangements necessary. Species can be observed in one's backyard,[253] one's coming and going,[1146] or the research projects with which one is associated.[215]

B. Sporulation Characteristics

Sporulation characteristics were determined for agaric fruit bodies in Estonia by Kalamees.[618] A glass slide placed on a metal plate under a fruit body was used to determine intensity of sporulation; a film of spores visible to the microscope (+), to the naked eye (+ +), and a thick layer (+ + +) were observed daily at 9 a.m., and cap and stipe diameters were recorded. Apparently, fruit bodies started to sporulate before they reached maximum size, and sporulation continued for several days to several weeks depending on species and on climatic conditions. The majority of fruit bodies under observation were damaged by slugs and snails, and some were damaged by rodents.

Shantz and Piemeisel[1027] noted fruiting of fairy rings in relation to stage of development in the annual cycle.

Nuss[838] developed a technique for studying sporulation of polypore fruit bodies in which a glass slide was placed under a sporulating sporophore to catch the spores. Spore release was conditioned by temperature and humidity, and during the early stages of spore discharge, the morphology of the trapped spores indicated that they were as yet immature.

C. Relation of Fruiting to Growth and Source of Nutrition

In macrofungi in nature, observations of development from fruiting to sporulation and spore discharge to spore germination and growth of the mycelium in the same fungus have been rather casual. It has been noted for some species that fruiting begins early in the growth of a fungus upon the substratum; in other cases fruiting is supposed to occur only when the nutrient supplies in the substratum have been exhausted. Little has been said about the amount of nutrients present in a particular substratum and used by a fungus, which are required by a fungus before fruiting can be initiated, or the amount required by a fungus in order to produce a certain amount of fruiting structure.

In certain situations a fungus will not fruit as long as adequate nutrient supplies are available for continued growth.[247,257] On this basis, the fruiting of cultures of fungi isolated from natural materials is promoted in some instances by providing what may seem to be a starvation diet.[1082]

III. ENHANCEMENT OF FUNGAL GROWTH BY PROVISION OF CERTAIN CHEMICALS

In the study of the effect of fertilizers it has been found that certain wild populations are responsive to chemical treatment. In England[563] the abundant fruiting of certain fungi previously known in the region by only a single or a few fruit bodies has been noted. In Japan, Sagara[995] and Sagara and Hamada[997] have noted a similar reaction, especially in the case of ammonophilous fungi, to the application of chemical fertilizers such as urea. In certain areas the results may be noticeable after only one application, the populations not being similarly affected after another application of the same fertilizer.

Sagara[996] has found *Hebeloma vinosophyllum* Hongo, a fungus which fruits luxuriantly after the soil has been fertilized with urea or other ammonium-producing materials, in soils in which had been shallowly buried carcasses of a cat and a dog. He suggested that "the present finding may lead to speculation that a victim of homicide buried in soil could be located using the fruiting bodies of this or some other fungi, assuming the burial was shallow."

IV. CIRCADIAN RHYTHM

Rhythms in various areas of growth, fructification, sporulation, and spore discharge have been noted throughout the fungi.

In a study of the rice false smut fungus *Ustilaginoidea virens*, Sreeramulu and Vital,[1095] using a Hirst spore trap, found that there was a 4-week period corresponding to the heading of the rice during which most spores were released. During this period, on dry days, a definite diurnal periodicity was noted with the daily maximum occurring at 2200 hr. Rainfall reduced the incidence of spore-fall. The highest hourly estimate of spores was 328,200 spores per cubic meter of air.

The pattern of spore discharge in a number of Hymenomycetes was studied in Kansas, Colorado, and Michigan[488] using a Kramer-Collins spore sampler. It was suggested that relative humidity, availability of adequate soil moisture, temperature, and light are factors which may affect spore discharge. In small forms of such genera as *Mycena*, *Inocybe*, and *Psilocybe*, where basidiocarps were growing in a relatively protected microenvironment, spore discharge was continuous with little variation throughout the 24-hr period. Species of *Cortinarius*, *Lactarius*, *Collybia*, *Laccaria*, *Crepidotus*, *Panaeolus*, and *Oudemansiella* developed a period of maximum spore discharge during the midpart of the night and a minimum during the daytime. In species of *Leccinum*, *Tylopilus*, *Suillus*, and *Boletus*, spore discharge peaks often developed earlier in the day and continued for only a short time into the night. In species of *Ganoderma* and *Poria* studied, a double peak pattern was found with one occurring about 0600 and the other about 1800 hr.

A series of definitions dealing with circadian periodicity by Pennycook and Newhook[890] include:

1. Periodicity — the recurrent temporal pattern of an activity or process; a result of the interaction of exogenous and/or endogenous factors
2. Rhythm — a periodically repeated fluctuation controlled by an innate time-measuring system; all are endogenous
3. Diurnal — having a daytime maximum
4. Diel or daily — preferred terms for the general sense "24-hourly"
5. Circadian — restricted exclusively to endogenous rhythms with a free-running period of 20 to 28 hr
6. Zeitgeber — a periodic exogenous factor acting as a forcing oscillation
7. Entrainment — consists of the adjustment (within limits) of the period of the rhythm to that of the Zeitgeber and the synchronization of all individual rhythms to a precise phase relationship to the Zeitgeber; occurs by means of a differential phase shifting of the circadian rhythm
8. Free-running period—when the Zeitgeber is removed, a circadian rhythm will display its free-running period as a sustained or a damped oscillation

The circadian rhythm of ballistospore discharge in the apple phylloplane populations of *Sporobolomyces roseus* and *Tilletiopsis minor* were studied[890] under a 12:12 light to dark regime and constant relative humidity. The ballistospores were discharged almost exclusively at night. *Tilletiopsis minor* ballistospore discharge periodicity retained a constant phase relationship with the light cycle regardless of variation in experimental procedure. *T. minor* ballistospore discharge remained periodic during a 50-hr interval under constant low illumination of 0.5 lux. It was suggested that all available evidence supports the hypothesis that the periodicity of basidiospore discharge is due to a circadian rhythm which is entrained by the environmental light/dark cycle. The expression of periodicity can be modified by relative humidity changes.

Gilliam[444] found in a study of *Marasmius rotula* using Kramer-Collins spore trapping equipment that in this species spore discharge does not follow the circadian rhythm found in other agarics tested but follows a rain-dependent pattern varying in amount and duration with the amount of rainfall.

V. FUNGAL SUCCESSION: ELIMINATION OF HABITAT

Garrett[433] has said that fungal succession proceeds to an end point, the elimination of a habitat. The actual progression of the series of fungi required to completely eliminate a leaf, twig, or log has rarely been described. The appearance of the succession may be broken into when one collects the bracket fungus from the log, the mushroom from the soil, or the aquatic Hyphomycete from a leaf, but unless the entire mycelium of a species involved is removed or destroyed, the succession will continue until the substratum used by its enzyme system is eliminated.

Kendrick and Burges[639] followed the succession of fungi and other microorganisms on pine litter. Following preliminary action by certain fungi, invertebrates took over following which other fungi continued the process of decomposition.

From the point of view of the individual species, the succession ends when the supply of nutrients has been exhausted even though it has produced breakdown products used to advantage by later organisms in the series. From the point of view of the community of organisms inhabiting a specific habitat, as one nutrient source is exhausted another becomes available, and the community resulting from these activities can be interpreted as a series of individuals competing for one nutrient or waiting for another to become available.

The kind of wood subjected to decomposition, spruce or beech, in an experiment conducted in Czechoslovakia,[1144] as well as the direction and character of the decomposition, are dependent on the availability of the substrate to the organisms available to attack it. Loss of dry weight in blocks of wood was used to determine effectiveness of fungi used in an experiment on succession of fungi on wood.

VI. UBIQUITY

On the basis of basic nutrition[254] and basic genetic patterns, the widespread distribution of a number of common species of fungi has been demonstrated by a series of collections, isolates, or strains of fungi. An individual collector, by early schooling and experience, may think of isolated geographic or political units as habitats actually isolated in space and time. As one becomes familiar with collections and isolates from wider and wider areas and increases the range of his travels, he becomes more familiar with a range of local biotas. Depending on techniques of identification, lengthy lists of synonyms may be developed, species or form species may be separated out, or species aggregates may be recognized. These comments agree with those of Vallentyne[1175] who reached similar conclusions on the basis of certain biophysical aspects of the environment.

VII. LONGEVITY AND SURVIVABILITY OF FUNGI

According to Sussman,[1123]

> The fact that survival, especially in nature, depends on the interaction of numerous interesting environmental and organismal elements means that studies of the effects of isolated variables oversimplify the issues. Furthermore, conclusions derived from laboratory cultures must be tempered by recognition of the extreme heterogeneity of microbial populations, and their mutability. Consequently, the accuracy of the extrapolation from such studies in natural situations, is difficult to evaluate. Nevertheless, under any circumstances there is the biological equivalent of the "uncertainty principle" such that any manipulation is likely to disturb a natural situation. Therefore, such upsets must be reckoned with as part of biological research, and it is through laboratory studies in combination with those in the field, that an understanding of survival of fungi in nature will be obtained.

Longevity implies survivability. The mycelium of fairy rings in the short-grass prairies of eastern Colorado has been demonstrated;[1027] these organisms have been known to survive for several hundred years, continuing to produce fruit bodies throughout the period. The wood decay fungus *Agaricum officinalis* is known to produce fruit bodies reaching 60 years (or growth seasons) in age. The fungus must have survived in the wood that long or possibly 1 or 2 years longer.

Chapter 19

USES FOR FUNGI

I. INTRODUCTION

It has been implied above that there are several uses to which fungi can be put for man's benefit. The earliest of these was probably the culture of mushrooms and the collection of wild mushrooms for food sources. Possibly older than this was the development and manufacture of fermentation products from natural yeasts and from mucoraceous fungi.[82,83,541] More recently, as fermentation processes have become more elaborate and refined, additional fungi have been found useful in the production of various types of commodities[927,1072] such as penicillins and other antibiotics, food[1140] and feed materials, drugs, and chemicals.[717,988] When other sources failed, fungi have been employed in the production of fats for use in food supplements. Fungi have been surveyed[1276] for fat production, and fat production techniques have been improved.[823] The potentialities of fungi for biosynthesis of many products has been discussed.[462,471]

While development of the pure culture technique has taken place in the last century,[323,717,718] it is only in the last 70 years that techniques have become refined to the extent that certain species can be employed in the assay[696] of various types of compounds. Techniques have been developed to assay vitamins and other highly specialized substances. Fungi are sensitive to the amount of elements, for instance, molybdenum, in soil and in solutions.

The development of sewage treatment facilities has led, quite by accident, to further use for fungal populations. Both on the trickling filter[228] and in the activated sludge tank[274] fungi are one type of organism which contributes to the treatment process. Recent work[249] has indicated that the probable extent of this type of activity, as well as the whole problem of the biological treatment of sewage, is full of problems based on questions concerning how this activity takes place, what contribution each group of organisms makes, and related inquiries.

II. FUNGUS FOODS AND POISONS

Fungi have served for food for a long time, probably since before the dawn of history.[933] It has been reported that fungus poisons have been used to eliminate enemies and as stimulants. Cowell[286] emphasized the early recognition of the poisonous nature of "venereous meats" in a survey of edible and poisonous fungi, emphasizing the fact that few general practitioners, especially in England, are equipped to cope with problems of identification of the fungi their patients may have eaten.

The culture of mushrooms is one of the more specialized phases of horticultural science. A number of books and pamphlets has been written on this subject. To increase production without loss of flavor, the use of liquid culture has been proposed.[579] A number of edible fungi has been adapted to liquid culture.[1120] The culture of truffles has become an industry of some importance in Europe where it is necessary to develop mycorrhizae with oaks in forest plantations. As yet, techniques and media for growing morels and truffles in artificial culture have not been perfected.

Use of poisonous fungi as intoxicants has been practiced for a long time,[933] and other fungi which are not deadly poisonous, but whose edibility is questioned by connoisseurs, are used for similar purposes. The intoxicating effect of such fungi may not be felt for long periods, and immunity to the poison is built up gradually by increasing dosages. The edible inky cap *Coprinus atramentarius* is reported from France[578] to have produced intoxication in at least one case.

Many animals also depend on fungi as food sources. Workers in Austria[1012] have shown that definite relations can be established between fungi and beetles which use them as food. Here the same degrees of dependence on fungi as food sources have been noted as are shown between fungi and burned areas mentioned above. Slugs and rodents use fungi as food sources, as has been noted in Canada,[150] in the study of the hypogaeous fungi of Mount Shasta,[216] and in *Endogone*,[320] a fleshy fruited hypogaeous member of the Zygomycotina. Observations in tropical Africa[531] have shown that termites are able to grow certain agarics as food sources. Miller and Halls[796] have shown that a number of species of fleshy fungi are used as food by deer and other mammals in the southeastern U.S.

The relations between species of the fruit fly *Drosophila* and yeasts growing in habitats from which both the flies and the yeasts have been taken in the Yosemite region of California have been carefully studied.[181,325,908]

That there is an important dependence between nematodes and fungi has been noted in a study made in Holland.[787] Here it was noted that, as in the case of beetles, there is a dependence between the nematodes and the fleshy fungi on which they subsist, in which some species of nematodes require the mushroom as a habitat while others do not. In Russia[764] a method for determining predictable fruiting periods of edible fleshy fruited fungi has been proposed.

A definitive survey of the various uses of fungi by man and the processes developed in producing their by-products has been developed by Prescott and Dunn.[927] Bacteria, yeasts, and molds are treated in separate sections. Fermentation products, vitamins, and enzyme production by yeasts are surveyed. The fungi produce a wide range of organic acids, as well as fats, enzymes, and antibiotics; they also transform steroids.

Gray[465,466] has surveyed the direct use of fungi as food for man and animals, the use of fungi in food processing, and the potential uses of fungi as food or food supplements. He has also indicated the toxic substances in fungi and their effects on man and animals. Additional information is given on the kinds of wild fungi available for use by man as food, mushroom poisoning, ethnomycology, and hallucinogenic fungi, and the methods of preparing fungi for use as human foods. It is noted that comparatively small amounts of fungi are used as food in regard to the amounts of fungi available and the amounts of fungus materials which can be produced with the present technology and with technologies which can be developed in the future.

One of the more recent of these developing technologies was described by Church and colleagues[198] as a result of studies conducted in Minneapolis on the use of corn-processing wastes in the development of fungus protein through the activity of several species of filamentous fungi assigned to the Fungi Imperfecti. In these studies 21 strains of 11 fungal species were screened. *Trichoderma viride* reduced the 5-day biochemical oxygen demand (BOD_5) from 4000 mg/ℓ to 50 mg/ℓ during the continuous fermentation process. Sterile conditions, used only in the first stage of transfer of mycelium to the process, were not required beyond that point. Only about 0.5 to 0.7 lb of O_2 was required per pound BOD_5 used. Mycelium was removed from the process by coarse filtration. Palatability of the mycelium to rats was excellent, and digestability in weanling rats was excellent. Upon expanding from a laboratory scale fermentor to an 11,000-gal scale fermentation system, corn-canning wastes were used. An oxidation ditch with a rotary brush aerator was tested. pH was controlled to about 3.7, and ammonium nitrogen and inorganic phosphate were added. Residence time averaged 18 to 22 hr. About 96% BOD_5, 88% COD (chemical oxygen demand), and 93% TOC (total organic carbon) were removed, essentially zero levels of ammonium nitrogen and inorganic phosphate could be attained in the effluent stream, and organic phosphate levels were reduced by 80%. Mycelium yields on a dry-weight basis were equal to about 50% of the BOD_5 of the feed, and the nitrogen content of the mycelium was equavalent to about 50% protein.

Updegraff and co-workers[1171] have developed a process for treating coffee processing wastes in Central America. Of several fungi tested, *Aspergillus oryzae* gave the best results and was used in developing the program. Good protein production has been achieved.

Ladisch and co-workers[669] have noted that corn (maize) crop residues are a potential source of single-cell protein which can be used as supplements for animal feed produced from the 67 to 167 million tons of corn crop residues produced annually in the U.S. and presently wasted.

A. Mycotoxins

A symposium based on some of the earlier studies of toxin-producing fungi was edited by Wogan.[1271] Major subjects considered were fungi of cereal grains and their products, toxin production in cereal grain causing toxic alimentary aleukia in man, and aspects of aflatoxicosis in farm animals.

In the sections on mycotoxins of the first U.S.-Japan Conference on Toxic Microorganisms, the proceedings of which were edited by Herzberg,[540] the major subject areas considered included the mycotoxin problem, the identification of mycotoxins, the production of mycotoxins, the biological effects of mycotoxins, their formation and characterization, their presence in food and feeds, and the differentiation and irradiation of mycotoxins. Results were reported for different mycotoxins produced by 21 species of fungi. Toxic compunds produced by species of *Fusarium* are of serious concern in the production of rice. Japanese scientists reported a broad series of studies on the chemistry and pathology of these toxins. Some success was shown in the U.S. in attempts to minimize aflatoxin formation under field conditions in crops of peanuts, cotton, and soybeans. Advances were reported in the isolation, identification, chemistry, biosynthesis, destruction, pharmacology, pathology, and other aspects of toxins and on our understanding of the fungi that cause them. Mycotoxins represent serious potential hazards in certain crops and commercial foods but so far have not been demonstrated in commercially fermented foods.

In an effort to present to as wide an audience of farm technicians as possible information concerning the cause of mycotoxins and methods for the control of the fungi that cause them, a symposium-like series of papers has been presented.[97] There are records of mass poisoning of human populations by mycotoxins. Myxotoxin-induced liver cancer is known in Africa, and studies in Thailand implicate mycotoxins in child fatalities. That molds and mycotoxins are serious farm problems is indicated by the fact that a midwestern feedlot has lost 25 to 100 steers daily, 493 of 1000 have been lost in 48 hr, and over 250,000 turkeys have been lost during a recent 3-year period. Mycotoxins thus force changes in management practice. Young animals are more susceptible to the action of mycotoxins. Lack of appetite and weight loss are the first warning signs, and liver damage is frequent. Mold-contaminated toxic batches of food are blamed. Fungi are thought to be the cause of major damage in stored grain; moisture content, temperature, and other factors are considered to be of critical importance in the development of heating and other damage. More analysis techniques are needed for determining the presence of mycotoxins. The newest method is indicative of mycotoxin presence only; three principal methods are available, and there are chemical tests for toxins in peanuts, corn, and other products. High-moisture grain is effectively preserved with organic acids, and this kind of treated grain may be particularly beneficial to dairy cattle. Treated grain stores and handles easily. That high-moisture corn has increased food value is now a proven fact — the feed needed per pound of grain is 10% lower when high-moisture corn is fed. Thus, early and rapid harvest is now possible. Because feeding moldy grain is risky, organic grain preservatives offer unique strategies. Proper application is essential, and treated grain can be easily stored. No detrimental effects on animal health have been noted in any species fed acid-treated grain. In fact, a 5 to 9% increase in rate of gain has been observed where acid-treated corn is fed. Various corn handling systems are considered in depth, since the corn handling system should provide the greatest feeding value for the lowest cost and must minimize risks from molds. Research at the Food and Drug Administration is being stepped up, especially in analytical techniques for mycotoxin detection development. Necessary attributes for chemicals necessary to prevent fungal growth are described by Environmental Protection Agency personnel. Practical use evaluation tests are necessary. The outlook on preservation is reported as "very favorable".

The toxins produced by higher fungi have been listed,[515] their sources described, and the toxic effects produced by them summarized. These toxins include Amanita toxins, ibotenic acid, related constituents with neuralgic effects, muscarine, gastrointestinal irritation, and mushroom allergies.

Ciegler[199] noted that mycotoxins may be produced by a variety of fungi in food and foodstuffs that, on consumption, cause disease in man and animals; symptoms may be either acute or chronic. While acute symptoms may be more dramatic, chronic effects may be more important since they are insidious, difficult to detect, and can unknowingly cause economic losses in livestock as a result of feed efficiency and weight gains. Symptoms in mycotoxicoses tend to be nonspecific; hence, they are difficult to diagnose. Many of the mycotoxins are hepatotoxins, but other organs of the body can be, and often are, affected. The problem is worldwide, not confined to any geographic area, and is extremely complex since mycotoxins can be produced on grains in the field, during harvest and processing, and during storage of any given food or feed. The tendency of these toxins to occur in comparatively low concentrations complicates detection and analysis; the problem may be further exacerbated by the potential occurrence of mixed toxins and of toxins bound to the substrate in which they may be produced, which render them difcult to detect.

According to Moreau,[808] the dosages of mycotoxins should be carefully considered but not overemphasized. Lactating cows are not affected to a great extent by aflatoxin B_1; however, when that mycotoxin occurs in cattle feed, it may be metabolized and found in cow's milk as aflatoxin M, which is hepatotoxic and carcinogenic and could be dangerous for human health, especially that of small children. While many molds develop in powdered milk, none of them are considered to be toxinogenic. Some strains of *Penicillium camemberti* of Camembert cheese could be toxic. Many mycotoxins have been identified in the products of *Penicillium roquefortii*, but the major toxin, known as PR-toxin, could not be produced in the normal conditions of Roquefort cheese manufacture.

Effects of aflatoxin B_1, a toxic metabolite of *Aspergillus flavus* and *A. parasiticus*, on insects have been studied in laboratory experiments[670] with *Drosophila melanogaster*. First instar larvae were placed on a diet including aflatoxin B_1, and it was found that percentage of larvae entering the pupal stage, percentage of pupae emerging as adults, time for maturation, sex, and size of the pupal case, adult body, and wings were all reduced.

Studies with the Syrian hamster *Mesocricetus auratus*[715] have shown that, in this animal, aflatoxin B_1 — administered chronically in the diet at the rate of 1 ppm —

and subcutaneously implanted 17-beta-estradiol, evaluated at 98-, 161-, 210-, and 257-day periods, yielded liver and kidney damage when administered individually, and this damage was increased when these substances were administered concomitantly.

Aflatoxin B_1 was administered[1137] intravaginally in K-4 jelly in female rats prior to mating. There was a decrease in percent fertilization with increased concentration of aflatoxin B_1 (0.25 to 2.5 μg/100 g body weight). Fetuses showed neither gross malformations nor any significant change in body weight from the controls. There was no change in the mean number of offspring per litter.

The activity of sterigmatocystin in three rodents, *Mesocrinitus auratus* (the Syrian or golden hamster), *Meriones unguiculatus* (the Mongolian gerbil), and *Mus musculus* (the mouse), was tested.[714] Dietary levels used were 0, 10, 25, 50, and 150 ppm. For young gerbils receiving 150 ppm, there was 100% lethality in 5 days. At 50 ppm, mice showed an initial loss of weight followed by gradual recovery. At 50 ppm, there appeared to be a retardation of growth rate in all three animals. There was kidney and liver damage in all three animals, varying in type and intensity, at 10 to 50 ppm.

III. FUNGI IN CULTIVATION BY MAN

Fungi have been used by man in the production of alcoholic beverages since before the dawn of recorded history. Such use is continued to the present day.[1074] Other traditional uses for fungi include soybean fermentation, baking, and the mold ripening of cheese. Currently there are two additional uses for fungi: mushroom culture and industrial fermentation. Among mushroom culture are paddy straw mushroom culture, shiitake culture, and truffle cultivation. Industrial fermentations are directed toward production of antibiotics, proteins, fats, and vitamins, the production of organic acids such as citric acid, itaconic acid, d-lactic acid, gluconic acid, and fumaric acid, and the modification of sterols.

A. Fermented Foods

In certain Asian countries, including India, Pakistan, Nepal, Sikkim, and Tibet, large quantities of fermented foods are consumed. These have been listed by English and vernacular names, fungi responsible for the fermentation, and native and laboratory techniques used for producing these foods.[82,83]

In developing laboratory techniques[541] for producing oriental fermented foods, it is necessary to have a supply of fungal spores available for use. Desirable inoculum characteristics include: (1) production of spores in large quantities, (2) uniform viability and genetic stability over a period of several months, (3) high percentage of spore germination in a short time, (4) correct proportion of strains when mixed strains are used, (5) ready dispersibility of spores in fermentation substrate, and (6) freedom from contaminating organisms. Methods of preparing inocula of different fungi on different substrate were given, as well as of storage of inocula in such a state as to ensure 100% viability.

Yeast derived from a strain of *Torulopsis utilis* was grown using continuous type feeding methods in spent sulphite liquor at commercial levels in a pulp mill in Wisconsin.[1245,1246] Chemical analyses and vitamin assays made on samples of dry yeast showed it to be uniform in chemical characteristics and in the content of various components of the vitamin B complex. Wood yeast[968,969] was found to be equal to a high quality soybean meal as a source of protein for the production of milk by cows. Under the conditions of an experiment,[965] *Torula* yeast was satisfactory as the only protein supplement for laying pullets producing hatching eggs and for growing chicks.

Since World War II when yeast[707] was used as an important food in Germany, there have been periodic attempts to revive interest in yeast as food. The production attempt in the Caribbean failed because of production difficulties. Certain economic conditions do not favor the use of food yeast largely because of political considerations. There are many possibilities for the industrial exploitation of yeast beside the use as food, in fermenation, and in baking. Some of these may lead to a more general use of by-product yeast on a large scale for nutritional puposes.

The submerged culture[711] growth of the morel mushrooms, *Morchella crassipes, M. esculenta,* and *M. hortensis,* was investigated using cheese whey, pumpkin canning wastes, and corn (*Zea mays*) canning wastes as substrata. *M. hortensis* was selected as culture of choice on the basis of the more efficient conversion of glucose and lactose components of the wastes to protein. The protein content of the mycelium of *M. hortensis* was approximately 35% when grown in cheese whey and pumpkin wastes when supplemented with nitrogen and phosphorus.

In the course of a project[463,464,467,473] designed to explore the Fungi Imperfecti as potential synthesizers of edible protein, 175 isolates were screened using a shake flask technique, and their economic coefficients were determined using crude glucose as a carbon source and ammonium nitrate as the inorganic nitrogen source. The maximum incubation period was arbitrarily set at 4 days. Approximately 7% of the isolates were found to have an economic coefficient of 2.00 or less. The economic coefficient is the unit weight of carbohydrate required to produce one unit weight of dried fungus tissue; the economic coefficient of 45% of the isolates was within the range of 2.01 to 3.00. Yield, protein, and lipid content could be increased by proper alterations of the physical and chemical environment, and the process proved to be feasible. The use of cheap and abundant carbohydrate sources such as sweet potatoes,[468] corn, rice,[472] cassava flour and root, citrus molasses, corn molasses, beet molasses, whole and shredded sugar beets,[470] wood fiber, paper pulp, cocoa husks, manioc,[469] and cane juice[474] have been investigated. The best results, based on total crude protein per liter of medium, were obtained with sweet potatoes and beet molasses. With sugar beets the time required to complete the process was only 30 hr. In most cases the end product was odorless and tasteless, and the calorific value was high. One result of this work was the development of a patent.[463]

The use of pear-canning wastes[3] for production of alcohol was explored using a strain of *Saccharomyces cerevisiae.* Batches of ground pear wastes were fermented in 30 hr with a 90% efficiency of converting sugar into alcohol. However, when supplemented by the addition of small amounts of nitrogen in the form of ammonium sulfate, urea, or ammonium carbonate, pear-canning wastes could be fermented continuously in a 10-hr cycle. It was noted that the disposal of cannery wastes by alcoholic fermentation must be based upon the utilization of the whole waste, and it was found that use of the juice alone could be extrapolated to the whole waste.

It has been found that in an activated sludge process treating corn-canning wastes,[792] it was possible to produce merchantable quantities of fodder yeast while removing the BOD_5 of this type of industrial waste.

IV. PRODUCTION OF ORGANIC ACIDS AND OTHER PRODUCTS

Texts describing the chemical activites of fungi, both from the point of view of their requirements and their ability to produce useful as well as harmful metabolites, include those of Foster,[395] Prescott and Dunn,[927] Rainbow and Rose,[932] and Rose.[980]

Using cultures of *Rhizopus oryzae*[976,977] and calcium base spent sulphite liquor, the feasibility of using a fungus fermentation to convert sugars in spent sulphite liquor to fumaric acid was studied. Based on techniques used for successful fermentation carried out in stationary culture and laboratory fermenters, production of fumaric acid was demonstrated in both growth and replacement phases. Based on total amount of sugar available, overall fumaric acid yield of 40% by weight was obtained, with 70% sugar removal.

Lysine is an amino acid in short supply in many foods, particularly those of cereal origin, and lysine is the primary limiting amino acid in wheat protein when it is used as a food. Cultures of over 600 fungi were screened[963] for the production of extracellular free lysine in submerged agitated culture. Of these isolates, 63 produced free lysine sufficient to be detected by turbidometry, and 12 of these produced 5 to 15 gamma lysine per milliliter of culture filtrate. With two strains of *Ustilgo maydis,* yields were increased to 200 to 300 gamma lysine per milliliter of culture mixture by adjustment of culture conditions. Using two strains of *Ustilago maydis* and one of a species of *Gliocladium,* extracellular lysine production[342] has reached 400μg/mℓ in an experimental system in which lysine production was concurrent with the formation of most of the extracellular organic nitrogen.

An extracellular dextranase[1041,1042] was purified approximately 76-fold from cell-free culture filtrates of *Fusarium moniliforme.* The purified dextranase was of the endo type, and isomaltose was identified as the primary end product of dextran hydrolysis. The *F. moniliforme* dextranese was found to differ from the commercial preparation by its greater relative activity against glucose isolated from *Streptococcus mutans.* Culture conditions and isolation procedures were described.

V. ANTIBIOTIC PRODUCTION

A survey of antibiotics produced by fungi was presented by Broadbent[137] in tabular form giving the producing organism, the organism or organisms against which the antibiotic is effective, and the name of the compound where this was known.

Various aspects of the discovery, development, production, chemistry, and improvement of penicillin and cephalosporin were reviewed by Abraham.[1]

Laboratoy evaluation at the manufacturer's level of various antibiotics has led to the discovery and development of new antifungal and antibacterial preparations. Methods used in determining the activities of various substances against different disease-producing agents desribed by Grundberg and Titsworth[485] have resulted in the recognition of such antifungal agents as 2-dimethyl-amino-6-(beta-diethylamino-ethyloxy) benzothiazole dihydrochloride (asterol), which is active against the dermatophytes; the antibiotic saramycetin, which is active against systemic mycoses; and flucytosine, which is active against yeasts such as *Candida albicans* and *Cryptococcus neoformans.*

The isolation of the penicillin nucleus,[366] 6-amino-penicillanic acid, has facilitated the preparation of literally thousands of new penicillins by the addition of novel side chains at the 6-amino position. As a result, about 30 semisynthetic penicillins have become available and are now in clinical use. Careful selection of those penicillins to be used is of paramount importance.

Some of the laboratory procedures used in the preclinical evaluation of cephalosporins have been summarized.[928] The in vitro and in vivo characteristics of several known cephalosporins and two groups of experimental compounds have been described and compared.

Chapter 20

TECHNIQUES

I. INTRODUCTION

Different segments of a population of fungi in any one area are observed by different techniques. The most obvious species, the macromycetes, those with fleshy or woody fruit bodies readily visible to the naked eye, appear usually in late summer or early autumn but represent responses to different habitats within the area. Among the fleshy species are those which grow in the soil, on the litter, as myorrhizae, or on rotting wood; among the woody fruit bodies are those of fungi growing in heartwood, sapwood, or on the ground in association with buried wood or living roots, among other habitats. Each of these habitat types may represent different biological adaptations, and while each of these appears to be distinct, in fact there may be a certain amount of antagonism or mutual tolerance associated with their proximities. In addition, there are other populations active at the same time and in the same areas including soil molds, aquatic fungi which are adapted to living in soils, and fungus species restricted to specific habitats such as dung, dead herbaceous stems, tree bark, and other habitats within the area. Most of the techniques used in the search for all kinds of fungi are described in the *Mycology Guidebook*.[1111]

In a comprehensive manual edited by Booth,[119] the methodology of the study of a wide variety of fungal groups is given in 24 chapters by 22 workers in as many areas of mycology. Chapters summarize general methods, culture media, techniques for preparation of material for microscopic observation, for isolation of fungi, for preservation of fungi, for purification and maintenance of yeasts, and techniques for study of such special groups as Phycomycetes, Heterobasidiomycetes, Homobasidiomycetes, lichens, aquatic fungi, and soil fungi. Techniques are given for air sampling, sampling for fungi pathogenic for man and animals, including those causing diseases of keratinized tissues and those causing subcutaneous and deep seated mycoses, methods for genetic studies, autoradiographic techniques, fluorescent light techniques, electron microscopy, chemical tests in agaricology, immunological techniques, effects on fungi of visible and ultraviolet light, and production and use of fungal protoplasts. Most, if not all, of these techniques have application in one or another study involving the synecology or autecology of fungi.

II. AQUATIC FUNGI

A. Fresh Water

In the study of aquatic fungi, including chytrids, various types of baits[285] have been used, but at present the most generally accepted bait for saprolegniaceous fungi is hemp seed. In Germany[557] such additional seeds as rice, corn, and rye were used. These seeds are handled in two basic ways: materials are brought to the laboratory and placed in culture dishes with the hemp seeds,[509] or the seeds are placed in containers, such as teaballs[258] or nylon cloth sacks,[557] and allowed to remain in the test body of water for various periods of time. Test periods of 2 to 6 days have given various degrees of effectivity. Upon completion of the exposure time, the seeds are brought to the laboratory in a still moist condition, preferably in a container of water from the sampled site, and re-exposed with fresh hemp seed, in water from the sampled site which has been sterilized or in tap water or distilled water in which activated carbon has been boiled and filtered off. When the usual halo of fungus growth appears around the hemp seed, it can be studied during its development. As the fungus develops it can be studied under the microscope, and various techniques have been described for preservation of such material. In addition to hemp seeds, rosaceous fruits, meats, and whole pitted dates have been used in trapping such fungi. These are exposed in hardware cloth or other wire containers, buoyed with cork stoppers, and skewered on wire dowells inserted in one of the corks. These are tied[258] to a permanent emplacement on the shore using 30-lb nylon fishing line. The baits are returned to the laboratory after exposure as noted above. Additional sampling baits include cellophane, human hair, pieces of human skin peeled from sunburn blisters, other keratin containing materials, and materials reasonably uncontaminated with the fungi for which one is searching. Such baits are exposed in the laboratory in culture dishes to which the sample is added, with water in which the spores of the fungus being sought can swim.

It has been reported that where hemp seeds are unavailable turnip seeds have been substituted. Rooney and McKnight[979] used a series of baits exposed in the lake they were studying. These included twigs of river birch, weeping white birch, poplar, hackberry, and pine; fruits such as apples and rose hips; chitinous and keratinaceous materials including human hair, snake skin, and insect exuviae; and other things including cellophane, filamentous algae, pine pollen, and pine needles. Willoughby[1257] used cornmeal agar in which a sample of water to be tested was incorporated just before hardening, following which pieces of the solidified agar were placed in a petri dish with water. Each hypha developing on the margin of the agar was considered to have arisen from growth from one zoospore. Emerson,[357] Emerson and Weston,[360] and others have used rosaceous fruits which were enclosed in containers and submerged in the water to be tested; this technique recovers a somewhat more specialized segment of the population, usually, than the floating of hempseed and other baits.

B. Marine Water

One of the standard devices for the baiting of marine fungi is the exposure of blocks of wood of various types, singly or in various combinations of units bolted together. Crow and colleagues[287] found that a membrane adsorption technique could be used for the isolation of bacteria and fungi from coastal surface films. The technique yielded larger numbers of a more diverse population than had been previously recognized.

C. Polluted Water

For the study of fungi in sewage, sewage treatment systems, and polluted waters, a laboratory guide has been prepared by Cooke,[234] of which a somewhat updated version has been prepared.[257] In this guide the medium used most commonly for primary isolation is neopeptone-dextrose-rose bengal-aureomycin agar. Comparisons have been made with different materials in the formulation of this medium which is basically Emmons' American Sabouraud agar. The peptone products neopeptone and polypeptone have been compared, as have these two peptones and dextrose with or without the inorganic supplements recommended by Waksman.[1196] Results with the same samples were so similar that (Table 1) it was decided to use either peptone source without the added salts. Presumably, sufficient amounts of required atoms or ions are present as impurities in other components of the medium so that their addition is unnecessary.[240]

In addition to colonies obtained on agar in the above technique, yeasts can be obtained in liquid culture using yeast nitrogen base (YNB-Difco) and dextrose.[234] Yeasts and filamentous fungi are also isolated on membrane filters which are incubated on

TABLE 1

Total, Percent of Total, Average Number of Colonies, and Range of Number of Colonies in Samples from Sewage Treatment Plant Processes on Four Media

	Rose bengal agar with			
Category	Neopeptone (no salts)	Polypeptone (no salts)	Neopeptone (salts)	Polypeptone (salts)
Total colonies	1112	1051	1094	1067
Percent of total	25.7	24.3	25.3	24.7
Average number of colonies	55.6	52.6	54.7	53.4
Range	14—128	16—170	14—139	15—178

Note: Mention of products is for identification only.

From Cooke, W. B., *Mycopathol. Mycol. Appl.*, 34, 313, 1968. With permission.

nutrient-soaked pads or on agar media. It may be noted that in none of these techniques is every member of a population obtained; some species grow better in liquid media, some on solid media, and some do well on either. Factors of competition or sůppression of segments of a population which are imperfectly understood may also be involved in successful recoveries of populations.

For identification of the yeast cultures obtained, the techniques outlined in Lodder[719] are recommended. They may be supplemented as follows. In the testing of numerous cultures for carbon source utilization, a multipoint replicator may be used. One type is illustrated in Figure 1. Agar prepared with YNB-Difco is placed in flat-bottomed petri dishes to which have been added 2% solutions of the carbon sources to be tested, except raffinose which is tested using a 4% solution. The medium is permitted to harden and the surface to dry. To each of 25 8 × 30 mℓ shell vials is added enough sterile distilled water to fill them ⅔ full. They are capped with aluminum foil and autoclaved. When cool, a loop of cells from each of 25 pure cultures of yeasts is added. The vials are arranged in known order in the vial holder. The replicator has been autoclaved and has been placed in the rest block without contaminating the needles. The replicator is removed from the rest block, and the needles are inserted in the vials and placed on the agar plate block so that the tip of each needle touches the agar surface leaving its inoculum. The process is repeated until all the plates, each containing a different carbon source, is inoculated. The inoculator is autoclaved in preparation for the next series. Four series can be inoculated on 1 day. Growth on the inoculated plates is read after 3, 6, and 14 days of incubation at room conditions of light and temperature. For testing different temprature effects, additional plates may be prepared, one using glucose for 5°C in a refrigerator or constant temperature chamber, if available, and one using a medium adaptable to warm temperature incubation[237] for use in a 37 or 45°C incubator, or both.

III. TERRESTRIAL FUNGI

A. Macromycetes

Most of the fungi studed in relation to synecological projects using macroscopic fruit bodies have large, definitely organized fruit bodies which may be gelatinous, fleshy, or woody. The search for such fungi is made relatively simple by their macro-

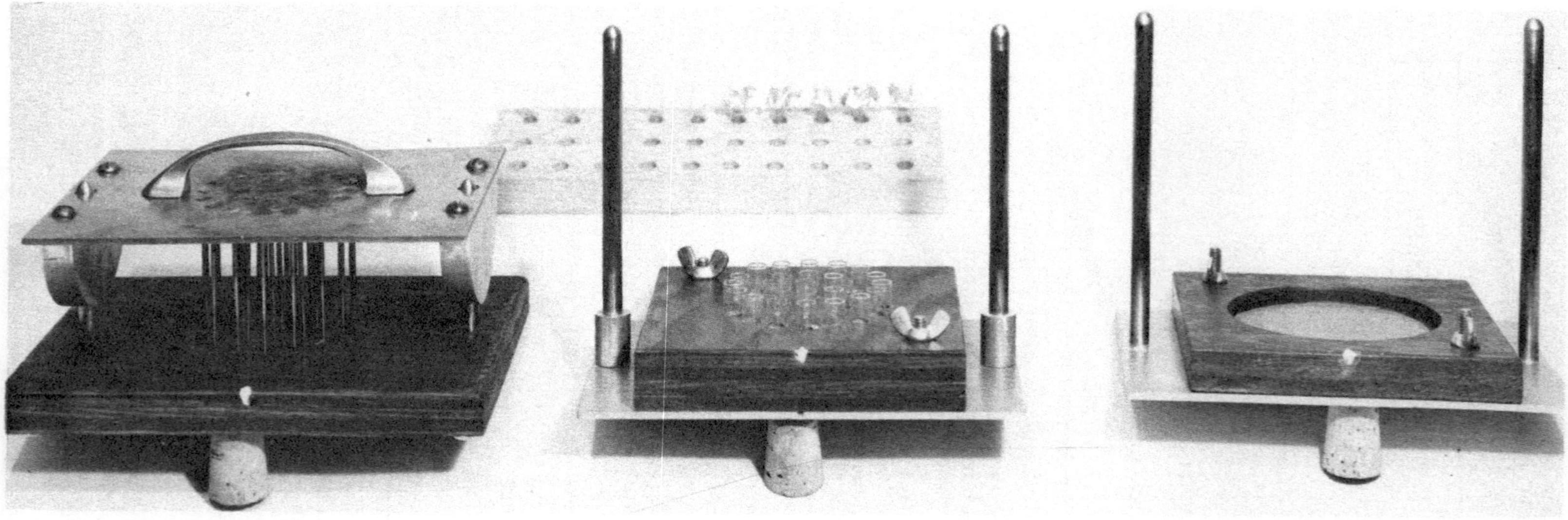

Figure 1. Equipment used for replication of yeast cultures on a variety of media. Rear: A block of oak wood, relatively resistant to repeated autoclaving, with spaces for 30 9 × 30 mm shell vials. Left front: A wood block used for storage of the replicator; holes in the wood receive the replicator needles; replicator in place made of stainless steel, with a strap handle of the same material, on the top side, with 25 "wire needles" 6 cm long soldered into holes. Holes in the side pieces coincide with the poles in the three stands. Center front: Wooden block mounted as in piece on right; 25 holes are bored at equal depths in block to accommodate 9 × 30 mm shell vials, 16 in outer row, 8 in inner row, 1 in center, to match position of needles in replicator. Right front: Wooden block mounted on stainless steel plate to which guide posts are welded; wooden block has recessed area to hold petri dish in position during inoculation. Notches in front edge of all instruments indicate starting point of numbering of each culture and match red glass pencil mark on bottom of petri dish in which agar with various test compounds has solidified and surface dried before inoculation. (From Cooke, W. B., *J. Elisha Mitchell Sci. Soc.*, 84, 219, 1968. With permission.)

scopic appearance. The mycological collector will remove the entire fruit body including the base of the stipe which may be buried in the ground or wood in or on which it grows. In areas for which no comprehensive list of fleshy fungi is available, and for persons just becoming familiar with mycology, the following technique is outlined. In the field as complete a description as possible of the fruit body is taken in field notes, including color, shape, dimensions, markings on surface of cap or pileus and stipe, presence of veil or volva, shape of gills or tubes, their method of attachment to the stipe, their color, and any other information which may appear to be necessary for characterization of the fruit body leading to an accurate identification. A spore print is made on white paper and other papers as desired. The specimen is dried by any of a number of processes but as soon as possible after the collection is made in such a way that the insect larvae which are usually associated with it will not destroy it before it can be studied further. After drying, the fruit body is preserved in paper packets or boxes, if these are available, so that the fruit body cannot be crushed or eaten by insects which may be expected to accompany unfumigated herbarium facilities.

Woody fruit bodies associated with wood rots need to be dried thoroughly. However, in the case of decaying wood with which the fruit body of the fungus causing the decay is not present, specimens of rotten wood may be returned to the laboratory and the fungus cultured from the wood using techniques described in the literature of forest pathology by Nobles,[834,835] Davidson and Vaughn,[300] and others. In connection with the detection in the field or laboratory of the presence of laccase in the fruit bodies of wood decay fungi, a test using syringaldazine has been devised.[499] A test has been reported[441] for extracellular phenol oxidase in field tests of wood decay fungi with gum guaiac.

For the study of many fungi, especially those whose fruit bodies cannot be collected by hand, various types of techniques have been developed for obtaining species or strains of species in pure culture. Most of the large fleshy or woody mushrooms or bracket-type fungi can be cultured with proper care. This is done from tissues of the fruit bodies, spores from spore prints, or from mycelia in wood or other material in which the fungus is causing decay. Handling these fungi with aseptic techniques is important, for the surface of the fruit bodies is an excellent habitat for yeasts[30] and bacteria as well as filamentous fungi, and the surface wood is an excellent habitat for molds which can penetrate deep into decaying wood as secondary or tertiary users of the nutrients available,[747] as well as other organisms such as yeasts and bacteria. Potentially contaminated surfaces can be removed, and relatively pure culture material can be obtained, by inserting sterilized chisel-forceps into the decayed wood from which cultures are desired and singeing the surfaces of the removed material with an open flame.

B. Micromycetes

In the cases of small fungi occurring on litter, decaying vegetable and animal matter of all kinds, dung, and similar materials, fruit bodies and receptacles are usually preserved in air dry condition in paper packets. Leaf spots, rusts, smuts, and similar materials occurring on living plant parts are pressed and packeted. Collection of materials from which to recover other fungi is described below.

IV. SOIL FUNGI

The most concise statement of the soil mold plate count technique is that published by Brierley and colleagues[136] in England. This has been modified several times by changing the numbers of the plates used, the amount of material sampled, certain nutrients in the medium, and the method of completing the dilution; however, the basic technique remains the same. These workers found that, within a certain ratio of material to be sampled to optimum growth, the medium developed by Waksman and Fred[1199] in a 9-cm petri dish using eight replicates and at certain dilutions, plates with discrete colonies which can be read and studied with little difficulty can be obtained. The Waksman medium requires that the reaction be maintained low enough so that bacteria are not troublesome. A reaction of 4.1 gave good results but did not retard growth of rapidly spreading fungi. An early compilation of many types of culture media[690] has been replaced by more concise summaries[69,321] of commercially prepared unipurpose and multipurpose dehydrated media. With the availability of antibiotics, it was found[758] that, by omission of acidification and by addition of rose bengal and streptomycin, bacteria could be eliminated or reduced to a minimum, and spreading fungi could be slowed down in their growth rate so that plates could be adequately counted and colonies picked for future study or use. This medium was further modified[214] since Aureomycin® (chlortetracycline or tetracycline) was preferable to streptomycin for sewage because of the numbers of bacteria present. Sources of organic nitrogen other than peptone (a hydrolysate of beef), such as phytone[69] or soytone[321] (hydrolysates of soybeans), could be used to good advantage.

As an aid to workers in the study of soil fungal populations, a guidebook has been prepared.[881] While a rather exacting protocol has been described, it has been said[879] that the worker should not be rigidly bound by any one set of instructions but should modify those instructions, as well as any suggested technique, to the extent required by the circumstances of the materials to be studied. The basic outline for an ideal project should include a statement of objectives and a description of the habitat, including the selection of the site, the sampling area, and the environmental factors to be measured such as soil, biotic, and meterological factors. Among the considerations of sampling are the collection of samples in the field, the sampling units, and statistical procedures such as types of distribution, assessment of the variability and reproducibility of the samples, and comparison of means of two batches of samples and of samples from more than two soil units. Among the general principles of sampling are planning and procedures with simple and stratified random samples. To determine the form and arrangement of microorganisms in soil, a number of techniques are available such as the direct examination of soil particles by light and electron microscopy; the direct examination of the soil profile; sectioning of soil and leaf litter; capillary and nylon mesh techniques for mycelial distribution; and examination of root-microbe interaction by the Fahraeus slide technique and root observation boxes. For the isolation of microorganisms, there are direct methods of isolation of fungi, using the Warcup method, and of other organisms, and such indirect methods as the plating of untreated samples and of pretreated samples in which growth is precluded or allowed. The biomass of fungi in the soil is measured. The expression of the results may be made on the basis of the weight of a soil constituent, the volume of soil, the specific surface area of a soil and its constituents, or the surface area of the ground. Problems in the application of these techniques to soil biomass determinations are discussed in relation to conversion of cell and spore counts and of mycelial lengths. Techniques of measurement and counting include agar film, stained smears, dilution plate count, and extinction dilution methods. The partial sterilization reinoculation method of biomass measurement is described. Among chemical estimations of biomass is the muriatic acid content of the soil. Microbial activity in the soil is measured by metabolic activity in the field and, in the laboratory, by gas exchange, enzyme assay, thermal measurements, and rates of substrate disappearance or metabolic change. Growth rate is measured in sterile soil, in natural environments, and by determination of competitive saprophytic ability. The growth of microorganisms in model systems is studied by soil sterilization, use of other particulate matter, artificial roots in the form of colloidon membranes, and continuous culture techniques. Finally microbial activity in the soil can be considered by the use of the inferrential assessment of metabolic activities and growth as they occur in the soil.

Techniques used in the study of the microfungal succession on leaves in the litter have been surveyed by Lindsey.[709] Beech and Davenport[91] have summarized the techniques for quantitative surveys of yeast cells on leaves and fruits of such plants as apples and grapes, important sources of these organisms in fermentation processes.

A tube has been developed in England[190,191] in place of soil for the study of soil organisms. It is a test tube, culture tube, or centrifuge tube, with openings on the sides. The tube is filled with sterile soil and placed in the ground in soils whose microbiological characteristics are to be examined. The organims developing in the new soil migrate from the test soil and can be studied by direct observation or by plating. In addition, recolonization of sterilized soil by microorganisms can be studied by similarly prepared tubes.[369]

Among the more thorough descriptions of methods of studying soil fungus populations are those given by Parkinson and colleagues in a recent International Biological Programme Handbook.[881] Techniques are given for developing all types of populations which may be present using a variety of standard and recently developed methods. In addition, for plant pathogens certain techniques have been suggested by Johnson and Curl.[601] Parkinson[879] has summarized much of the material described above and has given cogent reasons for developing improvements in some of the more sophisticated techniques. A symposium[984] dealing with modern methods in the study of soil microorganisms has been published, giving a number of techniques based on studies conducted in several areas with special reference to the biology of various biomes considered in the International Biological Programme.

Using soil washing techniques, including modification of the Parkinson technique, Gams,[419] Gams and Domsch,[421,422] and Gams et al.[423] found that *Mycelium radicis atrovirens* is a common soil fungus in forest soils, that more fungus mycelium was associated with mineral elements than with organic particles, that there was little seasonal variation in fungal populations in the soils of the area studied, and that different populations were more the result of earlier types of cultivation rather than the physical and chemical nature of the soil itself. Using buried nylon cloth strips; Gams[419] found that fungi grew beneath the frozen layer of soil in the winter in the Austrian Tyrol and that the nylon strip gave better qualitative than quantitative results in a winter study. However, Gams[417] found that narrow nylon strips attached to microscope slides and buried in the soil yielded information concerning actively growing fungi in the soil.

In England[1206,1207] a "soil plate" technique has been developed in which small quantities of soil materials are placed in petri dishes to which agar is added in which the fungi developing from the soil sample can be studied. It can be determined by microscopic observation of the plates whether the fungus colonies originate from spores, chlamydospores, or pieces of mycelium in chunks of soil which could be agglomerates of soil particles and organic matter or organic matter alone. In this way, a more adequate idea of what is happening in the soil and whether the fungi in question are actively participating in reduction of organic matter in the soil at the time of sampling can be obtained. Such observations can more adequately be used to determine the qualitative and quantitative fungus populations of the soil than the direct soil plate count technique previously used.

It has been said[427] that in one technique we observe what cannot be seen; in the other we are able to identify what otherwise cannot be studied. With a little familiarity with the populations under study, the species composition can be readily determined regardless of whether the fungi are on clear agar or on agar containing dyes or pigmented materials such as rose bengal, crystal violet, or ox gall. A comparison[1055] has been made between direct and plate count techniques for quantitative estimates of soil fungi. Other workers have shown that soil plating techniques must be carefully handled to avoid large statistical errors.

A technique for study of soil fungi[1260] utilized a round plastic scalp massager whose teeth were sterilized with a solution such as 0.02% quaternary ammonium salt (such as Hyamine® 1622). The massager, with as many teeth removed as desirable, is touched to a sample of soil, then to an agar surface in a petri dish. The massager is sterilized between inoculations with different soils. The sterilizing agent is removed in three changes in sterile distilled water, and as many plates may be prepared as needed in relatively rapid succession from one soil sample. Several types of media can be inoculated in succession from the same sample without resterilization. With modification, the technique can be made to fit a variety of conditions and available materials.

In a text on micropedology, direct observation of soil samples by a special microscope has been described.[662] In addition to the use of a special microscope, examination of soils has been carried out by a modification of the Rossi-Cholodny technique. A review[218] has covered several modifications of the use of glass slides in the study of soil populations by direct observation. In work with cellulose-decomposing fungi,[1051] potato-dextrose agar acidified with lactic acid, as well as other media, has been used to advantage. In studies[261] on fungi in stored mattresses, the rose bengal-Aureomycin® agar mentioned above, has been used in tandem with the medium including actidione mentioned below. Other techniques have been described[505] for study of active mycelia on living roots and other surfaces in the soil. Durbin [344] has surveyed those techniques for isolation and observation of soil fungi published up to 1971.

Szabo and colleagues,[1129] in studying the humus horizon of a mull-type rendzina microscopically, have described a number of types of bacterial, actinomycete, and fungal colonies which may be of homogeneous or heterogeneous composition. These colony types may be used in characterizing the populations of the area being studied.

V. AIR FUNGI

Various techniques have been developed for the entrapment and study of fungus spores in the air. Most of the techniques rely on the fall of spores from the air by gravity.[805] Various media have been developed[1126] for the study of colonies of fungi whose spores have fallen onto the agar plates holding the media. Methods of sampling air populations[633] at high altitudes from airplanes have been described in Canada. Here, plates containing agar or covered with adhesive material are exposed from the plane in definite time periods. In England,[549] an automatic volumetric spore trap and, in Kansas,[851] a slit sampler have been developed which reportedly entrap spores which can be counted or studied with better results than with earlier types of samplers. A New Zealand worker [805] has presented results of comparative researches in which the gravity and impact methods of sampling air populations have been used.

Gregory and Monteith[479] have edited the papers presented in a symposium on airborne microbes, including a number of papers based on sampling methods, the physics of airborne particles and their deposition, members of populations, and activities of populations once successfully deposited.

Following a statement of general principles, Davies[302] has described three major types of air samplers. Among the impacters are described the slit sampler, the Cascade impacter, the automatic volumetric spore trap or Hirst trap, the Andersen sampler, and the rotorod. Among the impingers are described the general features of impingers and their advantages and disadvantages in contrast with impacters, the Porton impinger, the preimpinger, and the multistage liquid impinger. In "fall-out" or sedimentary sampling are described the "gravity slide" and the settle plate. Several other methods of air sampling are also described including filters, the ram-jet impacter for sampling from aircraft and cyclones, large-volume air samplers, and biological air samplers. A section is devoted to the use of selective media in air sampling, and a section on visual recognition of fungal spores and pollens is accompanied by line drawings of the more common types encountered.

VI. PLANT AND ANIMAL PATHOGENIC FUNGI

A critical review[1159] and catalogue of selective media for the isolation of all kinds of pathogenic fungi, plant or animal, has been published. Johnson and Curl[601] have suggested a wide range of techniques for studying plant pathogenic fungi.

For isolation of human and animal pathogenic fungi from natural materials, in surveys for their occurrence as natural members of a population, two basic techniques are in current use. For fungi causing systemic mycoses, a portion of a sample of the subject habitat is suspended in water or saline,[681] and the suspension shaken and allowed to settle. The supernatant is used for injection of susceptible animals, commonly mice.[679] In this method, 1 mℓ of the supernatant is injected, usually intraperitoneally, into a mouse. After 2 to 4 weeks the mouse is sacrificed, and portions of such internal organs as liver and spleen are streaked and left on neopeptone-dextrose agar slants. After 1 to 2 weeks or more of incubation at room conditions of light and temperature, the tubes are examined for the presence of the suspected fungus. In this way, such fungi as *Histoplasma capsulatum* and *Coccidioides immitis* have been found in soil and substratum samples. The second basic technique involves plating samples of suspected materails such as soils, scrapings from shoes, shower stalls,[20] and other types of materails in plates in which a Sabouraud-type medium is fortified with actidione, penicillin, and streptomycin,[20] or in which an enriched medium is fortified with ox gall, gentian violet, and streptomycin.[712] Cycloheximide[434] has also been suggested as a useful antibiotic in the isolation of human pathogens. Penicillin and streptomycin reduce the bacterial contaminants, although Aureomycin® was found to be an acceptable substitute, while actidione reduces the numbers of filamentous fungi which crowd such plates preventing observation and study of the slower growing and less definitive pathogens. In this way various dermatophytes have been found in areas where their presence was suspected but otherwise not demonstrated.

VII. SPECIAL MEDIA

To get good growth of fungi in the laboratory it has been suggested that — rather than depend on synthetic media, such as those described above whose formulae include known concentrations of known chemical nutrients, and on semisynthetic media, including potato-dextrose agar — natural materials such as pieces of healthy, diseased, or rotted plant parts[1082] can be used. Such materials need not be autoclaved but may be sterilized dry at room temperature by small amounts of highly toxic volatile materials such as ethylene oxide or propylene oxide of which 1 mℓ can be added to a screw-capped quart glass jar kept capped for 24 hr.[497]

In the study of different products to determine their mold populations, special media have been developed. For instance, in the study of citrus fruit juice preparations for microbiological content it was found[822] that most available media were inadequate and that a better medium containing citric acid or citrus concentrates could be prepared. This medium was later [1107] developed as a dehydrated agar medium which could be used in any laboratory.

Various techniques have been developed to study fungi on materials in which they were suspected of causing deterioration. For the study of fungi on cloth, the papers mentioned above under this subject include techniques for study. Other materials are also subject to deterioration. Study of fungi on leather[303] is particularly difficult. In addition to getting fungi to grow on such substrates in pure culture, it is important to know how to evaluate[1164] this information.

VIII. SPECIAL TECHNIQUES

A further refinement of technique, the molecular or membrane filter, has had increasing use in the study of fungi, especially yeasts. With this technique, a suspension of the sample being tested is filtered and washed through the membrane filter; the materials, including spores and other fungus fragments to be tested, remain on the filter which is transferred to a nutrient soaked pad in a suitable receptacle, or to the surface of an agar plate on which the spores are allowed to germinate and grow,[298] or are fixed and stained.[451] In the study of the passage of *Histoplasma capsulatum* spores through water treatment processes,[781] the membrane filter technique was more useful than the agar plate technique. That broths may have somewhat different concentrations of nutrients than solid media, even though they are prepared in the same way, because of adsorption on the filter or absorption in the filter pad, may play some part in the final results. This was a problem, partly because of the pH levels used in the development of the membrane filter for use in sanitary bacteriology.[200]

A wide vaiety of methods has been developed for the use of workers in many fields, including plant pathology, medical mycology, fungus genetics, and others, for the isolation of single spores and the development of cultures from single hyphal tips. The techniques developed have been thoroughly reviewed.[544,545] A common and popular technique is still the flooding of an agar plate with a dilute suspension of spores and the cutting out of agar blocks bearing single spores or the lifting of single spores from the agar to another agar plate.

To purify cultures of fungi that either have become contaminated accidentally or are mixtures in which one or more species of importance to a study are to be separated, the van Tieghem cell technique has been developed.[48,939] A van Tieghem cell, a glass ring, is placed on the agar, and the mycelium of the desired fungus is allowed to grow under it and outside its limiting area. The tips of the desired outgrowing hyphae are removed to clean agar plates, and the colony is maintained in pure culture.

Among the techniques available for developing a fungus colony on a slide and mounting the preparation with or without the agar is one described by Cooke[234] and one described by Duddington and Dixon.[341]

IX. CULTURE COLLECTIONS

A committee of the American Society of Bacteriologists[522] has presented general statements concerning methods of maintaining culture collections of industrially important microorganisms. Such techniques apply to bacteria, yeasts, and filamentous fungi and are becoming standardized as the genetic implications of heterokaryosis and parasexuality in the Fungi Imperfecti are becoming better understood.

Perspectives and problems of those dealing with culture collections of fungi were considered in a symposium held in Canada and edited by Martin.[759] Principal subjects covered included the role and organization of the culture collection, some fundamental aspects of cell preservation, and an assessment of the methods in use for the preservation of live cultures.

In the development of culture collections in which are stored living strains of most of the fungi which have been isolated from nature, the larger repositories have developed lists or catalogues of available cultures. These lists are updated periodically and include ones from the U.S., the Netherlands, Great Britain, and Japan. Other lists are published occasionally and may be restricted to special interest groups of fungi. These lists include data concerning maintenance, source of isolates, and sometimes use of the listed fungi. Reese and colleagues[954] have described the U.S. Army Quartermaster culture collection, now housed at the University of Massachusetts, and Fennell[380] has described the conservation of fungus cultures. Mazur[770] has considered the theory, problems, and practice of cold temperature preservation of fungus cultures.

Several methods have been devised for the maintenance of fungal spores and cultures in as compact and viable a condition as possible. Such methods arose primarily from a desire to retain the large numbers of fungus cultures in collections developed in monographic studies of fungus species and genera, in a need for material on which to make comparative studies of series of isolates based on product analyses and fermentative capabilities, and in collections developed during and after World War II as a result of studies of cellulolytic activity and especially as these collections increased in size. The principal techniques which have been described include lyophilization,[380,511] preservation under mineral oil,[21,1102] preservation in sterile soil pans,[68] and preservation at subfreezing temperatures.[783] Holding cultures in the laboratory at room temperature, in the experience of the writer, necessitates transfer at intervals of 3 to 4 months, depending on the relative humidity of the air. Keeping cultures in refrigerators when tubes are stoppered with cotton requires transfer once a year for maintaining viable cultures and somewhat less often when tubes are covered with paraffin or screw caps to prevent drying. The longevity of spores kept under these conditions has been described, based on samples removed after varying periods of time up to 5 years. However, it has been noted that the accumulation of CO_2 in tubes which are essentially sealed can be detrimental to the cultures.[760]

A. Water Storage

It has been found that some human pathogenic fungi survive well in water storage,[184] surviving by storage in sterile distilled water for at least a year and possibly indefinitely if the evaporated water is replaced annually. This experience has been confirmed[774] with a different series of human pathogens. Using 417 isolates of 147 species belonging to 66 genera of filamentous fungi, yeasts, and aerobic Actinomycetales, cultures were stored in sterile distilled water at room temperature over periods ranging from 12 to 60 months, in four independent experiments. Of these cultures, 93% survived storage in sterile distilled water. Important factors influencing storage in water over the longer periods of time were the selection of good sporulating cultures and sufficient inoculum consisting of spores and hyphae. The technique was found to be simple, inexpensive, and reliable.

In relation to water storage of ectomycorrhizal and plant pathogenic fungi,[750] cultures were grown on agar in petri dishes, and mycelial discs 8 mm in diameter were removed to test tubes with sterile distilled water and stored at 5°C. Of the symbiotic fungi, 64, 61, and 41 isolates of the original 64 isolates of 14 species survived 1, 2, and 3 years in storage, respectively; of the pathogenic species, only 19, 10, and 8, of the 27 isolates of 15 species survived 1, 2 and 3 years. Time in pure culture before water storage did not appear to affect viability. After 3 years of storage, three symbionts and one pathogen were found to have retained their original growth rate and root infecting abilities on pine seedlings. The same four isolates maintained on agar at 5°C and subcultured every 4 to 6 months grew more slowly and did not infect so many feeder roots of pine as the water-stored isolates.

X. BIOMASS

Techniques for measuring biomass of soil fungi are given in detail by Parkinson and colleagues.[881] An additional technique has been described.[397] The estimation of the biomass of live fungi in leaf litter was carried out by phase contrast microscopy. Observed hyphae were divided into three categories based on amount of cell content present. The categories are live, active; live, inactive; and dead. The technique was tested on soil and leaf litter samples and found comparable with other techniques.

XI. MICROCLIMATE

Once the numbers and members of a population are known,[981] it may be possible to correlate the population of one area with that of another adjacent or distant area or another similar or dissimilar plant association. There is no certainty that the same series of fungi found one year will be found the next. Some species may not appear in two or more successive seasons, and the fruiting of others is thought to be very erratic. In part, this may be attributed to seasonal conditions. Averages of climatic factors may lead one to suppose that conditions are uniform over certain areas. However, a check of weather records reported for an area may show that there is considerable fluctuation from day to day, and this fluctuation is not consistent. For very few, if any, areas are microclimatic data available which would help to determine or forecast potential fruiting possibilities of a habitat under consideration. Relative humidity, available moisture, and temperature are important characteristics, but far too little is known about how these are received in any area in relation to the fruiting capabilities of the fungal population. For interpretation of macromycete fruiting patterns few, if any, microclimatic data are available on even a short term basis for meaningful use.

XII. FLUORESCENCE

A number of years ago in Boston two doctors found that, by use of UV light, photography of fungi and bacteria could be improved.[368] This information lay fallow for a long time, except that UV light was used under certain rar e conditions for the study of occasional organisms. With improvement of equipment and development of a certain amount of interest in the subject, it has been shown more recently[556] in Vienna that the ultraviolet light technique, especially when secondary fluorescence is induced with stains, can be used in mycology with good results. Certain anatomical and morphological characteristics not readily demonstrated otherwise can be brought out with stains which produce a secondary fluorescence. Primary fluorescence has functioned in a very limited way in the microscopic analysis of fungus structure.

Isolates of *Aspergillus flavus*[1015] cultured in the soil in the vicinity of buried slides were detected with fluorescent antibody techniques. All isolates of *A. flavus* were detectable on contact slides after staining with the labeled antibody. Most of the fungi obtained there, including *A. flavus,* reacted weakly with the labeled antibody. Its specificity and its adaptability to direct examination of soils suggest that the method may be visual.

XIII. AUTORADIOGRAPHY

An autoradiographic procedure has been developed[1193] to detect metabolically active hyphae in leaf litter. Experiments with *Sclerotium rolfsii* and with hypae on the surfaces of beech leaf litter showed that active hyphae could be detected after exposure for 2 hr to 10 μCi/mℓ of uniformly labeled ^{14}C-glucose followed by film exposure for 2 days. Controls indicated negligible background radioactivity and no passive uptake of radionuclide by hyphae. Liquid photographic emulsion was superior to stripping film because closer contact could be obtained between the hyphae and the emulsion. The use of photographic emulsion was facilitiated by removing hyphae from leaf litter surfaces in polystyrene peels and then covering those peels with emulsion.

XIV. COMPUTERIZED FERMENTATION

The use of computers in the fermentation industry has been advocated[305,580] as a method of revitalizing the industry. Computers have been used in a direct digital control scheme as a more economical substitute for analog controls. Computers have been used previously in data acquisition and analysis schemes. It is suggested that probable uses of computers will be for on-line analysis of cell biomass concentration, growth rate, and substrate control through component material balancing. It is thought that fermentation companies, in the near future, will begin to use a system of computers in which microprocessors will control fermentation functions; these microprocessors will interface with a microcomputer for management control; in return, the minicomputer will interface for data transfer and storage with a large frame computer handling many company tasks.

XV. STERILIZATION AND ASEPSIS

While aseptic processing and packaging of food is an accomplished fact,[678] the importance of environmental air is often overlooked. Potentially contaminated air and the degree and type of airborne contaminants are to be considered in the planning of an aseptic operation. Procedures are available for sampling and counting of airborne particulates and for enumeration of bacterial colonies and of fungal spores in the air.

The U.S. Pharmacopoedia and National Formulary are recognized as official drug compendia. Standards for sterility tests[849] including specimen taking, media, incubation, adventitious contamination, and validation are considered in detail. Statistical design, microbial limit tests, prederivatives, and pyrogen (fever-producing) tests are discussed briefly. Internationalization of the compendia is considered desirable.

Halleck[493] has noted that innovative and tailorized sterilization cycles have been developed that maintain the quality characteristics required in the sterilized products. A recommended guideline for the role of sterilization in research in new sterile product development programs is presented.

XVI. SOME STATISTICAL METHODS

Few workers have been able to adapt statistical methods to the study of macromycete populations, usually because the methods of population study have been of a qualitative survey type rather than a quantitative type. In a study of the latter type,[215] the unanticipated seasonal climatic factors which appeared to govern the distribution of fungi throughout the growing season were so erratic that statistical analyses were impractical. However, in England, Parker-Rhodes[870-872,875-877] has found conditions to be ideal for developing statistical analyses of various factors in fungal distribution. Because problems differ and have not been treated in the same way, the resulting papers will be treated separately here.

A theory of anemophilous dissemination of fungi, especially of Basidiomycetes, was presented.[870] The mean radius of dissemination depends on the structure of the basidiocarp, the height of the pileus in the case of agarics, the behavior of spores, and the habitat. The present theory applies only to grassland species. "The circumscription ratio is a single-valued function of the mean radius of dissemination, and governs monotonocally the coefficient of inbreeding and all parameters dependent thereon."[870] The results of plotting circumscription ratios, of which the mean radii of dissemination are abscissae, indicate that a considerable variation in the general level of heterozygosity is conceivable as a result of purely anemophilous dispersal. It is suggested that if dissemination of small agarics is purely anemophilous, Skokholm Island populations could show incipient speciation.

In presenting the mycobiota of Skokholm Island in tabular form, some floristic and ecological considerations are given.[871] The ecological status of each species is described as either prevalent (present more than two times in any one year) or subvalent. In the latter case, the species may be either preposite (represented by only one record in any one year) or apposite (with at least two records in one year). Habitats, habits, season classes, and distribution were tabulated. Statistical analyses have yielded information concerning sampling efficiency in relation to individuals and species which may be affected by psychological factors inherent in the observer, the total number of species in the area under study, the various habitats under study, and the proportion of species fruiting in any one season.

A statistical analysis of the fungal populations on Skokholm island is presented[872] under the heading "population densities". Estimate is made of the fungal populations, constancy of the populations is discussed, total populations are summarized, and the population distribution on the island is described. The most abundant species in each habitat, together with the communities, hosts, or substrate on which they occur, are listed according to five principal groups: humicolous, quisquilicolous, coprophilous, metacoprophilous, and lignicolous.

Correlation of the fungal species with the plant associations of Skokholm Island is discussed[875] statistically. The statistical concept of fidelity is defined and the statistical concepts of "preference" and "aversion" are described. One type of fungus may demonstrate a preference for a grassland association and an aversion to a bog community. Certain species have been able to develop in an area to the exclusion of all others. These are considered saturants, a term contrasting with the subvalent species mentioned above.

Experiences in England[876] and reports of workers in Colorado[1027] suggest a statistical basis for the study of fairy rings and related phenomena. Parker-Rhodes developed this idea on a geometrical model and considered the rate of growth of impingent groups, interspecific competition, the problem of nonuniform fields, factors of effective reproduction rate, and the longevity of fairy rings. It is concluded that the fairy ring is highly inefficient and that the communities in which it arises are open. These are conditions that are likely to promote the evolution of species of nearly identical ecological requirements.

The fungus foray is an institution in the activities of mycological societies on both sides of the Atlantic Ocean and in Japan. The foray brings together a group of fungus collectors once a year for several days or a week to collect and study intensively the fungi of the area chosen. In this way very much has been learned of fungus "floristics". To determine the efficiency of the foray as a means of obtaining floristic information, a test was made of records from one group of forays in England[879] covering a 5-year period. Here the statistician was limited, since no experimental design was developed before or during the foray. The records showed that records made on individual occasions yielded only about 25% of the total possible species. This percentage is attributed in large part to the erratic fruiting habits of the fungus species rather than to the inefficiency of the collectors. Thus, long periods of time are required in which to obtain an adequate knowledge of a fungus population. The statistical formulae by which the conclusions were obtained may be applied with caution to comparable data from forays from other regions. Forays are as reliable a method of floristic study as can be expected, but the results could be improved if the forays were smaller and were more frequent.

XVII. IMPORTANCE VALUES

A set of terms describing the habitats of organisms in polluted waters was prepared in Germany in 1908,[657] and this so called saprobic system was elaborated[656] to make it more readily available. These terms were not readily usable for the organism and its role in the polluted habitat, and a different set of terms was introduced.[220,221] The prefix lyma- (filth) was used with a set of four suffixes (-biont, -phile, -xene, and -phobe) representing four grades of adaptability of an organism to a habitat. Since the prefix sapro- had been used earlier, Fjerdingstad in Denmark[384] preferred to use sapro- instead of lyma- with the four suffixes. However, since all active fungi in the pollution environment are saprobes, saying that a fungus is a saprobiont or a saprophile is redundant, and saproxene and saprophobe are contradictory when applied to the fungi. It is interesting to note that *Fusarium aquaeductuum* and *Leptomitus lacteus* are considered[384,385] to be saprobiont in Denmark, although elsewhere they are known from other habitats also. A worker in Czechoslovakia[1058] has refined and elaborated the Kolkwitz and Marsson system used for describing saprobity. Techniques for measuring saprobity were described by Sladecek.[1059] Methods and techniques for studying fungi in polluted water supplies are given in recent volumes designed to aid the field and laboratory worker in the U.S.,[934] and similar publications are available at least in Central Europe[35] and in Germany.[490]

Palmer[861] introduced a method for determining the pollution tolerance of algae, based on reports by 165 authors, where the genus and species most often referred to as significant are given a composite rating for use in rating water samples with high organic loadings. Based in part on this, a method of listing fungi by importance values has been developed for fungi.[247,255,257] In this method the number of times a colony of a species of fungus has appeared on a primary isolation plate from samples of materials from sewage treatment systems and streams carrying organic enrichment of various kinds is totaled. These values have been termed the "observation number". On an arbitrary basis, when the observation number is over 10 the fungus is a lymaphile, between 2 and 10 a lymaxene, and below 2 a lymaphobe. Lymabionts are restricted to sewage-polluted habitats and presently include only one species with a rather low observation number.

Chapter 21

SUMMARY STATEMENT

For every enzyme system resulting in the production of a natural compound, simple or complex, there is an enzyme system whose action will result in the complete or nearly complete reduction of that naturally produced compound to its component elements, ions, or simpler compounds.

Major habitats for fungal growth may be listed starting with the young bud, its progress to maturity, opening, unfolding of leaves and flowers, the maturing of flowers and initiation of fruits, the maturing of fruits, their ripening, harvesting, and decay, the development of leaves, their maturing, senescence, and death, leaf fall, and decay; the development of twigs, their bark and wood, and their branches.

Leaf fall may deposit leaves on soil or in water. Processes of decomposition on soil proceed through marked stages by numerous, more or less specialized organisms. Decomposition in water is carried out by more specialized organisms whose spores are adapted to water transport and deposition.

The skin of man and of animals, including appendages, provides the surface on which numerous types of organisms grow. Yeasts and possibly less fastidious organisms are found on such surfaces even though the surface has been washed within hours of the search. Keratinolytic organisms may be found on such surfaces fairly regularly. Such surfaces are continuous with other body surfaces such as nostrils, lungs, mouth, throat, respiratory, alimentary, and reproductive canals where other organisms may be found as fortuitous inhabitants which, with relatively slight upsets in host systems, may become pathogenic.

If it is assumed that the fungi and the green algae had a nearly common origin in time in Precambrian time about 1 billion years ago, it must be assumed that this occurred in water and that the water was probably fresh water. These fungi may have been mastigiomycetous, with flagellated zoospores. These fungi may have been accidentally or obligately associated with developing chlorophyllous plants which found a habitat on land, swampy, moist, or dry. The obligately associated types may have developed in habitats in the plant related to the moister portions of the area in which the three-dimensional plant developed, the roots or root-like structures, and as such formed the endotrophic mycorrhizae noted in the earliest vascular floras and still existing today as a most important feature of certain members of the vascular floras of the world.

Fungus tissue is not easily fossilized even in amber which is the medium of preservation of many chitinous insects. In certain rare cases a perennial polypore has been preserved. The decay pattern of at least one wood decay fungus has been preserved. The apparent pattern of leaf disease fungi and of mycelium in wood have been preserved, but so far the instances are so isolated that little more than inference from circumstantial evidence can be used as a substitute for the fossil record. The lack of preserved wood in areas where forests have probably stood points to the existence of wood decay fungi; the continued existence of trees known to be dependent, at least under certain circumstances, on ectomycorrhizae; the fact that plant and animal disease probably predates man's recognition of their existence; the existence of fungi whose flesh or whose by-products have been used by man, other mammals, and insects, as part of a sort of biological consciousness that such things exist and are usable parts of the environment, points to the great age of fungi regardless of the ecological niche to which they may be apparently confined today.

The exact point of time, measured in terms even of millenia, cannot presently be determined when the fungi left the water for the land. Morphological characters have been used abundantly to derive systems of relationship between groups of terrestrial fungi and between terrestrial or amastigiomycetous and aquatic or mastigiomycetous fungi. Tentative attempts have been made to relate these fungi by several biochemical methods. At least one physiological method has been proposed within a very small but important group of fungi.

In order to produce a fructificaton which can be analysed for morphological characteristics to determine relationships traditionally, a fungus must have an enzyme system with which to extract from its environment those elements, ions, and molecules which it must enzymatically process for production of assimilative, storage, and reproductive tissues. The interplay of these processes must be adequately understood before all aspects of interrelationships can be appreciated. Until then, all saprobe-pathogen, saprobe-weak parasite-obligate parasite, saprobe-perthogen-pathogen, or other relationships must be inferred on the basis of obvious present-day relationships.

Some fungi can compete with others in the same space for the same nutrients from the same substratum. Some fungi can occupy a single substrate, extracting usable nutrients until this source is exhausted, before moving to another nutrient source or going into a resting state until another source becomes available. Other fungi can degrade a nutrient source, extract certain degradation products, leaving other products for other organisms, concomitantly or successively, which may repeat or continue the process to the exhaustion of the substratum.

Lignin and molecules containing certain heavy metal elements or ions are among the more difficult natural occurring molecules to be degraded. These molecules, or remnants containing them, may remain in the litter, humus, or humic acids for indefinite periods of time, reaching to as much as, or more than, 1500 years. In the case of deposits of organic materials which remain intact, such as coal, the disintegration is halted by the adverse conditions for the life of the disintegrating organisms. Petrifaction results when, under special conditions, organic molecules are replaced by inorganic molecules, or their interstices are penetrated by inorganic molecules under conditions where decomposition of remaining organic molecules is impossible.

Where two or more fungi occupy the same territory, the combination is usually fortuitous, except possibly in the case of mycoparasitism where the mutual occupation is more apparent than real, enzymatically speaking. On or in the forest floor, only those fungi will form the association which can adapt, enzymatically, to the substrates available, and only those fungi will survive which have the higher inoculum potential and the greater competitive ability. The degrees to which these factors may be demonstrated vary with the fungus. It would be convenient for the observer if all species present fruited regularly, in the proper relation to the assimilative mycelium with a given number of sporophores in each fruiting period, etc., but these conditions never happen. Climatic and edaphic factors seem to prevent regular fruiting, some species do not fruit every year or season, numbers of sporophores produced vary from season to season, and sporophores may not be produced regularly in relation to the location of the assimilative tissues. Occurring with the larger fruiting fungi, as competitors, as degraders of other substrata, or in other relationships, may be a small to large number of microfungi (in the sense of the size of the fructification). Total production of fungal tissue for such an association can be estimated on a season by season basis but must be reported with reservations based on the vagaries mentioned. The surveys resulting in such information must be intensive enough to smooth out curves based on the individual's inability to maintain an intensive surveillance program.

Among the more useful fortuitous assemblages of fungi are those taking advantage of the habitat provided by the operator of the sewage treatment plant. While the specific habitat has been in existence for less than 100 years, the responding populations have become fairly well established, varying only as one species is favored over another, one species is "pushed out" by another, or a group of others. Within the habitat the fungus rarely, if ever, fruits, but its productivity is considerable and can be measured fairly easily.

There is very little information available about how much nutrient material must be available to a fungus for the production of X amount of assimilative tissue or fruit body tissue. It is assumed, of course, that if a fungus is able to grow and fruit in a habitat, enough nutrients are available in that habitat to satisfy its requirements. But how much is enough for the current and the succeeding generations, whether or not the mycelium is perennial, and whether or not the fructification is perennial, are relatively unknown quantities.

Productivity and biomass are determined in at least two ways. The more common is the determination of fresh weight of carpophores of fungi as they occur in the field or the length of hyphae in a viable mycelium in the litter. If conversions are made to dry weight, thus eliminating the very variable value of the amount of water present in these structures, it must be assumed that any one unit chosen is typical or representative of the whole population of that species present in the sampled location. In the case of the mycelium, it must be assumed that the dried and weighed sample is representative of the observed sample, even though the two may have been grown on different substrata in the one case, or on or in a medium from which harvest can be effected without significant damage.

From pole to pole, from the International Date Line to Greenwich Meridian and back to the International Date Line, there are fungi wherever there is dead organic matter from whatever source. There is at least one fungus, usually more, decomposing the dead remains of any single vascular plant. Most vascular plants may be attacked by at least one species of pathogenic fungus exhibiting one or more of the various types of pathogenesis. Control of pathogenic fungi on essential crop plants grown in uniculture has been effected by the use of certain chemicals which have proved to be environmentally hazardous. Replacement chemicals and processes, while less quickly effective, lead to experimental demonstrations of techniques which may be as effective and less environmentally harmful.

The fungi will continue to be the instruments of removal of dead organic matter in every instance in which their capabilities are permitted to be effective.

REFERENCES

1. **Abraham, E. P.**, Some aspects of the development of the penicillins and cephalosporins, *Dev. Ind. Microbiol.*, 15, 1, 1974.
2. **Adams, J. M., Seidler, D., Terry, D., Smart, K. I., and Cabelli, V. J.**, The Indigenous Bacterial and *Mycotic Flora of the Air and Its Influence on the Detection of BW Aerosols*, Tech. Rep. No. 279, U. S. Army Chemical Corps Research and Development Command, Dugway Proving Grounds, Utah, 1961, 1.
3. **Adams, S. L.**, Utilization of cannery fruit waste by continuous fermentation, *Wash. State Inst. Technol. Bull.*, 207, 1, 1950.
4. **Adler, J. P.**, Viruses of yeast, *Dev. Ind. Microbiol.*, 16, 152, 1975.
5. **Ahearn, D. G. and Meyers, S. P.**, Fungal degradation of oil in the marine environment, in *Recent Advances in Aquatic Mycology*, Jones, E. B. G., Ed., Elek Science, London, 1976, 125.
6. **Ahearn, D. G., Wilson, L. A., Julian, A. J., Reinhardt, D. J., and Ajello, L.**, Microbial growth in eye cosmetics: contamination during use, *Dev. Ind. Microbiol.*, 15, 211, 1974.
7. **Ainsworth, G. C.**, Host-parasite relationships, *J. Gen. Microbiol.*, 12, 352, 1955.
8. **Ainsworth, G. C.**, Fungal parasites of vertebrates, in *The Fungi, An Advanced Treatise*, Vol. 3, Ainsworth, G. C. and Sussman, A. S., Eds., Academic Press, New York, 1968, 211.
9. **Ainsworth, G. C.**, The numbers of fungi, in *The Fungi, An Advanced Treatise*, Vol. 3, Ainsworth, G. C. and Sussman, A. S., Eds., Academic Press, New York, 1968, 505.
10. **Ainsworth, G. C.**, *Introduction to the History of Mycology*, Cambridge University Press, Cambridge, 1976.
11. **Ainsworth, G. C., Sparrow, F. K., and Sussman, A. S., Eds.**, *The Fungi, An Advanced Treatise*, Vol. 4a, Academic Press, New York, 1973.
12. **Ainsworth, G. C., Sparrow, F. K., and Sussman, A. S., Eds.**, *The Fungi, An Advanced Treatise*, Vol. 4b, Academic Press, New York, 1973.
13. **Ainsworth, G. C. and Sussman, A. S., Eds.**, *The Fungi, An Advanced Treatise*, Vol. 1, Academic Press, New York, 1965.
14. **Ainsworth, G. C. and Sussman, A. S., Eds.**, *The Fungi, An Advanced Treatise*, Vol. 2, Academic Press, New York, 1966.
15. **Ainsworth, G. C. and Sussman, A. S., Eds.**, *The Fungi, An Advanced Treatise*, Vol. 3, Academic Press, New York, 1968.
16. **Ajello, L.**, Soil as a natural reservoir for human pathogenic fungi, *Science*, 123, 876, 1956.
17. **Ajello, L.**, Natural history of the dermatophytes and related fungi, *Mycopathol. Mycol. Appl.*, 53, 93, 1974.
18. **Ajello, L., Ed.**, *Coccidioidomycosis*, University of Arizona Press, Tucson, 1967.
19. **Ajello, L., Dean, D. F., and Irwin, R. S.**, The zygomycete *Saksaena vasiformis* as a pathogen of humans with a critical review of the etiology of zygomycosis, *Mycologia*, 68, 52, 1976.
20. **Ajello, L. and Getz, M. E.**, Recovery of dermatophytes from shoes and shower stalls, *J. Invest. Dermatol.*, 22, 17, 1954.
21. **Ajello, L., Grant, V. Q.,and Gutzke, M. A.**, Use of mineral oil in the maintenance of cultures of fungi pathogenic for humans, *Arch. Dermatol. Syphilol.*, 63, 747, 1951.
22. **Ajello, L. and Padhye, A.**, Keratinophilic fungi of the Galapagos Islands, *Mykosen*, 17, 239, 1974.
23. **Albritton, E. C., Ed.**, *Standard Values in Nutrition and Metabolism*, W. B. Saunders, Philadelphia, 1954.
24. **Alderman, D. J.**, Fungal diseases of marine animals, in *Recent Advances in Aquatic Mycology*, Jones, E. B. G., Ed., Elek Science, London, 1976, 223.
25. **Aldon, E. F.**, Endomycorrhizae enhance survival and growth of fourwing saltbush on coal mine spills, *U.S. For. Serv. Res. Note RM*, 294, 1, 1975.
26. **Alexopoulos, C. J.**, *Introductory Mycology*, 2nd Ed., John Wiley & Sons, New York, 1962.
27. **Allen, P. J.**, Metabolic aspects of spore germination in fungi, *Annu. Rev. Phytopathol.*, 3, 313, 1965.
28. **Amadjian, V.**, *The Lichen Symbiosis*, Blaisdell, Waltham, Mass., 1967.
29. **Amann, A. and Gottleib, D.**, Paper chromatography of antifungal antibiotics, *Appl. Microbiol.*, 3, 181, 1955.
30. **Anderson, K. W. and Skinner, C. E.**, Yeasts in decomposing fleshy fungi, *Mycologia*, 39, 165, 1947.
31. **Anderson, P. J. and Rankin, W. H.**, *Endothia* canker of chestnut, *Cornell Univ. Bull.*, 347, 531, 1941.
32. **Andrews, H. N. and Lenz, L. W.**, Fossil polypores from Idaho, *Ann. Mo. Bot. Gard.*, 34, 113, 1947.

33. **Anon.**, Efficiency of microorganisms in sewage, in Water Pollution Research, Report of the Director, England, 1955, 55.
34. **Anon.**, Forty-second Report of the Director of Water Examination for the Years 1965 and 1966, Series III, Bacteriology Section, Metropolitan Water Board, London, 1966, 11.
35. **Anon.**, Standard Methods for the Water Quality Examination for the Member Countries of the Council for Mutual Economic Assistance, The Ministry of Forestry and Water Management in co-operation with the Hydraulic Research Institute, Prague, 1968.
36. **Apinis, A. E.**, Thermophilous fungi of coastal grasslands, in *Soil Organisms*, Doeksen, J. and van der Drift, J., Eds., North-Holland, Amsterdam, 1963, 427.
37. **Apinis, A. E.**, Concerning occurrence of Phycomycetes in alluvial soils of certain pastures, marshes, and swamps, *Nova Hedwigia Z. Kryptogamenkd.*, 8, 103, 1964.
38. **Apinis, A. E.**, Biocoenotic relationships of grassland soil fungi, *Mitt. Florist. Soziol. Arbeitsgem.*, 14, 68, 1969.
39. **Apinis, A. E.**, Das Verhalten der Pilze in bestimmten Grasland-Gesellschaften, in *Gesellschaftens-morphologie*, Dr. W. Junk, Den Haag, 1970, 172.
40. **Apinis, A. E.**, Thermophilous fungi in certain grasslands, *Mycopathol. Mycol. Appl.*, 48, 63, 1972.
41. **Apinis, A. E.**, Uber die Grenzbereiche der Pilzvegetation, in *Tatsachen und Probleme der Grenzen in der Vegetation*, J. Cramer, Lehre, 1974, 139.
42. **Apinis, A. E.**, Concerning degradation of barley grain by fungi, *Biodegradation and Humification*, Kilbertus, G., Reisinger, O., Mourey, A., and Cancela da Fonseca, J. A., Eds., Rapport du Ier Colloque International — Nancy, 1974, Université de Nancy 1, Nancy, France, 1975, 19.
43. **Apinis, A. E.**, Microfungi colonizing nodes and internodes of aerial standing culms of *Phragmites communis* Trin., *Nova Hedwigia Z. Kryptogamenkd.*, 26, 495, 1975.
44. **Apinis, A. E., Chesters, C. G. C., and Taligoola, H. K.**, Colonization of *Phragmites communis* leaves by fungi, *Nova Hedwigia Z. Kryptogamenkd.*, 23, 313, 1972.
45. **Apinis, A. E., Chesters, C. G. C., and Taligoola, H. K.**, Microfungi colonizing submerged standing culms of *Phragmites communis* Trin., *Nova Hedwigia Z. Kryptogamenkd.*, 23, 473, 1972.
46. **Apinis, A. E. and Taligoola, H. K.**, Biodegradation of *Phragmites communis* Trin. by fungi, *Biodegradation and Humification*, Kilbertus, G., Reisinger, O., Mourey, A., and Cancela da Fonseca, J. A., Eds., Rapport du Ier Colloque International — Nancy, 1974, Université de Nancy 1, Nancy, France, 1975, 19.
47. **Ardini, J.**, Factors affecting the germination of orchid seed, *Bot. Rev.*, 33, 1, 1967.
48. **Ark, P. A. and Dickey, R. S.**, A modification of the van Tieghem cell for purification of contaminated fungus cultures, *Phytopathology*, 40, 389, 1950.
49. **Armstrong, D. N. and Peterson, M. L.**, Microbiological Studies of Compost Plant Dust, Program Element 1D2063, National Environmental Research Center, Research and Monitoring, U.S. Environmental Protection Agency, Cincinnati, 1972, 1.
50. **Åronson, J. M.**, The cell wall, in *The Fungi, An Advanced Treatise*, Vol. 1, Ainsworth, G. C. and Sussman, A. S., Eds., Academic Press, New York, 1965.
51. **Arsvoll, K.**, Fungi causing winter damage on cultivated grasses in Norway, *Medlinger fra Norges Landbrukshøgskde*, 54(9), 1, 1975.
52. **Arthur, J. C.**, *Manual of the Rusts in the United States and Canada*, Purdue Research Foundation, Lafayette, Ind., 1934.
53. **von Arx, J. A.**, Beitrage zur Kenntnis der Gattung *Mycosphaerella*, *Sydowia, Ann. Mycol.*, Ser. II, 3, 7, 1949.
54. **von Arx, J. A.**, *The Genera of Fungi Sporulating in Pure Culture*, J. Cramer, Lehre, 1970.
55. **Ashida, J.**, Adaptation of fungi to metal toxicants, *Annu. Rev. Phytopathol.*, 3, 153, 1965.
56. **Atkinson, G. F.**, The influence of mushrooms on the growth of some plants, *Cornell Univ. A. E. S. Bull.*, 240, 217, 1906.
57. **Atkinson, R. G.**, The application of a nutritional grouping method to soil fungi, *Can. J.Bot.*, 33, 281, 1955.
58. **Ausemus, E. H.**, Breeding for disease resistance in wheat, oats, barley, and flax, *Bot. Rev.*, 9, 207, 1943.
59. **Austwick, P. K. C.**, Pathogenicity, in *The Genus Aspergillus*, Raper, K. B. and Fennell, D. J., Eds., Williams & Wilkins, Baltimore, 1965, 82.
60. **Austwick, P. K. C.**, Effects of adjustment to the environment on fungal form, in *The Fungi, An Advanced Treatise*, Vol. 3, Ainsworth, G. C. and Sussman, A. S., Eds., Academic Press, New York, 1968, 419.
61. **Backus, M. P. and Stauffer, J. F.**, The production and selection of a family of strains in *Penicillium chrysogenum*, *Mycologia*, 47, 429, 1955.
62. **Bärlocher, F. and Kendrick, B.**, Hyphomycetes as intermediaries of energy flow in streams, in *Recent Advances in Aquatic Mycology*, Jones, E. B. G., Ed., Elek Science, London, 1976, 435.

63. **Baker, G. E.**, Heterocaryosis in *Penicillium notatum*, *Bull. Torrey Bot. Club*, 71, 367, 1944.
64. **Baker, G. E.**, Inadvertent distribution of fungi, *Can. J. Microbiol.*, 12, 109, 1966.
65. **Baker, G. E. and Meeker, G. A.**, Ecosystems, mycologists and the distribution of fungi in the Central Pacific, *Pac. Sci.*, 26, 418, 1972.
66. **Baker, J. H.**, Yeasts, molds, and bacteria from an acid peat on Signy Island, in *Antarctic Ecology*, Holdgate, M. W., Ed., Academic Press, New York, 1970, 717.
67. **Baker, K. F. and Snyder, W. C., Eds.**, *Ecology of Soil-borne Plant Pathogens; Prelude to Biological Control*, University of California Press, Berkeley, 1965.
68. **Bakerspiegel, A.**, Soil as a storage medium for fungi, *Mycologia*, 45, 596, 1953.
69. Baltimore Biological Laboratory, *Culture Media, Materials and Apparatus for the Microbiological Laboratory*, 4th ed., Baltimore Biological Laboratory, Inc., Baltimore, 1956.
70. **Bandoni, R. J.**, Mycological observations on the aqueous films covering decaying leaves and other litter, *Trans. Mycol. Soc. Jpn.*, 15, 309, 1974.
71. **Bandoni, R. J.**, Significance of the tetraradiate spore in dispersal of terrestrial fungi, *Rep. Tottori Mycol. Inst.*, 12, 105, 1975.
72. **Bandoni, R. J.**, Surface-active spore slimes, *Can. J. Bot.*, 53, 2543, 1975.
73. **Barghoorn, E. S. and Linder, D. H.**, Marine fungi: their taxonomy and biology, *Farlowia*, 1, 395, 1944.
74. **Barnett, J. A.**, Biochemical differentiation of taxa with special reference to the yeasts, in *The Fungi, An Advanced Treatise*, Vol. 3, Ainsworth, G. C. and Sussman, A. S., Eds., Academic Press, New York, 1968, 557.
75. **Barras, S. J.**, Release of fungi from mycangia of southern pine beetles observed under a scanning electron microscope, *Z. Angew. Entomol.*, 79, 173, 1975.
76. **Barras, S. J. and Perry, T.**, Fungal symbionts in the prothoracic mycangium of *Dendroctonus frontalis* (Coleopt. Scolytidae), *Z. Angew. Entomol.*, 75, 95, 1972.
77. **Barras, S. J. and Perry, T. J.**, Interrelationships Among Microorganisms, Bark or Ambrosia Beetles, and Woody Host Tissue, An Annotated Bibliography, 1965—1974, Gen. Tech. Rep. No. S010, U.S. Forest Service, Pineville, Louisiana, 1975.
78. **Barron, G. L. and Dierkes, Y.**, Naematophagous fungi: *Hohenbuehelia*, the perfect state of *Nematoctonus*, *Can. J. Bot.*, 55, 3054, 1977.
79. **Bartnicki-Garcia, S.**, Cell wall chemistry, morphogenesis and taxonomy of fungi, *Annu. Rev. Microbiol.*, 22, 87, 1968.
80. **Batra, L. R.**, Ectosymbiosis between ambrosia fungi and beetles. I *Indian J. Mycol. Plant Pathol.*, 2, 165, 1972.
81. **Batra, L. R., Batra, S.W. T., and Bohart, G. E.**, The mycoflora of domesticated and wild bees (Apoidae), *Mycopathol. Mycol. Appl.*, 49, 13, 1973.
82. **Batra, L. R and Millner, P. D.**, Some Asian fermented foods and beverages and associated fungi, *Mycologia*, 66, 942, 1974.
83. **Batra, L. R. and Millner, P. D.**, Asian fermented foods and beverages, *Dev. Ind. Microbiol.*, 17, 117, 1976.
84. **Batra, S. W. T. and Batra, L. R.**, The fungus gardens of insects, *Sci. Am.*, 217, 112, 1967.
85. **Bawden, F. C.**, The role of plant hosts in microbial ecology, in *Microbial Ecology*, Williams, R. E. O. and Spicer, E. C., Eds., Cambridge University Press, Cambridge, 1957, 299.
86. **Baxter, D. V.**, *Pathology in Forest Practice*, 2nd ed., John Wiley & Sons, New York, 1952.
87. **Baxter, D. V.**, Fungus development from field to forest in spruce plantings of 1904 and later, *Pap. Mich. Acad. Sci. Arts Lett.*, 49, 3, 1964.
88. **Baxter, D. V. and Middleton, J. T.**, Geofungi in forest successions following retreat of the Alaskan glaciers, in *Recent Advances in Botany*, Vol. 2, Bailey, D. L., Ed., University of Toronto Press, Toronto, 1961, 1514.
89. **Beck, E. C.**, The application of serological methods to the differentiation of closely related smut fungi, *Can. J. Res. Sect. C*, 16, 391, 1938.
90. **Becker, J. G. and Shaw C. G.**, Fungi in domestic sewage treatment plants, *Appl. Microbiol.*, 3, 173, 1955.
91. **Beech, F. W. and Davenport, R. R**, A survey of methods for the quantitative examination of the yeast flora of apple and grape leaves, in *Ecology of Leaf Surface Microorganisms*, Preece, T. F. and Dickinson, C. H., Eds., Academic Press, New York, 1971, 139.
92. **Benjamin, C. R., Haynes, W. C., and Hesseltine, C. W.**, Microorganisms: what they are, where they grow, what they do, *U.S. Dept. Agric. Misc. Publ.*, 955, 1, 1964.
93. **Benjamin, C. R. and Slot, A.**, Fungi of Haiti, *Sydowia, Ann. Mycol.*, Ser. II, 23, 125, 1969.
94. **Benjamin, R. K.**, Laboulbeniomycetes, in *The Fungi, An Advanced Treatise*, Vol. 4a, Ainsworth, G. C. and Sussman, A. S., Eds., Academic Press, New York, 1973, 223.

95. **Ben-Meier-Glueck, Y.**, Fungi isolated from the air of groves and packing sheds and from the skin of shamouti oranges, *Rec. Agric. Res. Stn.*, 35, 1952.
96. **Benoit, R. E. and Hall, C. L., Jr.**, The microbiology of some dry valley soils of Victoria Land, Antarctica, in *Antarctic Ecology*, Haldgate, M. W., Ed., Academic Press, New York, 1970, 697.
97. **Berg, G. L.**, Ed., Master manual on molds and mycotoxins, *Farm Technol. Agri-Fieldman* (Suppl.), 28(5-Fall) 1a, 1972.
98. **Bergman, K.**, *Phycomyces, Bacteriol. Rev.*, 33, 99, 1969.
99. **Berkeley, G. H.**, Root rots of certain non-cereal crops, *Bot. Rev.*, 10, 67, 1944.
100. **Berner, N. H. and Ahearn, D. G.**, Observations on the growth and survival of *Cladosporium resinae* in jet fuel, *Dev. Ind. Microbiol.*, 18, 705, 1977.
101. **Bessey, C. E.**, An enormous puffball, *Am. Nat.*, 18, 530, 1884.
102. **Bessey, E. A.**, *Morphology and Taxonomy of the Fungi*, Blakiston, Philadelphia, 1950.
103. **Billings, W. D.**, The environmental complex in relation to plant growth and distribution, *Q. Rev. Biol.*, 27, 251, 1952.
104. **Billings, W. D.**, *Plants and the Ecosystem*, Wadsworth, Belmont, Calif., 1965.
105. **Bingham, R. T., Hoff, R. J., and McDonald, G. I.**, Disease resistance in forest trees, *Annu. Rev. Phytopathol.*, 9, 433, 1971.
106. **Bisby, G. R.**, Geographical distribution of fungi, *Bot. Rev.*, 9, 466, 1943.
107. **Bisby, G. R. and Ainsworth, G. C.**, The numbers of fungi, *Trans. Br. Mycol. Soc.*, 26, 16, 1943.
108. **Bistis, G.**, Studies on the genetics of *Ascobolus stercorarius* (Bull.) Schroet., *Bull. Torrey Bot. Club*, 83, 35, 1956.
109. **Björkman, E.**, The ecological significance of the ectotrophic mycorrhizal associates in forest trees, *Sven. Bot. Tidskr.*, 43, 223, 1949.
110. **Björkman, E.**, Soil antibiotics acting against the root-rot fungus (*Fomes annosus* Fr.), *Physiol. Plant.*, 1, 1, 1949.
111. **Blank, F.**, On the cell walls of dimorphic fungi causing systemic infections, *Can. J. Microbiol.*, 1, 1, 1954.
112. **Bohus, G.**, On the drought-resistance of fungi, *Ann. Hist. Nat. Mus. Nat. Hung.*, 7, 263, 1956.
113. **Bohus, G. and Babos, M.**, Coenology of terricolous macroscopic fungi of deciduous forests: contribution to our knowledge of their behaviour in Hungary, *Bot. Jahrb.*, 80, 1, 1960.
114. **Bohus, G. and Babos, M.**, Mycocoenological investigation of acidophilous deciduous forests in Hungary, *Bot. Jahrb.*, 87, 404, 1967.
115. **Boidin, J. and Abadie, F.**, Les levures des liqueurs tannantes vegetales: leur action sur les tanins pyrogalliques, *Bull. Soc. Mycol. Fr.*, 70, 353, 1954.
116. **Bollen, W. C. and Lu, K. C.**, Douglas-Fir Bark Tanin Decomposition in Two Forest Soils, Res. Paper No. PNW 85, U.S. Forest Service, Portland, Oregon, 1969.
117. **Boone, D. M. and Keitt G. W.**, *Venturia inaequalis* (Cke.) Wint. VIII. Inheritance of color mutant characters, *Am. J. Bot.*, 43, 226, 1956.
118. **Booth, C.**, *The Genus Fusarium*, Commonwealth Mycological Institute, Kew, 1971.
119. **Booth, C.**, Ed., *Methods in Microbiology*, Vol. 4, Academic Press, New York, 1971.
120. **Boughey, A. S.**, The ecology of fungi which cause plant diseases, *Trans. Br. Mycol. Soc.*, 32, 179, 1949.
121. **Bourquin, A. W., Kiefer, L. A., Berner, N. H., Crow, S., and Ahearn, D. G.**, Inhibition of estuarine microorganisms by polychlorinated biphenyls, *Dev. Ind. Microbiol.*, 16, 256, 1975.
122. **Boyce, J. S.**, *Forest Pathology*, McGraw-Hill, New York, 1938.
123. **Bracken, A.**, *The Chemistry of Micro-organisms*, Pitman, New York, 1955.
124. **Bradley, H. and Swedberg, C.**, *Keeping Food Safe*, Doubleday, Garden City, N.Y., 1975.
125. **Brandes, G. A.**, Advances in fungicide utilization, *Annu. Rev. Phytopathol.*, 9, 363, 1971.
126. **Brandsberg, J. W.**, Fungi found in association with *Histoplasma capsulatum* in a naturally contaminated site in Clarksburg, Maryland, USA, *Sabouraudia*, 6, 246, 1968.
127. **Brandsberg, J. W., Weeks, R. J., Hill, W. B., and Piggott, W. R.**, A study of fungi found in association with *Histoplasma capsulatum*: three bird roosts in southeastern Missouri, USA, *Mycopathol. Mycol. Appl.*, 38, 71, 1969.
128. **Brauns, F. E.**, *The Chemistry of Lignin*, Academic Press, New York, 1952.
129. **Brauns, F. E. and Brauns, D. A.**, *Chemistry of Lignin, Supplement Volume Covering the Literature for 1949—1960*, Academic Press, New York, 1960.
130. **Bresinsky, A. and Dichtel, B.**, Bericht der Arbeitsgemeinschaft zur Kartierung von Grosspilzen in der DDR (1), *Z. Pilzkd.*, 37, 75, 1971.
131. **Brewer, D.**, Studies on the slime accumulation in pulp and paper mills. I. Some fungi isolated from mills in New Brunswick and Newfoundland, *Can. J. Bot.*, 36, 941, 1958.
132. **Brewer, D.**, Studies on slime accumulations in pulp and paper mills. II. Physiological studies of *Phialophora fastigiata* and *P. richardsiae*, *Can. J. Bot.*, 37, 339, 1959.

133. **Brewer, D.**, Studies on slime accumulations in pulp and paper mills. III. The stimulation of growth by "white water", *Can. J. Bot.*, 37, 517, 1959.
134. **Brian, P. W.**, Antibiotics production by fungi, *Bot. Rev.*, 17, 357, 1951.
135. **Brian, P. W.**, The ecological significance of antibiotic production, in *Microbial Ecology*, Williams, R. E. O. and Spicer, E. C., Eds., Cambridge University Press, Cambridge, 1957, 168.
136. **Brierley, W. B., Jewson, S. T., and Brierley, M.**, The quantitative study of soil fungi, *Proc. 1st Int. Cong. Soil Sci.*, 3, 48, 1928.
137. **Broadbent, D.**, Antibiotics produced by fungi, *Bot. Rev.*, 32, 219, 1966.
138. **Broadfoot, W. M. and Pierre, W. H.**, Forest soil studies. I. Relation of rate of decomposition of tree leaves to their acid-base balance and other chemical properties, *Soil Sci.*, 48, 329, 1939.
139. **Brock, T. D.**, *Principles of Microbial Ecology*, Prentice-Hall, Englewood Cliffs, N.J., 1966.
140. **Brock, T. D.**, Microbial growth under extreme conditions, in *Microbial Growth*, Cambridge University Press, Cambridge, 1969, 15.
141. **Brock, T. D.**, *Biology of Microorganisms*, Prentice-Hall, Englewood Cliffs, N.J., 1970.
142. **Brodie, H. J.**, The splash-cup mechanism in plants, *Can. J. Bot.*, 29, 224, 1951.
143. **Brodie, H. J.**, *The Birds' Nest Fungi*, Univerity of Toronto Press, Toronto, 1975.
144. **Brook, P. J.**, Effect of light on ascospore discharge by five fungi with bitunicate asci, *New Phytol.*, 74, 85, 1975.
145. **Brower, G. and Gaddis, L.**, Filamentous waste treatment systems at low pH, *J. Water Pollut. Control Fed.*, 41, R61, 1969.
146. **Brown, W.**, The physiology of host-parasite relations, *Bot. Rev.*, 2, 236, 1936.
147. **Buchner, P.**, *Endosymbiosis of Animals with Plant Microorganisms*, (trans. by Bertha Mueller with the collaboration of F. H. Foeckler of *Endosymbiose der Tiere mit pflanzlichen Mikroorganismen*), Interscience, New York, 1965.
148. **Buechner, H. A., Seabury, J. H., Campbell, C. C., Georg, L. K., Kaufman, L., and Kaplan, W.**, The current status of serologic, immunologic and skin tests in the diagnosis of pulmonary mycoses, *Chest*, 63, 259, 1973.
149. **Buller, A. H. R.**, *Researches on Fungi*, Vol. 1, Longmans, Green & Company, London, 1909.
150. **Buller, A. H. R.**, *Researches on Fungi*, Vol. 2, Longmans, Green & Company, London, 1922.
151. **Buller, A. H. R.**, *Researches on Fungi*, Vol. 3, Longmans, Green & Company, London, 1924.
152. **Buller, A. H. R.**, *Researches on Fungi*, Vol. 4, Longmans, Green & Company, London, 1931.
153. **Buller, A. H. R.**, *Researches on Fungi*, Vol. 5, Longmans, Green & Company, London, 1933.
154. **Buller, A. H. R.**, *Researches on Fungi*, Vol. 6, Longmans, Green & Company, London, 1934.
155. **Buller, A. H. R.**, *Researches on Fungi*, Vol. 7, University of Toronto Press, Toronto, 1950.
156. **Buller, A. H. R.**, The diploid cell and the diploidization process in plants and animals, with special reference to the higher fungi, *Bot. Rev.*, 7, 335, 1941.
157. **Burges, A.**, Soil fungi and root infection, a review, *Broteria Ser. Cienc. Nat.*, Ser. 3, 8, 64, 1939.
158. **Burges, A.**, The downward movement of fungus spores in sandy soil, *Trans. Br. Mycol. Soc.*, 33, 142, 1950.
159. **Burges, A.**, The nature and distribution of humic acid, *Sci. Proc. R. Dublin Soc. Ser. A*, 1, 53, 1960.
160. **Burges, A.**, The microbiology of a podzol profile, in *Soil Organisms*, Doeksen, J. and van der Drift, J., Eds., North-Holland, Amsterdam, 1963, 151.
161. **Burges, A. and Latter, P.**, Decomposition of humic acid by fungi, *Nature (London)*, 186, 404, 1960.
162. **Burkholder, F. R. and Sinnott, E. W.**, Morphogenesis of fungus colonies in submerged shaken cultures, *Am. J. Bot.*, 32, 424, 1946.
163. **Burnett, J. W.**, Oxygen consumption during sexual reproduction of some Mucoraceae, *New Phytol.*, 52, 58, 1953.
164. **Burnett, J. H.**, The mating systems of fungi I, *New Phytol.*, 55, 50, 1956.
165. **Burnett, J. H.**, *Fundamentals of Mycology*, Edward Arnold, London, 1968.
166. **Burnett, J. H.**, *Mycogenetics. An Introduction to the General Genetics of Fungi*, John Wiley & Sons, New York, 1975.
167. **Byrne, A. R., Ravnik, V., and Kosta, L.**, Trace element concentrations in higher fungi, *Sci. Total Environ.*, 6, 65, 1976.
168. **Cain, R. F.**, Studies of coprophilous Sphaeriales in Ontario, *Univ. of Toronto Stud. Biol. Ser.*, 38, 3, 1934.
169. **Cain, R. F.**, Evolution of the fungi, *Mycologia*, 64, 1, 1972.
170. **Cameron, R. E., King, J., and David, C. N.**, Mircobiology, ecology and microclimatology of soil sites in dry valleys of southern Victoria Land, Antarctica, in *Antarctic Ecology*, Holdgate, M. W., Ed., Academic Press, New York, 1970, 702.
171. **Canter, H. M**, The importance of fungal parasitism in limnology, *Verh. Int. Verein. Theor. Angew. Limnol.*, 10, 107, 1949.

172. **Canter, H. M. and Lund, J. W. G.**, Studies on plankton parasites. I. Fluctuation in the numbers of *Asterionella formosa* Hass. in relation to fungal epidemics, *New Phytol.*, 47, 238, 1948.
173. **Canter, H. M. and Lund, J. W. G.**, Studies on plankton parasites. III. Examples of the interaction between parasitism and other factors determining the growth of diatoms, *Ann. Bot. (London)*, 15, 359, 1951.
174. **Canter, H. W. and Lund, J. W. G.**, Studies on plankton parasites. II. The parasitism of diatoms with special reference to the English lake district, *Trans. Br. Mycol. Soc.*, 36, 13, 1953.
175. **Canter, H. W. and Lund, J. W. G.**, The parasitism of planktonic desmids by fungi, *Oesterr. Bot. Z.*, 116, 351, 1969.
176. **Cantino, E. C.**, Physiology and phylogeny in the water molds, a reevaluation, *Q. Rev. Biol.*, 30, 138, 1955.
177. **Cantino, E. C.**, Morphogenesis in aquatic fungi, in *The Fungi, An Advanced Treatise*, Vol. 2, Ainsworth, G. C. and Sussman, A. S., Eds., Academic Press, New York 1966, 283.
178. **Carantino, S., Poisson, J., and Cahagnier, E.**, La microflore des graines de feverole: son evolution au cours d'essais de stockage au function des conditions hydriques realisées, *Rev. Mycol.*, 40, 223, 1976.
179. **Carl, E. A.**, Control of plant diseases by crop rotation, *Bot. Rev.*, 29, 413, 1969.
180. **Carmichael, J. W.**, *Geotrichum candidum, Mycologia*, 49, 820, 1957.
181. **Carson, H. L., Knapp, E. P., and Phaff, H. J.**, Studies on the ecology of *Drosophila* in the Yosemite region of California. III. The yeast flora of the natural breeding sites of some species of *Drosophila*, *Ecology*, 37, 538, 1956.
182. **Cartwright, K. St. G. and Findlay, N. P. K.**, *Decay of Timber and its Prevention*, (American reprint), Chemical Publishing Co., Brooklyn, 1950.
183. **Castellani, A.**, Fungi and fungous diseases, *Arch. Dermatol. Syphilol.*, 16, 383, 1927; 17, 76, 1928.
184. **Castellani, A.**, The cultivation of pathogenic fungi in sterile distilled water, *Commentarii*, 1(30), 1, 1963.
185. **Catchside, D. G.**, *The Genetics of Micro-organisms*, Sir Isaac Pitman & Sons, London, 1951.
186. **Chahal, D. S., Swan, J. E., and Young, M. M.**, Protein and cellulase production by *Chaetomium cellulolyticum* grown in wheat straw, *Dev. Ind. Microbiol.*, 18, 433, 1977.
187. **Chamberlain, A. C.**, Deposition of particles to natural surfaces, in *Airborne Microbes*, Cambridge University Press, Cambridge, 1967, 138.
188. **Chambers, C. C.**, The germicidal efficiency of silver, iodine, and quaternary ammonium compounds, in Disinfection and Chemical Oxidation in Water and Waste Treatment, Proc. 3rd Sanitary Engineering Conf., Urbana, Ill., 1961, 29.
189. **Charles, H. P. and Knight, B. C. J. G.**, *Organization and Control in Prokaryotic and Eukaryotic Cells*, Cambridge University Press, Cambridge, 1970.
190. **Chesters, C. G. C.**, A contribution to the study of the fungi in the soil, *Trans. Br. Mycol. Soc.*, 30, 100, 1937.
191. **Chesters, C. G. C.**, Concerning fungi inhabiting soil, *Trans. Br. Mycol. Soc.*, 32, 197, 1949.
192. **Chestukhni, V. Ya**, An ecological analysis of the decomposition of plant residues in young pine plantations, *Pochnovedenia (Pedology)*, 1948, 87, 1948; *Biol. Abstr.*, 23, 16283, 1948.
193. **Christensen, C. M.**, *The Molds and Man. An Introduction to the Fungi*, University of Minnesota Press, Minneapolis, 1951.
194. **Christensen, C. M.**, Deterioration of stored grains by fungi, *Bot. Rev.*, 23, 108, 1957.
195. **Christensen, J. J. and Rodenhiser, H. A.**, Physiologic specialization and genetics of the smut fungi, *Bot. Rev.*, 6, 389, 1940.
196. **Christensen, M., Wittingham, W. F., and Novak, R. O.**, The soil microfungi of wet mesic forests in southern Wisconsin, *Mycologia*, 54, 374, 1962.
197. **Christensen, M. and Wittingham, W. F.**, The soil fungi of open bogs and conifer swamps in Wisconsin, *Mycologia*, 57, 882, 1965.
198. **Church, B. D., Nash, H. A., and Brosz, A. W.**, The use of Fungi Imperfecti in treating food processing wastes, *Dev. Ind. Microbiol.*, 13, 30, 1972.
198a. **Chute, H. L., O'Meara, D. C., and Barden, E. S.**, A bibliography of avian mycoses, *Maine Agric. Exp. Stn. Misc. Publ.*, No. 655, 1962.
199. **Ciegler, A.**, Mycotoxins: occurrence, chemistry, biological activity, *Lloydia*, 38, 21, 1975.
200. **Clark, H. F., Geldreich, E. E., Jeter, H. L., and Kabler, P. W.**, The membrane filter in sanitary bacteriology, *Public Health Rep.*, 66, 951, 1951.
201. **Cochran, V. M.**, *Physiology of Fungi*, John Wiley & Sons, New York, 1958.
202. **Cohen, R., Roth, F. J., Delgado, E., Ahearn, D. G., and Kelser, M. H.**, Fungal flora of the normal human small and large intestine, *N. Engl. J. Med.*, 280, 638, 1969.
203. **Coker, W. C. and Mathews, V. D.**, *Saprolegniales, North Am. Flora*, 2,15, 1937.

204. **Coley-Smith, J. R. and Cooke, R. C.**, Survival and germination of fungal sclerotia, *Annu. Rev. Phytopathol.*, 9, 69, 1971.
205. **Colwell, R. R., Walker, J. D., Conrad, B. F., and Seesman, P. A.**, Microbiological studies of Atlantic Ocean water and sediments from potential off-shore drilling sites, *Dev. Ind. Microbiol.*, 17, 269, 1976.
206. Commonwealth Mycological Institute, Distribution Maps of Plant Diseases, Commonwealth Mycological Institute, Kew, 1942 to 1978.
207. **Conant, N. F., Smith, D. T., Baker, R. D., and Callaway, J. L.**, *Manual of Clinical Mycology*, 3rd ed., W. B. Saunders, Philadelphia, 1971.
208. **Connell, G. H. and Skinner, C. E.**, The external surface of the human body as a habitat for nonfermenting nonpigmented yeasts, *J. Bacteriol.*, 66, 627, 1953.
209. **Cooke, R.**, *The Biology of Symbiotic Fungi*, John Wiley & Sons, New York, 1977.
210. **Cooke, W. B.**, A survey of literature on fungus sociology and ecology. I, *Ecology*, 29, 376, 1948.
211. **Cooke, W. B.**, *Oxyporus nobilissimus* and the genus *Oxyporus* in North America, *Mycologia*, 41, 442, 1949.
212. **Cooke, W. B.**, Ecological life history outlines for fungi, *Ecology*, 32, 736, 1951.
213. **Cooke, W. B.**, A survey of the literature on fungus sociology and ecology. II, *Ecology*, 34, 211 1953.
214. **Cooke, W. B.**, Fungi in polluted water and sewage. I. Literature Review; II. Isolation technique; III. Fungi in a small polluted stream, *Sewage Ind. Wastes*, 26, 539, 1954.
215. **Cooke, W. B.**, Fungi, lichens and mosses in relation to vascular plant communities in eastern Washington and adjacent Idaho, *Ecol. Monogr.*, 25, 119, 1955.
216. **Cooke, W. B.**, Fungi of Mount Shasta, 1936—1951, *Sydowia, Ann. Mycol.*, Ser. II, 9, 94, 1955.
217. **Cooke, W. B.**, Subalpine fungi and snowbanks, *Ecology*, 36, 124, 1955.
218. **Cooke, W. B.**, Colonization of artificial bare areas by microorganisms, *Bot. Rev.*, 22, 613, 1956.
219. **Cooke, W. B.**, Potential plant pathogenic fungi in sewage and polluted water, *Plant Dis. Rep.*, 40, 681, 1956.
220. **Cooke, W. B.**, Checklist of fungi found in sewage and polluted water, *Sydowia, Ann. Mycol.*, Ser. II, Beiheft 1, 146, 1957.
221. **Cooke, W. B.**, The use and value of fungi as biological indicators of pollution, in *Trans. Semin. Biol. Problems in Water Pollution*, Sanitary Engineering Center, Cincinnati, 1957, 84.
222. **Cooke, W. B.**, Natural and induced degradation of lignin, *Tappi*, 40, 301, 1957.
223. **Cooke, W. B.**, Nutritional requirements of nine common sewage fungi, *Sewage Ind. Wastes*, 29, 1243, 1957.
224. **Cooke, W. B.**, Continuous sampling of trickling filter populations. I. Procedures, *Sewage Ind. Wastes*, 30, 21, 1958.
225. **Cooke, W. B.**, The ecology of the fungi, *Bot. Rev.*, 24, 341, 1958.
226. **Cooke, W. B.**, An ecological life history of *Aureobasidium pullulans* (de Bary) Arnaud, *Mycopathol. Mycol. Appl.*, 12, 1, 1959.
227. **Cooke, W. B.**, Fungi in polluted water and sewage. IV. The occurrence of fungi in a trickling-filter-type sewage treatment plant, *Purdue Univ. Eng. Bull.*, 43, 26, 1959.
228. **Cooke, W. B.**, Trickling filter ecology, *Ecology*, 40, 273, 1960.
229. **Cooke, W. B.**, The cyphellaceous fungi, a study in the Porotheleaceae, *Sydowia Ann. Mycol.*, Ser. II, Beiheft 4, 1, 1961.
230. **Cooke, W. B.**, Pollution effects on the fungus population of a stream, *Ecology*, 42, 1, 1961.
231. **Cooke, W. B.**, Some effects of spray disposal of spent sulphite liquor on soil mold populations, *Purdue Univ. Eng. Bull.*, 45(2), 35, 1961.
232. **Cooke, W. B.**, A taxonomic study in the "black yeasts", *Mycopathol. Mycol. Appl.*, 17, 1, 1962.
233. **Cooke, W. B.**, Species of *Fusarium* isolated from a waste stabilization pond system, *Mycopathol. Mycol. Appl.*, 18, 225, 1962.
234. **Cooke, W. B.**, A Laboratory Guide to Fungi in Polluted Waters, Sewage and Sewage Treatment Systems, Publ. No. 999-WP-1, U.S. Public Health Service, Cincinnati, Ohio, 1963.
235. **Cooke, W. B.**, Fungi associated with spent sulphite liquor disposal in a natural sand bed, *Tappi*, 46, 573, 1962.
236. **Cooke, W. BX.**, The enumeration of yeast populations in a sewage treatment plant, *Mycologia*, 57, 696, 1965.
237. **Cooke, W. B.**, Growth of yeasts at 37°C, *Mycopathol. Mycol. Appl.*, 25, 195, 1965.
238. **Cooke, W. B.**, The occurrence of fungi in acid mine-drainage, *Proc. Purdue Univ. Eng. Bull.*, 51, 258, 1966.
239. **Cooke, W. B.**, Fungal populations in relation to pollution of the Bear River, Utah-Idaho, *Utah Acad. Proc.*, 44, 298, 1967.
240. **Cooke, W. B.**, Carbon/nitrogen relationships of fungus culture media, *Mycopathol. Mycol. Appl.*, 34, 305, 1968.

241. **Cooke, W. B.**, Studies in the genus *Prototheca*. I. Literature review, *J. Elisha Mitchell Sci. Soc.*, 84, 213, 1968.
242. **Cooke, W. B.**, Studies in the genus *Prototheca*. II. Taxonomy, *J. Elisha Mitchell Sci. Soc.*, 84, 217, 1968.
243. **Cooke, W. B.**, Some fungi of the Cache la Poudre River, Colorado, *Mycopathol. Mycol. Appl.*, 35, 361, 1968.
244. **Cooke, W. B.**, A bibliography of histoplasmosis, *Mycopathol. Mycol. Appl.*, 39, 1, 1969.
245. **Cooke, W. B.**, Fungi in soils over which digested sewage sludge has been spread, *Mycopathol. Mycol. Appl.*, 39, 209, 1969.
246. **Cooke, W. B.**, Fungi associated with the activated sludge process of sewage treatment at the Lebanon, Ohio, sewage treatment plant, *Ohio J. Sci.*, 70, 129, 1970.
247. **Cooke, W. B.**, Our Mouldy Earth, Res. Contract Ser. Publ. No. OWR O, U.S. Department of the Interior, Cincinnati, 1970.
248. **Cooke, W. B.**, Fungi in the Lebanon sewage treatment plants and in Turtle Creek, Warren Co., Ohio, *Mycopathol. Mycol. Appl.*, 42, 89, 1970.
249. **Cooke, W. B.**, The role of fungi in waste treatment, *CRC Crit. Rev. Environ. Control*, 1, 581, 1970.
250. **Cooke, W. B.**, The effect of sludge additives on soil fungus populations, *Mycopathol. Mycol. Appl.*, 44, 205, 1971.
251. **Cooke, W. B.**, Fungi in burned and unburned chaparral soils, *Sydowia Ann. Mycol.*, Ser. II, 24, 164, 1971.
252. **Cooke, W. B.**, Zonal distribution patterns in the fungi, *Mycopathol. Mycol. Appl.*, 48, 7, 1972.
253. **Cooke, W. B.**, Backyard fungi, *Ohio J. Sci.*, 73, 88, 1973.
254. **Cooke, W. B.**, The ubiquity of , *Rep. Tottori Mycol. Inst.*, 12, 193, 1975.
255. **Cooke, W. B.**, Fungi in sewage, in *Recent Advances in Aquatic Mycology*, Jones, E. B. G., Ed., Elek Science, London, 1976, 389.
256. **Cooke, W. B.**, Fungi in and near streams carrying acid mine-drainage, *Ohio J. Sci.*, 76, 231, 1976.
257. **Cooke, W. B.**, *The Fungi of "Our Mouldy Earth"*, in press.
258. **Cooke, W. B. and Bartsch, A. F.**, Aquatic fungi in water with high waste loads, *Sewage Ind. Wastes*, 31, 1316, 1959.
259. **Cooke, W. B. and Brazis, A. R.**, Occurrence of molds and yeasts in dairy products, *Mycopathol. Mycol. Appl.*, 35, 281, 1960.
260. **Cooke, W. B. and Busch, K. A.**, Activity of cellulose decomposing fungi isolated from sewage polluted water, *Sewage Ind. Wastes*, 29, 210, 1957.
261. **Cooke, W. B. and Foter, M. J.**, Fungi in used bedding materials, *Appl. Microbiol.*, 6, 169, 1958.
262. **Cooke, W. B. and Hawksworth, D. L.**, A preliminary list of the families proposed for the fungi (including the lichens), *Commonw. Mycol. Inst. Mycol. Pap.*, No. 121, 1970.
263. **Cooke, W. B. and Hirsch, A.**, Continuous sampling of trickling filter populations. II. Populations, *Sewage Ind. Wastes*, 30, 138, 1958.
264. **Cooke, W. B. and Kabler, P. W.**, The survival of *Histoplasma capsulatum* in water, *Lloydia*, 16, 252, 1953.
265. **Cooke, W. B. and Kabler, P. W.**, Isolation of potentially pathogenic fungi from polluted water and sewage, *Public Health Rep.*, 70, 689, 1955.
266. **Cooke, W. B. and Lawrence, D. B.**, Soil mold fungi isolated from recently glaciated soils in southeastern Alaska, *J. Ecol.*, 47, 529, 1959.
267. **Cooke, W. B. and Ludzack, F. J.**, Predacious fungus behavior in an activated sludge system, *Sewage Ind. Wastes*, 30, 1490, 1958.
268. **Cooke, W. B. and Matsuura, G. S.**, Physiological studies in the "black yeasts", *Mycopathol. Mycol. Appl.*, 21, 225, 1963.
269. **Cooke, W. B. and Matsuura, G. S.**, Removal of ABS from solutions by a common fungus of sewage, *Mycopathol. Mycol. Appl.*, 19, 287, 1963.
270. **Cooke, W. B. and Matsuura, G. S.**, A study of yeast populations in a waste stabilization pond system, *Protoplasma*, 57, 163, 1963.
271. **Cooke, W. B. and Matsuura, G. S.**, Distribution of fungi in a waste stabilization pond system, *Ecology*, 50, 689, 1969.
272. **Cooke, W. B., Moore, W. A., and Kabler, P. W.**, B.O.D. satisfaction by fungi, *Sewage Ind. Wastes*, 28, 1075, 1956.
273. **Cooke, W. B., Phaff, H. J., Miller, M. W., Shifrine, M., and Knapp, E. P.**, Yeasts in polluted water and sewage, *Mycologia*, 52, 210, 1960.
274. **Cooke, W. B. and Pipes, W. O.**, The occurrence of fungi in activated sludge, *Mycopathol. Mycol. Appl.*, 40, 249, 1970.
275. **Cooley, J. S.**, Root diseases of deciduous fruit trees, *Bot. Rev.*, 12, 83, 1946.

276. **Cooney, D. G. and Emerson, R.,** *Thermophilic Fungi, An Account of their Biology, Activities and Classification,* W. H. Freeman, San Francisco, 1964.
277. **Copeland, H. F.,** The kingdoms of organisms, *Q. Rev. Biol.,* 13, 383, 1938.
278. **Copeland, H. F.,** *The Classification of Lower Organisms,* Pacific Books, Palo Alto, Calif., 1956.
279. **Corda, A. C. I.,** *Icones fungorum hucusque cognitarum,* 64 large plates, Vol. 1 to 4, J. G. Calve, Prag., Vol. 5 to 6, Fr. Ehrlich, Prag, 1837 to 1854.
280. **Corner, E. J. H.,** The classification of resupinate Thelephoraceae, Abstr. Papers, 10th Int. Bot. Congr., 1964, 52.
281. **Corner, E. J. H.,** *The Natural History of Palms,* Weidenfeld & Nicholson, London, 1966.
282. **Couch, J. N.,** *The Genus Septobasidium,* University of North Carolina Press, Chapel Hill, 1938.
283. **Couch, J. N.,** The structure and action of the cilia in some aquatic Phycomycetes, *Am. J. Bot.,* 28, 704, 1941.
284. **Couch, J. N.,** A new fungus on crab eggs, *J. Elisha Mitchell Sci. Soc.,* 58, 158, 1942.
285. **Couch, J. N.,** Technic for collection, isolation and culture of chytrids, *J. Elisha Mitchell Sci. Soc.,* 55, 208, 1939.
286. **Cowell, R.,** Venereous meats, *New Sci.,* 75, 766, 1977.
287. **Crow, S. A., Ahearn, D. G., Cook, W. L., and Bourquin, A. W.,** Densities of bacteria and fungi in coastal surface films as determined by a membrane absorption procedure, *Limnol. Oceanogr.,* 20, 644, 1975.
288. **Cummins, G. B.,** *Supplement to Arthur's Manual of the Rusts of the United States and Canada,* (isued with reprint of Arthur's book), Hafner Press, New York, 1A, 1962.
289. **Curtis, E. J. C.,** Sewage fungus: its nature and effects, *Water Res.,* 3, 289, 1969.
290. **Daniels, S. L.,** The adsorption of microorganisms onto solid surfaces: a review, *Dev. Ind. Microbiol.,* 13, 211, 1972.
291. **Darby, R. T. and Goddard, D. F.,** Studies on the respiration of the mycelium of the fungus *Myrothecium verrucaria, Am. J. Bot.,* 37, 379, 1950.
292. **Daubenmire, R. F.,** Temperature gradients near the soil surface with reference to techniques of measurement in forest ecology, *J For.,* 41, 601, 1943.
293. **Daubenmire, R. F.,** *Plants and Environment,* 3rd ed., John Wiley & Sons, New York, 1974.
294. **Daubenmire, R. and Prusso, D. C.,** Studies on the decomposition rates of tree litter, *Ecology,* 44, 589, 1963.
295. **Davenport, R. R.,** Distribution of yeasts and yeast-like organisms from aerial surfaces of developing apples and grapes, in *Microbiology of Aerial Plant Surfaces,* Dickinson, C. H. and Preece, T. F., Eds., Academic Press, New York, 1976, 325.
296. **Davey, C. B.,** Sawdust composts: their preparation and effect on plant growth, *Soil Sci. Soc. Am. Proc.,* 17, 59, 1953.
297. **Davey, C. B.,** Transformations of sawdust in the course of its decomposition under the influence of *Coprinus ephemereus, Soil Sci. Soc. Am. Proc.,* 19, 376, 1955.
298. **Davey, C. B. and Wilde, S. A.,** Determination of the number of soil microorganisms by the use of molecular membrane filters, *Ecology,* 36, 760, 1955.
299. **Davidson, R. W.,** Fungi causing stains in logs and lumber in the southern states including five new species, *J. Agric. Res. (Washington, D.C.),* 50, 789, 1935.
300. **Davidson, R. W. and Vaughn, D. B.,** Fungi Causing Decay of Living Oaks in the Eastern United States and Their Cultural Identification, Tech. Bull. No. 785, U.S. Department of Agriculture, Washington, D.C., 1942.
301. **Davies, J. S., Zajic, J. E., and Wellman, A. M.,** Fungal oxidation of gaseous alkanes, *Dev. Ind. Microbiol.,* 15, 256, 1974.
302. **Davies, R. R.,** Air sampling for fungi, pollen and bacteria, in *Methods in Microbiology, Vol. 4,* Norris, J. R., et al., Eds., Academic Press, New York, 1971, 367.
303. **Day, D.,** A method of growing fungi on leather, *J. Am. Leather Chem. Assoc.,* 44, 558, 1959.
304. **Delwicke, C. C.,** The cycling of carbon and nitrogen in the biosphere, in *Microbiology and Soil Fertility,* Gilmour, C. M. and Allen, O. N., Eds., Oregon State University Press, Corvallis, 1965, 29.
305. **Demain, A. L.** The health of the fermentation industry: a prescription for the future, *Dev. Ind. Microbiol.,* 18, 73, 1977.
306. **Demoulin, V.,** Phytogeography of the fungal genus *Lycoperdon* in relation to the opening of the Atlantic, *Nature (London),* 242, 123, 1972.
307. **Demoulin, V.,** The origin of Ascomycetes and Basidiomycetes: the case for a red algal ancestry, *Bot. Rev.,* 40, 315, 1974.
308. **Denison, W. C. and Carroll, G. C.,** The primitive Ascomycetes: a new look at an old problem, *Mycologia,* 58, 249, 1966.

309. **Detroy, R. W. and Still, P. E.,** Fungal metabolites and viral replication in *Penicillium stoloniferum*, *Dev. Ind. Microbiol.*, 16, 145, 1975.
310. **DeVay, J. E.,** Mutual relationships in fungi, *Annu. Rev. Microbiol.*, 10, 15, 1956.
311. **Deverall, B. J.,** The physical environment for fungal growth. I. Temperature, in *The Fungi, An Advanced Treatise*, Vol. 1, Ainsworth, G. C. and Sussman, A. S., Eds., Academic Press, New York, 1965, 543.
312. **Deverall, B. J.,** Psychrophiles, in *The Fungi, An Advanced Treatise*, Vol. 3, Ainsworth, G. C. and Sussman, A. S., Eds., Academic Press, New York, 1968, 129.
313. **Dick, M. W.,** The occurrence and distribution of Saprolegniaceae in certain soils of south-east England. II. Distribution within defined areas, *J. Ecol.*, 50, 119, 1962.
314. **Dick, M. W.,** The occurrence and distribution of Saprolegniaceae in certain soils of south-east England. III. Distribution in relation to pH and water content, J. *Ecol.*, 51, 75, 1963.
315. **Dick, M. W.,** The Saprolegniaceae of the environs of Blelham Tarn: sampling techniques and the estimation of propagule numbers, *J. Gen. Microbiol.*, 42, 257, 1966.
316. **Dick, M. W.,** The ecology of aquatic Phycomycetes, in *Recent Advances in Aquatic Mycology*, Jones, E. B. G., Ed., Elek Science, London, 1976, 513.
317. **Dickinson, C. H. and Preece, T. F., Eds.,** *Microbiology of Aerial Plant Surfaces*, Academic Press, New York and London, 1976.
318. **Dickinson, C. H.and Pugh, G. J. F., Eds.,** *Ecology of Plant Litter Decomposition*, Vol. 1 and 2, Academic Press, New York, 1974.
319. **Diehl, W. W.,** A basis for mycogeography, *J. Wash. Acad. Sci.*, 27, 244, 1937.
320. **Diehl, W. W.,** *Endogone* as animal food, *Science*, 90, 442, 1939.
321. Difco Laboratories, Difco Manual of Dehydrated Culture Media and Reagents for Microbiological and Clinical Laboratory Procedures, Difco Laboratories, Detroit, 1953.
322. **Dodge, B. O.,** Some problems in the genetics of the fungi, *Science*, 90, 379, 1939.
323. **Dodge, B. O.,** The fungi come into their own, *Mycologia*, 44, 273, 1952.
324. **Dodge, C. W.,** *Medical Mycology*, C. V. Mosby, St. Louis, 1935.
325. **Dobzhansky, T., Cooper, D. M., Phaff, H. J., Knapp, E. P., and Carson, H. L.,** Studies on the ecology of *Drosphila* in the Yosemite region of California. IV. Differential attraction of species of *Drosophila* to different species of yeasts, *Ecology*, 37, 544, 1956.
326. **Doeksen, J. and van der Drift, J., Eds.,** *Soil Organisms*, North-Holland, Amsterdam, 1963.
327. **Domsch, K. H. and Gams, W.,** Variability and potential of a soil fungus population to decompose pectin, xylan and carboxymethylcellulose, *Soil Biol. Biochem.*, 1, 29, 1969.
328. **Domsch, K. H. and Gams, W.,** Fungi in Agricultural Soils, (trans. by Hudson, P. S. from the German *Pilze aus Agrarböden*, Gustav Fischer Verlag, Stuttgart, 1970), Longman Group Ltd., London, 1972.
329. **Domsett, J. A. and Reid, J.,** Light microscope observations on the trapping of nematodes by *Dactylaria candida*, *Can. J. Bot.*, 55, 2956, 1977.
330. **Domsett, J. A. and Reid, J.,** Transmission and scanning electron microscope observations on the trapping of nematodes by *Dactylaria candida*, *Can. J. Bot.*, 55, 2963, 1977.
331. **Dotson, G. K., Dean, R. B., Cooke, W. B., and Kenner, B. A.,** Land spreading, a conserving and non-polluting method of disposal of oily wastes, *Proc. 5th Int. Water Pollution Control Conf.*, Vol. 1, Section II-36, Pergamon Press, New York, 1971, 1.
332. **Dowding, P. and Widden, P.,** Some relationships between fungi and their environment in tundra regions, in *Soil Organisms and Decomposition in Tundra*, Tundra Biome Steering Committee, Stockholm, 1974, 123.
333. **Driver, C. H. and Wheeler, H. E.,** A sexual hormone in *Glomerella*, *Mycologia*, 47, 311, 1955.
334. **Druett, H. A.,** The inhalation and retention of particulates in the human respiratory system, in *Airborne Microbes*, Gregory, P. H. and Monteith, J. L., Eds., Cambridge University Press, Cambridge, 1967, 165.
335. **Duddington, C. L.,** The ecology of predacious fungi. I. Preliminary survey, *Trans. Br. Mycol. Soc.*, 34, 322, 1951.
336. **Duddington, C. L.,** The predacious fungi: Zoopagales and Moniliales, *Biol. Rev.*, 31, 152, 1956.
337. **Duddington, C. L.,** *The Friendly Fungi*, Faber & Faber, London, 1957.
338. **Duddington, C. L.,** The predacious fungi and their place in microbial ecology, in *Microbial Ecology*, Williams, R. E. O. and Spicer, E. C., Ed., Cambridge University Press, Cambridge, 1957, 218.
339. **Duddington, C. L.,** Predacious fungi and soil nematodes, in *Soil organisms*, Doeksen, J. and van der Drift, J., Eds., North-Holland, Amsterdam, 1963, 298.
340. **Duddington, C. L.,** Fungal parasites of invertebrates: predacious fungi, in *The Fungi, An Advanced Treatise*, Vol. 3, Ainsworth, G. C. and Sussman, A. S., Eds., Academic Press, New York, 1968, 239.

341. **Duddington, C. L. and Dixon, S. M.**, Permanent preparations of fungi growing on agar, *Nature (London)*, 168, 38, 1951.
342. **Dulaney, D. L.**, Formation of extracellular lysine by *Ustilago maydis* and *Gliocladium* sp., *Can. J. Microbiol.*, 3, 467, 1957.
343. **Dunnington, E. A.**, Oyster parasite distribution studied in Maryland waters, *Maryland Tidewater News*, 12(9), 1, 1956.
344. **Durbin, R. D.**, Techniques for the observation and isolation of soil microorganisms, *Bot. Rev.*, 27, 522, 1961.
345. **Durrell, L. W.**, The composition and structure of walls of dark fungus spores, *Mycopathol. Mycol. Appl.*, 23, 339, 1964.
346. **Durrell, L. W.**, Hyphal invasion by *Trichoderma viride*, *Mycopathol. Mycol. Appl.*, 35, 138, 1968.
347. **Durrell, L. W. and Shields, L. M.**, Fungi isolated in culture from soils of the Nevada test site, *Mycologia*, 52, 636, 1960.
348. **Dye, M. H.**, Basidiocarp development and spore release by *Stereum purpureum* in the field, *N.Z. J. Agric. Res.*, 17, 93, 1974.
349. **Dye, M. H. and Vernon, T. R.**, Air-borne mould spores, *N.Z. J. Sci. Technol. Sect. B*, 34, 118, 1952.
350. **Dykstra, T. E.**, Production of disease-free seed, *Bot. Rev.*, 27, 115, 1961.
351. **Ehrlich, J.**, Etiological terminology, *Chron. Bot.*, 6, 248, 1941.
352. **Ehrlich, M. A. and Ehrlich, H. G.**, Fine-structure of the host-parasite interface in mycoparasitism, *Annu. Rev. Phytopathol.*, 9, 155, 1971.
353. **Ellis, D. H. and Griffiths, D. A.**, The location and analysis of melanins in the cell walls of some fungi, *Can. J. Microbiol.*, 20, 1379, 1974.
354. **Ellis, M. B., Ellis, E. A., and Ellis, J. P.**, British marsh and fen fungi I, *Trans. Br. Mycol. Soc.*, 24, 147, 1951.
355. **Emerson, R.**, An experimental study in the life cycles and taxonomy of *Allomyces*, *Lloydia*, 4, 77, 1941.
356. **Emerson, R.**, Current trends of experimental research on the aquatic Phcomycetes, *Annu. Rev. Microbiol.*, 1950, 169, 1950.
357. **Emerson, R.**, Mycological organization, *Mycologia*, 50, 589, 1958.
358. **Emerson, R.**, Thermophiles, in *The Fungi, An Advanced Treatise*, Vol. 3, Ainsworth, G. C. and Sussman, A. S., Eds., Academic Press, New York, 1968, 105.
359. **Emerson, R. and Held, A. A.**, *Aqualinderella fermentans* gen. et sp. n., a Phycomycete adapted to stagnant waters. II. Isolation, cultural characteristics and gas relations, *Am. J. Bot.*, 56, 1103, 1969.
360. **Emerson, R. and Weston, W. H.**, *Aqualinderella fermentans* gen. et sp. nov., a Phycomycete adapted to stagnant waters. I. Morphology and occurrence in nature, *Am. J. Bot.*, 54, 702, 1967.
361. **Emerson, S.**, The growth phase in *Neurospora* corresponding to the logarithmic phase in unicellular organisms, *J. Bacteriol.*, 60, 22, 1950.
362. **Emmons, C. W.**, Isolation of *Histoplasma capsulatum* from soil, *Public Health Rep.*, 64, 892, 1949.
363. **Emmons, C. W.**, Natural occurrence of opportunistic fungi, *Lab. Invest.*, 11, 1026, 1962.
364. **Emmons, C. W., Binford, C. H., Utz, J. P., and Kwon-Chung, K. J.**, *Medical Mycology*, 3rd ed., Lea & Febiger, Philadelphia, 1977.
365. **Engelbrecht, R. S., Foster, D. H., Greening, E. O., and Lee, S. H.**, New microbial indicators of wastewater chlorination efficiency *Environ. Prot. Technol. Ser.*, No. 670/2-73-082, 1974.
366. **English, A. R.**, Laboratory evaluation of new semisynthetic penicillins, *Dev. Ind. Microbiol.*, 16, 175, 1975.
367. **Eriksson, J.**, Studies in the Heterobasidiomycetes and the Homobasidiomycetes — Aphyllophorales of the Muddus National Park in north Sweden, *Symb. Bot. Ups.*, 16, 1, 1958.
368. **Ernst, H. C. and Wolbach, S. B.**, Ultra-violet photomicrography, *J. Med. Res.*, 14, 463, 1906.
369. **Evans, E.**, Soil recolonization tube for studying recolonization of sterilized soil by microorganisms, *Nature (London)*, 173, 1196, 1954.
370. **Eyerdam, W. J.**, A giant bracket fungus found in south eastern Alaska, *Sven. Bot. Tidskr.*, 46, 131, 1952.
371. **Falck, R.**, Die *Merulius*-Fäule des Bauholzes, in A. Moller: *Hausschwammforschungen in Amtlichen Auftrage*, 6, 1, 1912.
372. **Farley, J. D. and Riedel, R. M.**, Soil Fumigation, Bull. No. L-249, Cooperative Extension Service, Ohio State University, Columbus, 1976.
373. **Farr, M. L.**, Trans., *Einführung in die Mykologie*, (trans. of Alexopoulos, C. J., *Introductory Mycology*), Gustav Fischer Verlag, Stuttgart, 1966.
374. **Farrow, M. W.**, Tropical soil fungi, *Mycologia*, 46, 632, 1954.

375. **Favre, J.**, Les champignons superieurs de la zone alpine de Parc Nationale Suisse, Resultats des recherches scientifiques enterprises au Parc Nationale Suiss, Part 5, New Series 33, 1955, 1.
376. **Fawcett, H. S.**, Fungus and bacterial diseases of insects as factors in biological control, *Bot. Rev.*, 10, 327, 1944.
377. **Fayret, J. and Parguey-Leduc, A.**, Thermo-inhibition du dévelopment du saprophyte au cours de la maturation des perethèces de *Gnomonia leptostyla* (Fr.) Ces. and deNot., *Rev. Mycol.*, 40, 245, 1976. 1976.
378. **Fehér, D.**, *Die Mikrobiologie des Waldbodens*, J. Springer, Berlin, 1923.
379. **Fell, J. W.**, Yeasts in oceanic regions, in *Recent Advances in Aquatic Mycology*, Jones, E. B. G., Ed., Elek Science, London, 1976, 93.
380. **Fennell, D. I.**, Conservation of fungus cultures, *Bot. Rev.*, 26, 79, 1960.
381. **Fischer, G. W.**, The longevity of smut spores in herbarium specimens, *Phytopathology*, 26, 1118, 1936.
382. **Fischer, G. W.**, *Manual of North American Smut Fungi*, Ronald Press, New York, 1953.
383. **Fischer, G. W. and Holton, C. S.**, *Biology and Control of the Smut Fungi*, Ronald Press, New York, 1957.
384. **Fjerdingstad, E.**, Taxonomy and saprobic valency of benthic phytomicroorganisms, *Int. Rev. Gesamten Hydrobiol.*, 50, 475, 1965.
385. **Fjerdingstad, E.**, Microbial criteria of environmental qualities, *Annu. Rev. Microbiol.*, 25, 563, 1971.
386. **Flanagan, P. W. and Scarborough, M.**, Physiological groups of decomposer fungi on tundra plant remains, in *Soil Organisms and Decomposition in Tundra*, Tundra Biome Steering Committee, Stockholm, 1974, 159.
387. **Flanagan, P. W. and Veam, K.**, Relationships between respiration, weight loss, temperature and moisture in organic residues on tundra, in *Soil Organisms and Decomposition in Tundra*, Tundra Biome Steering Committee, Stockholm, 1974, 249.
388. **Flor, H. H.**, Current status of the gene-for-gene concept, *Annu. Rev. Phytopathol.*, 9, 275, 1971.
389. **Fogel, R.**, Insect Mycophagy: A Preliminary Bibliography, Gen. Tech. Rep. No. PNW-36, US. Forest Service, Portland, Oregon, 1975.
390. **Fogel, R.**, Ecological studies of hypogaeous fungi. II. Sporocarp phenology in a western Oregon Douglas fir stand, *Can. J. Bot.*, 54, 1152, 1976.
391. **Fogel, R. and Cromack, K., Jr.**, Effect of habitat and substrate quality on Douglas fir litter decomposition in western Oregon, *Can. J. Bot.*, 55, 1632, 1977.
392. **Foister, C. E.**, The relation of weather to fungus and bacterial diseases, *Bot. Rev.*, 1, 497, 1935.
393. **Foister, C. E.**, The relation of weather to fungus diseases of plants, *Bot. Rev.*, 12, 548, 1946.
394. **Foster, J. W.**, The heavy-metal nutrition of fungi, *Bot. Rev.*, 5, 207, 1939.
395. **Foster, J. W.**, *Chemical Activities of the Fungi*, Academic Press, New York, 1949.
396. **Fowler, W. B.**, Measurement of seasonal air temperatures near the soil surface, *J. Range Manage.*, 24, 158, 1971.
397. **Frankland, J. C.**, Estimation of live fungal biomass, *Soil Biol. Biochem.*, 7, 339, 1975.
398. **Frankland, J. C.**, Fungal decomposition of leaf litter in a deciduous woodland, *Biodegradation and Humification*, Kilbertus, G., Reisinger, O., Mourey, A., and Cancela da Fonseca, J. A., Eds., Rapport du Ier Colloque International—Nancy, 1974, Universite de Nancy 1, Nancy, France, 1975, 33.
399. **Franz, H.**, Biozoenotische und Synecologische Untersuchungen über die Bodenfauna und ihre beziehungen und mikro- und makroflora, in *Soil Organisms*, Doeksen, J. and van der Drift, J., Eds., North-Holland, Amsterdam, 1963, 345.
400. **Fraser, L.**, The distribution of sooty-mold fungi and its relation to certain aspects of their physiology, *Proc. Linn. Soc. N.S.W.*, 62, 25, 1937.
401. **Fries, E. M.**, *Systema Mycologicum*, Vol. 1 to 3, Johnson Reprint Corp., New York, New York, 1821 to 1832.
402. **Fries, N. E.**, Uber die Sporenkeimung bei einigen Gasteromyceten und mykorrhizanbildenen Hymenomyceten, *Ark. Mikrobiol.*, 12, 266, 1941.
403. **Fries, N. E.**, Untersuchungen über Sporenkeimung und Mycelentwicklung Bodenbewohnender Hymenomyceten, *Symb. Bot. Ups.*, 6, 1, 1943.
404. **Fries, N. E.**, The nutrition of fungi from the aspect of growth factor requirements, *Trans. Br. Mycol. Soc.*, 30, 118, 1948.
405. **Fry, B. A.**, *The Nitrogen Metabolism of Microorganisms*, Methuen, London, 1955.
406. **Fuller, M. S.**, The zoospore, hallmark of the aquatic fungi, *Mycologia*, 69, 1, 1977.
407. **Fullerton, R. A., Scott, D. J., and Graham, G. J.**, Effect of fungicides on the control of head smut of maize, and the relationship between infection level and crop yield, *N.Z. J. Agric. Res.*, 2, 177, 1974.
408. **Fulling, E. H.**, Plant life and the law of man. IV. Barberry, currant and gooseberry, and cedar control, *Bot. Rev.*, 9, 483, 1943.

409. **Furman, T. E. and Trappe, J. M.**, Phylogeny and ecology of mycotrophic achlorophyllous angiosperms, *Q. Rev. Bio.*, 46, 219, 1971.
410. **Gaertner, A.**, Vorkommen und Verteilung mariner niederer Pilze in der Nordsee und im Nordmeerbecken, *Veroeff. Inst. Meeresforsch. Bremerhaven* (Suppl.), 5, 123, 1974.
411. **Gaertner, A., Schaumann, K., and Ulken, A.**, Final discussion of the 2nd International Symposium on Marine Mycology, *Veroeff. Inst. Meeresforsch. Bremerhaven* (Suppl.), 5, 519, 1974.
412. **Gäumann, E. A.**, *Vergleichende Morphologie der Pilze*, Gustav Fischer Verlag, Jena, 1926.
413. **Gäumann, E. A.**, *Pflanzliche Infektionslehre*, Verlag Birkhäuser, Basel, 1946; Brierley, W. B., *Principles of Plant Infection*, (trans.), Hafner Press, New York.
414. **Gäumann, E. A.**, *Plant Infection Handbook* (authorized trans.), Hafner Press, New York, 1949.
415. **Gäumann, E. A.**, *The Fungi, A Description of their Morphological Features and Evolutionary Development*, (trans. by Wynd, F. L.), Hafner Press, Ne York, 1952.
416. **Gäumann, E. A. and Dodge, C. W.**, *Comparative Morphology of Fungi*, McGraw-Hill, New York, 1928.
417. **Gams, W.**, Isolierung von Hyphen aus dem Boden, *Sydowia Ann. Mycol.*, Ser. II, 13, 87, 1959.
418. **Gams, W.**, Studium cellulolytischer Bodenpilze mit Hilfe der Zellophanstreifen-Methode und mit Carboxymethyl-Zellulose, *Sydowia Ann. Mycol.*, Ser. II, 14, 295, 1960.
419. **Gams, W.**, Winterliches Pilzwachstum in Boden, *Sydowia, Ann. Mycol.*, Ser. II, 14, 288, 1960.
420. **Gams, W.**, *Mycelium radicis atrovirens* in forest soils, isolation from soil microhabitats and identification, in *Soil Organisms*, Doeksen, J. and van der Drift, J., Eds., North-Holland, Amsterdam, 1963, 176.
421. **Gams, W. and Domsch, K. H.**, Beitrage zur Anwendung der Bodenwaschtechnik fur die Isolierung von Bodenpilzen, *Ark. Mikrobiol.*, 58, 134, 1967.
422. **Gams, W. and Domsch, K. H.**, The spatial and seasonal distribution of microscopic fungi in arable soils, *Trans. Br. Mycol. Soc.*, 52, 301, 1969.
423. **Gams, W., Domsch, K. H., and Weber, E.**, Nachweis signifikand verschiedener Pilzpopulationen bei gleicher Bodennutzung, *Plant Soil*, 31, 439, 1969.
424. **Garner, J. H. B.**, Some notes on the study of bark fungi, *Can. J. Bot.*, 45, 540, 1967.
425. **Garrett, S. D.**, A century of root disease investigation, *Ann. Appl. Biol.*, 42, 211, 1945.
426. **Garrett, S. D.**, Function of the mycelial strands in substrate colonization by the cultivated mushroom, *Psalliota hortensis*, *Trans. Br. Mycol. Soc.*, 37, 51, 1954.
427. **Garrett, S. D.**, *Biology of Root-Infecting Fungi*, Cambridge University Press, Cambridge, 1956.
428. **Garrett, S. D.**, Rhizomorph behaviour in *Armillaria mellea* (Vahl.) Quel. II. Logistics of infection, *Ann. Bot. (London)*, 20, 193, 1956.
429. **Garrett, S. D.**, Effect of a soil microflora selected by carbon disulphide fumigation on survival of *Armillaria mellea* in woody host tissues, *Can. J. Microbiol.*, 3, 135, 1957.
430. **Garrett, S. D.**, Inoculum potential as a factor limiting lethal action by *Trichoderma viride* Fr. on *Armillaria mellea* (Fr.) Quél., *Trans. Br. Mycol. Soc.*, 41, 157, 1958.
431. **Garrett, S. D.**, Decomposition of cellulose in soil by *Rhizoctonia solani* Kühn., *Trans. Br. Mycol. Soc.*, 45, 115, 1962.
432. **Garrett, S. D.**, *Soil Fungi and Soil Fertility*, Pergamon Press, London, 1963.
433. **Garrett, S. D.**, *Pathogenic Root-Infecting Fungi*, Cambridge University Press, Cambridge, 1970.
434. **Georg, L. K., Ajello, L., and Papageorge, C.**, Use of cycloheximide in the selective isolation of fungi pathogenic to man, *J. Lab. Clin. Med.*, 44, 422, 1954.
435. **Gerdemann, J. W. and Trappe, J. M.**, The Endogonaceae of the Pacific Northwest, *Mycol. Mem.*, 5, 1, 1974.
436. **Ghilarov, M. S.**, On the interrelations between soil dwelling invertebrates and soil microorganisms, in *Soil Organisms*, Doeksen, J. and van der Drift, J., Eds., North-Holland, Amsterdam, 1963, 255.
437. **Gilbert, E.-J.**, *La mycologie sur la terrain*, Libraire le Francois, Paris, 1928.
438. **Gilbertson, R. L.**, *Fungi that Decay Ponderosa Pine*, University of Arizona Press, Tucson, 1974.
439. **Gilbertson, R. L., Martin, K. J., and Lindsey, J. P.**, Annotated Check List and Host Index for Arizona Wood-Rotting Fungi, Tech. Bull. No. 209, University of Arizona Experiment Station, Tucson, Arizona, 1974.
440. **Gilbertson, R. L.**, Notes on some corticioid lignicolous fungi associated with snowbanks in southern Arizona, *Persoonia*, 7, 171, 1973.
441. **Gilbertson, R. L., Lombard, F. F., and Canfield, E. R.**, Gum Guaiac in Field Tests for Extracellular Phenol Oxidases of Wood-Rotting Fungi and Other Basidiomycetes, Res. Paper No. 269, U.S. Forest Service, Madison, Wisconsin, 1975.
442. **Gilbertson, R. L., Burdsall, H. H., Jr., and Canfield, E. R.**, Fungi that decay mesquite in southern Arizona, *Myxotaxon*, 3, 487, 1976.
443. **Gilbertson, R. L. and Lindsay, J. P.**, Basidiomycetes that decay junipers in Arizona, *Great Basin Nat.*, 35, 288, 1975.

444. **Gilliam, M. S.**, Periodicity of spore release in *Marasmius rotula*, *Mich. Bot.*, 14, 83, 1975.
445. **Gilman, J. C.**, *A Manual of Soil Fungi*, 2nd ed., Iowa State College Press, Ames, 1957.
446. **Goddard, H. N.**, Can fungi living in agricultural soil assimilate free nitrogen?, *Bot. Gaz. (Chicago)*, 56, 249, 1913.
447. **Godeau, S.**, Synecological relations of biocoenose components in aerotanks, *Hidrobiologia*, 14, 323, 1973.
448. **Goidanich, G., et al.**, Ricerche sulla alterazione nulla conservazione della pasta di legno destinate al fabbricazione della carta, Ente Nazionale per la Cellulosa et al Carta, Rome, 1938.
449. **Golding, N. S and Miller, D. D.**, The optimum temperature of growth of 22 cultures of *Oospora lactis*, *J. Dairy Sci.*, 26, 251, 1943.
450. **Goodwin, T. W.**, Fungal carotinoids, *Bot. Rev.*, 16, 291, 1952.
451. **Gordon, M. A. and Culp, H. B., Jr.**, Detection of *Histoplasma capsulatum* and other fungus spores in the environment by means of the membrane filter, *Mycologia*, 45, 241, 1953.
452. **Gordon, W. L.**, The occurrence of *Fusarium* species in Canada. II. Prevalence and taxonomy of *Fusarium* species in cereal seed, *Can. J. Bot.*, 30, 209, 1952.
453. **Gottlieb, D.**, The physiology of spore germination in fungi, *Bot. Rev.*, 16, 229, 1950.
454. **Gould, S.**, Numbers to supplement binomial nomenclature, *Am. Sci.*, 42, 269, 1954.
455. **Gould, S. W.**, Family names of the plant kingdom, *Int. Plant Index*, 1, 1, 1962.
456. **Grainger, Mary**, Some chemical aspects of the fungi, *Naturalist*, p. 153, Autumn 1942.
457. **Graustein, W. C., Cromack, K., Jr., and Sollins, P.**, Calcium oxalate: occurrence in soils and effect on nutrient and geochemical cycles, *Science*, 198, 1252, 1977.
458. **Gray, K. R. and Biddlestone, A. J.**, Decomposition of urban wastes, in *Biology of Plant Litter Decomposition*, Dickinson, C. H. and Pugh, G. J., Eds., Academic Press, New York, 1974, 743.
459. **Gray, T. R. G. and Bell, F.**, The decomposition of chitin in an acid soil, in *Soil Organisms*, Doekson, J. and van der Drift, J., Eds., North-Holland, Amsterdam, 1963, 222.
460. **Gray, T. R. G. and Williams, S. T.**, *Soil Micro-organisms*, Oliver & Boyd, Edinburgh, 1971.
461. **Gray, W. D.**, *The Relation of Fungi to Human Affairs*, Henry Holt, New York, 1959.
462. **Gray, W. D.**, Protein and world population increase, in Proc. 13th. Ann. Meet. Agric. Res. Inst., Washington, D.C., 1964, 97.
463. **Gray, W. D.**, Process for the Production of Fungal Protein, U.S. Patent 3,151,038, 1964.
464. **Gray, W. D.**, Fungi as a potential source of edible protein, *Activities Rep.*, 17, 66, 1965.
465. **Gray, W. D.**, *The Use of Fungi as Food and in Food Processing*, Chemical Rubber Co., Cleveland, 1970.
466. **Gray, W. D.**, *The Use of Fungi as Food and in Food Processing*, 2nd ed., Chemical Rubber Co., Cleveland, 1973.
467. **Gray, W. D. and Abou-El-Seoud, M.**, Fungi Imperfecti as sources of protein, *Dev. Ind. Microbiol.*, 7, 221, 1965.
468. **Gray, W. D. and Abou-El-Seoud, M.**, Fungal protein for food and feeds. II. Whole sweet potato as a substrate, *Econ. Bot.*, 20, 119, 1966.
469. **Gray, W. D. and Abou-El-Seoud, M.**, Fungal protein for food and feeds. III. Manioc as a potential crude raw material for tropical areas, *Econ. Bot.*, 20, 251, 1966.
470. **Gray, W. D. and Abou-El-Seoud, M.**, Fungal protein for food and feeds. IV. Whole sugar beets or beet pulp as a substrate, *Econ. Bot.*, 20, 372, 1966.
471. **Gray, W. D. and Bushnell, W. R.**, Biosynthetic potentialities of higher fungi, *Mycologia*, 47, 646, 1955.
472. **Gray, W. D. and Karve, M. D.**, Fungal protein for food and feeds. V. Rice as a source of carbohydrate for the production of fungal protein, *Econ. Bot.*, 21, 110, 1967.
473. **Gray, W. D., Och, F. F., and Abou-El-Seoud, M.**, Fungi Imperfecti as a potential source of edible protein, *Dev. Ind. Microbiol.*, 5, 384, 1964.
474. **Gray, W. D. and Paugh, R.**, Fungal protein for food and feeds. VI. Direct use of cane juice, *Econ. Bot.*, 21, 273, 1967.
475. **Gregory, P. H.**, *The Microbiology of the Atmosphere*, Leonard Hill Books, London, 1961.
476. **Gregory, P. H.**, Dispersal, in *The Fungi, An Advanced Treatise*, Vol. 2, Ainsworth, G. C. and Sussman, A. S., Eds., Academic Press, New York, 1966, 709.
477. **Gregory, P. H. and Henden, D. R.**, Terminal velocity of basidiospores of the giant puffball (*Lycoperdon giganteum*), *Trans. Br. Mycol. Soc.*, 67, 399, 1976.
478. **Gregory, P. H., Hirst, J. M., and Last, F. T.**, Concentrations of the basidiospores of the dry rot fungus (*Merulius lacrymans*) in the air of buildings, *Acta Allergol.*, 6, 168, 1953.
479. **Gregory, P. H. and Monteith, J. L., Eds.**, *Airborne Microbes*, Cambridge University Press, Cambridge, 1967.

480. **Gremmen, J.**, A new, crystalline antibiotic substance produced by *Mollisia* species (Discomycetes), *Antonie van Leeuwenhoek J. Microbiol. Serol.*, 22, 58, 1956.
481. **Gremmen, J.**, Microfungi decomposing organic remains of pines, *Fungus*, 27, 34, 1957.
482. **Griffin, D. M.**, *Ecology of Soil Fungi*, Chapman & Hall, London, 1972.
483. **Griffin, H. D.**, The genus *Ceratocystis* in Ontario, *Can. J. Bot.*, 46, 689, 1968.
484. **Groves, J. W. and Skolko, A. J.**, Notes on seed-borne fungi. I. *Stemphylium*, *Can. J. Res. Sect. C*, 22, 190, 1944.
485. **Grundberg, E. and Titsworth, E.**, Laboratory evaluation of antifungal agents, *Dev. Ind. Microbiol.*, 16, 201, 1975.
486. **Guilliermond, A.**, Sexuality, development cycle and phylogeny of yeasts, *Bot. Rev.*, 6, 1, 1940.
487. **Guzmán, G.**, Some distributional relationships between Mexican and American mycoflora, *Mycologia*, 65, 1319, 1973.
488. **Haard, R. T. and Kramer, C. L.**, Periodicity of spore discharge in the Hymenomycetes, *Mycologia*, 62, 1145, 1970.
489. **Haas, H.**, Makromyzetenflora und Kernfäulebefall älteres Fichtenbestande auf der Schwäbischen Alb, *Mitt. Ver. Forstl. Stand. Forstp.*, 20, 50 1971.
490. **Haase, L. W.**, *Deutsche Einheitsverfahren zur Wasser-, Abwasser-, und Schlammuntersuchung Physikalische, chemische und bakteriologische Verfahren*, Verlag Chemie GMBH Weinheim/Bergstr., 1954.
491. **Hale, M. E., Jr.**, *The Biology of Lichens*, Edward Arnold, London, 1967.
492. **Hall, R.**, Molecular approaches to taxonomy of fungi, *Bot. Rev.*, 35, 285, 1969.
493. **Halleck, F. A.**, Industrial sterilization processes, *Dev. Ind. Microbiol.*, 18, 335, 1977.
494. **Hansen, H. N.**, The dual phenomenon in the Fungi Imperfecti, *Mycologia*, 30, 442, 1938.
495. **Hansen, H. N.**, Heterocaryosis and variability, *Phytopathology*, 32, 639, 1942.
496. **Hansen, H. N. and Smith, R. E.**, The mechanism of variation in imperfect fungi: *Botrytis cinerea*, *Phytopathology*, 22, 953, 1932.
497. **Hansen, H. N. and Snyder, W. C.**, Gaseous sterilization of biological materials for use as culture media, *Phytopathology*, 37, 369, 1947.
498. **Happ, G. M., Happ, C. M., and Barras, S. J.**, Bark beetle-fungal symbiosis. III. Ultrastructure of conidiogenesis in a *Sporothrix* ectosymbiont of the southern pine beetle, *Can. J. Bot.*, 53, 2702, 1975.
499. **Harkin, J. M., Larsen, M. J., and Bhat, J. R.**, Use of syringaldehyde for the detection of laccase in sporophores of wood rotting fungi, *Mycologia*, 66, 469, 1974.
500. **Harley, J. L.**, Mycorrhiza and soil ecology, *Biol. Rev.*, 23, 127, 1948.
501. **Harley, J. L.**, Mycorrhiza, in *Ecology of Soil-Borne Plant Pathogens*, Baker, K. F. and Snyder, W. C., Eds., University of California Press, Berkeley, 1965, 218.
502. **Harley, J. L.**, *The Biology of Mycorrhiza*, 2nd ed., Leonard Hill Books, London, 1969.
503. **Harley, J. L.**, Mycorrhiza, in *The Fungi, An Advanced Treatise*, Vol. 3, Ainsworth, G. C. and Sussman, A. S., Eds., Academic Press, New York, 1968, 139.
504. **Harley, J. L.**, Problems of mycotrophy, in *Endomycorrhizas*, Sanders, F. E., Mosse, B., and Tinker, P. B., Eds., Academic Press, New York, 1975, 1.
505. **Harley, J. L. and Waid, J. S.**, A method of studying active mycelia on living roots and other surfaces in the soil, *Trans. Br. Mycol. Soc.*, 38, 104, 1955.
506. **Harman, W. N.**, Benthic substrates, their effect on freshwater Mollusca, *Ecology*, 53, 271, 1972.
507. **Hartley, C.**, Fungicides for forest and shade trees and for forest products, *Bot. Rev.*, 16, 33, 1950.
508. **Harvey, J. V.**, A study of the water molds and Pythiums occurring in the soils of Chapel Hill, *J. Elisha Mitchell Sci. Soc.*, 41, 151, 1925.
509. **Harvey, J. V.**, Relationship of aquatic fungi to water pollution, *Sewage Ind. Wastes*, 24, 1159, 1952.
510. **Hasenclever, H. H. and Kocan, R. M.**, Serotypes in *Saccharomyces tellularis:* their relation to source of isolation, *Infect. Immun.*, 7, 610, 1973.
511. **Haskins, R. H. and Anastasiou, J.**, Comparisons of the survivals of *Aspergillus niger* spores lyophilized by various methods, *Mycologia*, 45, 523, 1953.
512. **Hatch, A. B. and Hatch, C. T.**, Some hymenomycetes forming mycorrhizae with *Pinus strobus* L., *J. Arnold Arbor. Harv. Univ.*, 14, 324, 1933.
513. **Hatch, W. R.**, Gametogenesis in *Allomyces*, *Ann. Bot. (London)*, 49, 623, 1935.
514. **Hatch, W. R.**, Conjugation and zygote germination in *Allomyces arbuscula*, *Ann. Bot. (London)*, 2, 583, 1938.
515. **Hatfield, G. W. and Brady, L. R.**, Toxins of higher fungi, *Lloydia* 38, 36, 1975.
516. **Hawker, L. E.**, *Physiology of Fungi*, University of London Press, London, 1950.
517. **Hawker, L. E.**, Ecological factors and the survival of fungi, in *Microbial Ecology*, Williams, R. E. O. and Spicer, E. C., Eds., Cambridge University Press, Cambridge, 1957, 238.
518. **Hawkes, H. A.**, Biological detection and assessment of river pollution, *Effluent Water Treat. J.*, 1963, 651, 1963.

519. **Hawkes, H. A.**, *The Ecology of Waste Water Treatment*, Pergamon Press, London and New York, distributed in the Western hemisphere by MacMillan Co., New York, 1963.
520. **Hawkes, H. A.**, The ecology of sewage bacteria beds, in *Ecology and the Industrial Society*, Goodman, G. T., Ed., Blackwell Scientific, Oxford, 1965, 119.
521. **Hawkes, H. A.**, Factors influencing the seasonal incidence of fungal growths in sewage bacteria beds, *Int. J. Air Water Pollut.*, 9, 693, 1965.
522. **Haynes, W. C., Wickerham, L. J., and Hesseltine, C. W.**, Maintenance of cultures of industrially important microorganisms, *Appl. Microbiol.*, 3, 361, 1955.
523. **Heal, O. W.**, Soil fungi as food for amoebae, in *Soil Organisms*, Doeksen, J. and van der Drift, J., Eds., North-Holland, Amsterdam, 1963, 289.
524. **Heald, F. D.**, *Introduction to Plant Pathology*, McGraw-Hill, New York, 1943.
525. **Heath, I. B.**, Ultrastructure of freshwater Phycomycetes, in *Recent Advances in Aquatic Mycology*, Jones, E. B. G., Ed., Elek Science, London, 1976, 603.
526. **Hedrick, L. R., Cook, W., and Woollett, L.**, Yeasts and molds in Lake Superior water and some of its tributaries, in *Proc. 11th Conf. Great Lakes Res.*, Ann Arbor, Michigan, 1968, 538.
527. **Hedrick, L. R. and Soyugenc, M.**, Yeasts recovered from mud samples from Lake Michigan, *Great Lakes Res. Div. Univ. Mich. Publ.*, 13, 69, 1965.
528. **Hedrick, L. R. and Soyugenc, M.**, Yeasts and molds in water and sediments of Lake Ontario, in *Proc. 10th Conf. Great Lakes Res.*, Ann Arbor, Michigan, 1967, 20.
529. **Hedrick, L. R., Soyugenc, M., du Pont, P., and Ambriosini, R.**, Yeasts in Lake Michigan and Lake Erie, *Great Lakes Res. Div.,Univ. Mich. Publ.*, 11, 77, 1964.
530. **Hedrick, L. R., Soyugenc, M., and Larsen, L.**, Yeasts in sediment core samples from Lake Michigan, *Great Lakes, Res. Div., Univ. Mich. Publ.*, 15, 27, 1966.
531. **Heim, R.**, Études descriptives et experimentales sur les Agarics termatophiles d'Afrique tropicales, *Mem. Acad. Sci. Inst. Fr.*, 64, 1, 1941.
532. **Heim, R.**, Phylogeny and natural classification of macro-fungi, *Trans. Br. Mycol. Soc.*, 30, 161, 1948.
533. **Henderson, M. E. K. and Farmer, V. G.**, Utilization by soil fungi of para-hydroxybenzaldehyde, ferulic acid, syringaldehyde and vanillin, *J. Gen. Microbiol.*, 12, 37, 1955.
534. **Hering, T. F.**, Growth of litter fungi in a forest soil, in *Soil Organisms*, Doeksen, J. and van der Drift, J., Eds., North-Holland, Amsterdam, 1963, 183.
535. **Hering, T. F.** The terricolous higher fungi of four Lake District woodlands, *Trans. Br. Mycol. Soc.*, 49, 368, 1966.
536. **Hering, T. F.**, Fungal decomposition of oak leaf litter, *Trans. Br. Mycol. Soc.*, 50, 267, 1967.
537. **Hering, T. F.**, Fungal associations in broad-leaved woodlands in north-west England, *Mycopathol. Mycol. Appl.*, 48, 15, 1972.
538. **Herrick, J. A.**, An unusually large *Fomes*, *Mycologia*, 45, 622, 1953.
539. **Hervey, A., Rogerson, C. T., and Leong, I.**, Studies on fungi cultivated by ants, *Brittonia*, 29, 226, 1977.
540. **Herzberg, M., Ed.**, Proc. 1st U.S. - Japan Conf. Toxic Micro-organisms, Mycotoxins — Botulism, U.S.-J.N.R. Joint Panels on Toxic Microorganisms and the U.S. Department of the Interior, Washington, D.C., 1970.
541. **Hesseltine, C. W., Swain, E. W., and Wang, H. L.**, Production of fungal spores as inocula for oriental fermented foods, *Dev. Ind. Microbiol.*, 17, 101, 1976.
542. **Higinbotham, N. L. and Powers, E. L.**, Some factors affecting the relative amounts of pigmentation in *Coccosporium* sp., *Am. J. Bot.*, 34, 483, 1947.
543. **Hilborn, M. T.**, The calorific value of decayed cordwood, *Phytopathology*, 26, 906, 1936.
544. **Hildebrand, E. M.**, Techniques for the isolation of single microorganisms, *Bot. Rev.*, 4, 627, 1938.
545. **Hildebrand, E. M.**, Techniques for the isolation of single microorganisms. II, *Bot. Rev.*, 16, 181, 1950.
546. **Hinds, T. E. and Krebil, R. G.**, Wounds and canker diseases in western aspen, *U.S. For. Serv. For. Pest. Leafl.*, p. 152, 1975.
547. **Hintikka, V.**, Psychrophilic Basidiomycetes decomposing forest litter under winter conditions, *Commun. Inst. For. Fenn.*, 59(2), 1, 1964.
548. **Hintikka, V.**, Studies on white-rot humus formed by higher fungi in forest soils, *Commun. Inst. For. Fenn.*, 69(2), 61, 1970.
549. **Hirst, J. M.**, An automatic volumetric spore trap, *Ann. Appl. Biol.*, 39, 257, 1952.
550. **Hirst, J. M. and Hurst, G. W.**, Long distance spore transport, in *Airborne Microbes*, Gregory, P. H. and Monteith, J. L., Eds., Cambridge University Press, Cambridge, 1967, 307.
551. **Ho, I. and Trappe, J. M.**, Translocation of ^{14}C from *Festuca* plants to their endomycorrhizal fungi, *Nature (London) New Biol.*, 244, 30, 1973.

552. **Ho, I. and Trappe, J. M.**, Nitrate reducing caacity of two vescicular-arbuscular fungi, *Mycologia*, 67, 886, 1975.
553. **Höfler, K.**, Über Pilzaspekt, *Vegetatio*, 5—6, 373, 1954.
554. **Höfler, K.**, Über Pilzsoziologie, *Verh. Zool. Bot. Ges. Wien*, 95, 58, 1955.
555. **Höfler, K.**, Pilzkäfer und Käferpilze, *Verh. Zool. Bot. Ges. Wien*, 100, 74, 1960.
556. **Höfler, K. and Packseider, E.**, Fluoreszenzmikroscopische Beobachtungen an höhren Pilze, *Oesterr. Bot. Z.*, 94, 99, 1947.
557. **Höhnk, W. and Bock, K. J.**, Ein Beitrag zur Okologie der saprophytischen Wasserpilze, *Vereff. Inst. Meeresforsch. Bremerhaven*, 3, 9, 1954.
558. **Holding, A. J., Heal, O. W., MacLean, S. F., Jr., and Flanagan, P. W.**, Eds., *Soil Organisms and Decomposition in Tundra*, Tundra Biome Steering Committee, Stockholm, 1974.
559. **Hollings, M. and Stone, O. M.**, Viruses that infect fungi, *Annu. Rev. Phytopathol.*, 9, 93, 1971.
560. **Holtman, D. F.**, Antibiotic products of fungi, *Bot. Rev.*, 13, 59, 1947.
561. **Holton, C. E.**, Physiologic specialization and genetics of the smut fungi. II, *Bot. Rev.*, 19, 187, 1953.
562. **de Hoog, G. S. and Hemanides-Nijhoff, E. L.**, The black yeasts and allied Hyphomycetes, *Stud. Mycol.*, 11, 1977.
563. **Hora, F. B.**, Productivity of toadstools in coniferous plantations natural and experimental, *Mycopathol. Mycol. Appl.*, 48, 35, 1972.
564. **Horsfall, J. G.**, *Fungicides and their Action*, Chronica Botanica, Waltham, Mass., 1945.
565. **Horsfall, J. G. and Dimond, A. E.**, *Plant Pathology, An Advanced Treatise*, Academic Press, Vol. 1 to 3, New York, 1959 to 1960.
566. **Hubert, E. E.**, *An Outline of Forest Pathlogy*, John Wiley & Sons, New York, 1931.
567. **Hudson, H. J.**, *Fungal Saprophytism*, Edward Arnold Ltd., London, 1972.
568. **Hudson, H. J. and Webster, J.**, Succession of fungi on decaying stems on *Agropyron repens*, *Trans. Br. Mycol. Soc.*, 41, 165, 1958.
569. **Hudson, P. S., Trans.**, *Fungi in Agricultural Soils*, (trans. of Domsch and Gams, *Pilze aus Agrarböden*), Longman Group, Ltd., London, 1972.
570. **Hueck, J.**, Myco-sociological methods of investigation, *Vegetatio*, 4, 84, 1952.
571. **Hueck, H. J.**, The biological deterioration of plastics, *Plastics*, 25, 419, 1960.
572. **Hütter, R. and DeMoss, J. A.**, Organization of the tryptophan pathway: a phylogenetic study of the fungi, *J. Bacteriol.*, 94, 1896, 1967.
573. **Hughes, G. C.**, Seasonal periodicity of the Saprolegniaceae in the south-eastern United States, *Trans. Br. Mycol. Soc.*, 45, 519, 1962.
574. **Hughes, G. C.**, Geographical distribution of the higher marine fungi, *Veroeff. Inst. Meeresforsch. Bremerhaven Suppl.*, 5, 419, 1974.
575. **Hughes, G. C.**, Studies of fungi in oceans and estuaries since 1961. I. Lignicolous, caulicolous and foliicolous species, *Oceanogr. Mar. Biol.*, 13, 69, 1975.
576. **Hughes, S. J.**, Conidiophores, conidia and classification, *Can. J. Bot.*, 31, 577, 1953.
577. **Hughes, S. J.**, Sooty moulds, *Mycologia*, 68, 693, 1976.
578. **Hugon, L.**, Une intoxication par *Coprinus atramentarius*, *Rev. Mycol.*, 3 (Suppl.), 48, 1938.
579. **Humfeld, H. and Sugihara, T. F.**, The nutrient requirements of *Agaricus campestris* grown in submerged culture, *Mycologia*, 44, 605, 1952.
580. **Humphrey, A. L.**, Computer-assisted fermentation developments, *Dev. Ind. Microbiol.*, 18, 58, 1977.
581. **Hunt, C. B. and Durrell, L. W.**, Distribution of fungi and algae, in Plant Ecology of Death Valley, California, Prof. Paper No. 509, U.S. Geological Survey, Washinton, D.C., 1966, 55.
582. **Hunt, H. W.**, A simulation model for decomposition in grasslands, *Ecology*, 58, 469, 1977.
583. **Hunt, J.**, Taxonomy of the genus *Ceratocystis*, *Lloydia*, 19, 1, 1956.
584. **Hunt, N. R.**, Destructive plant diseases not yet established in North America, *Bot. Rev.*, 12, 593, 1946.
585. **Hynes, H. B. N.**, *The Biology of Polluted Waters*, Liverpool University Press, Liverpool, 1960.
586. **Ingold, C. T.**, Fungi and water, *Trans. Br. Mycol. Soc.*, 37, 97, 1954.
587. **Ingold, C. T.**, Fungi, in *Vistas in Botany*, Pergamon Press, London, 1959.
588. **Ingold, C. T.**, Spore release, in *The Fungi, An Advanced Treatise*, Vol. 2, Ainsworth, G. C. and Sussman, A. S., Eds., Academic Press, New York, London, 1966, 679.
589. **Ingold, C. T.**, *Fungal Spores, Their Liberation and Dispersal*, Clarendon Press, Oxford, 1971.
590. **Ingold, C. T.**, *Biology of Fungi*, 3rd rev. ed., Hutchinson Biology Monographs,London, 1973.
591. **Ingold, C. T.**, Convergent evolution in aquatic fungi: the tetraradiate spore, *Biol. J. Linn. Soc.*, 7, 1, 1975.
592. **Ingram, M.**, *An Introduction to the Biology of Yeasts*, Pitman Publishing, New York, 1955.
593. **Ingram, M.**, Microorganisms resisting high concentrations of sugars or salts, in *Microbial Ecology*, Williams, R. E. O. and Spicer, E. C., Eds., Cambridge University Press, Cambridge, 1957, 91.

594. **Ingram, W. M., Cooke, W. B., and Hagerty, L. T.**, Snails associated with sewage treatment installations, *Sewage Ind. Wastes*, 30, 821, 1958.
595. **Isaac, P. C. G. and James, A.**, Some aspects of the biology of trickling filters, scientific papers from the Institute of Chemical Technology, Prague, *Techn. Water*, 8, 123, 1964.
596. **Ivarson, K. C. and Sowden, F. J.**, Decomposition of forest litters. I. Production of ammonia and nitrate nitrogen, changes in microbial population, and rate of decomposition, *Plant Soil*, 11, 237, 1959.
597. **Jackson, H. S.**, Life cycles and phylogeny in the higher fungi, *Trans. R. Soc. Can. Sect. 3*, 38, 1, 1944.
598. **James, N.**, Plate counts of bacteria and fungi in a saline soil, *Can. J. Microbiol.*, 5, 431, 1959.
599. **Janson, B. F. and Miller, R. L.**, Backyard Fruit Sprays for Insects and Diseases, Bull. No. L-1, Cooperative Extension Service, Ohio State University, Columbus, 1977.
600. **Jennison, M. W., Newcomb, M. D., and Henderson, B.**, Physology of the wood-rotting Basidiomycetes. I. Growth and nutrition in submerged culture in synthetic media, *Mycologia*, 47, 275, 1955.
601. **Johnson, L. F. and Curl, E. A.**, *Methods for Research on the Ecology of Soil-Borne Plant Pathogens*, Burgess Publishing, Minneapolis, 1972.
602. **Johnson, T. W., Jr.**, A species of *Olpidiopsis* parasitic on *Achlya glomerata*, *J. Elisha Mitchell Sci. Soc.*, 71, 58, 1955.
603. **Johnson, T. W., Jr.**, Inoculation studies with a polysporangiate *Rozella* parasitic in *Dictyuchus anomalus*, *Am. J. Bot.*, 42, 119, 1955.
604. **Johnson, T. W., Jr.**, *The Genus Achlya: Morphology and Taxonomy*, University of Michigan Press, Ann Arbor, 1956.
605. **Johnson, T. W., Jr.**, Notes on fungi from Mississippi. I. Aquatic Phycomycetes, *Am. Midl. Nat.*, 55, 184, 1956.
606. **Johnson, T. W., Jr.**, Saprobic marine fungi, in *The Fungi, An Advanced Treatise*, Vol. 3, Ainsworth, G. C. and Sussman, A. C., Eds., Academic Press, New York, 1968, 95.
607. **Johnson, T. W., Jr. and Meyers, S. P.**, Literature on halophilous and halolimnic fungi, *Bull. Mar. Sci. Gulf Caribb.*, 7, 330, 1957.
608. **Johnson, T. W., Jr. and Sparrow, F. K.**, *Fungi in Oceans and Estuaries*, J. Cramer, Weinheim, 1964.
609. **Johnson, T. and Newton, M.**, Specialization, hybridization, and mutation in the cereal rusts, *Bot. Rev.*, 12, 337, 1946.
610. **Joffe, A. Z.**, The mycoflora of a continuously cropped soil in Israel, with special reference to effects of manuring and fertilizing, *Mycologia*, 55, 271, 1963.
611. **Joffe, A. Z.**, The presence of aflatoxin in kernels from five years ground nuts crops and of *Aspergillus flavus* isolates from kernels and soils, *Plant Soil*, 33, 91, 1970.
612. **Jones, D. J.**, *Introduction to Microfossils*, Harper and Brothers, New York, 1956.
613. **Jones, E. B. G., Ed.**, *Recent Advances in Aquatic Mycology*, Elek Science, London, 1976.
614. **Kalamees, K.**, On fungal resources in certain east Estonian forest types, *Eesti NSV Tead. Akad. Toim. Fiu. Mat.*, 59, 203, 1960.
615. **Kalamees, K.**, Seasonal changes in the fungal cover, *Eesti NSV Tead. Akad. Toim. Biol.*, 15, 216, 1966.
616. **Kalamees, K.**, Mycocoenological methods based on investigations in the Estonian forests, *Acta Mycol.*, 4, 327, 1968.
617. **Kalamees, K.**, Peculiarities in the growth and development of fruit bodies in some fungal species, *Est. Contrib. Int. Biol. Programme Prog. Rep.*, 3, 47, 1971.
618. **Kalamees, K.**, Some methodological aspects of mycocoenology, *Est. Contrib. Int. Biol. Programme Prog. Rep.*, 2, 33, 1971.
619. **Kalamees, K.**, Theoretical aspects of mycocoenology, *Eesti NSV Tead. Acad. Toim. Biol.*, 20, 150, 1971.
620. **Kalamees, K. and Kollom, A.**, On the biological productivity of the Agaricales in forest ecosystems, *Est. Contrib. Int. Biol. Programme Prog. Rep.*, 3, 29, 1971.
621. **Kamat, M. N., Rao, V. G., and Subramoniam, V.**, Some observations on the nature of host-fungus relationships of market pathogens, *Rev. Mycol.*, 41, 171, 1977.
622. **Kanashiro, T.**, Lignocellulosic agricultural wastes degraded by *Pleurotus ostreatus*, *Dev. Ind. Microbiol.*, 18, 591, 1977.
623. **Kapica, L. and Clifford, A.**, Quantitative determination of yeasts in sputum, *Mycopathol. Mycol. Appl.*, 34, 27, 1968.
624. **Kaplan, W. M.**, Practical application of fluorescent antibody procedures in medical mycology, *Pan Am. Health Organ. Sci. Publ.*, 304, 178, 1974.

625. **Kaplan, W.**, Epidemiology of the principal systemic mycoses of man and lower animals and the ecology of their etiologic agents, *J. Am. Vet. Med. Assoc.*, 163, 1043, 1973.

626. **Kaplan, W., Arnstein, P., Ajello, L., Chandler, F., Watts, J., and Huklin, M.**, Fatal aspergillosis in imported parrots, *Mycopathologia*, 56, 25, 1975

627. **Karg, W.**, Die edaphischen Acarina in ihren Bieziegunger zur Mikroflora und ihre Eigung als Anzeiger für Prozesse der Bodenbildung, in *Soil Organisms*, Doeksen, J. and van der Drift, J., Eds., North-Holland, Amsterdam, 1963, 304.

628. **Karling, J. S.**, Keratinophilic chytrids. I. *Rhizophidium keratinophilum* n. sp., a saprophyte isolated on human hair and its parasite *Phlyctidium mycetophagum* n. sp., *Am. J. Bot.*, 33, 751, 946.

629. **Karling, J. S.**, Host reaction, host-parasite relationship, and taxonomic criteria in *Synchytrium*, *Mycologia*, 46, 293, 1954.

630. **Karling, J. S.**, *The Plasmodiophorales*, Hafner Press, New York, 1968.

631. **Kaufman, L.**, Serodiagnosis of fungal diseases, in *Manual of Clinical Immunology*, Rose, N. F. and Friedman, H., Eds., American Society for Microbiology and American Association for Immunologists, Washington, D.C., 1976, 363.

632. **Keener, P. D.**, Mycoflora of buds. I. Results of cultures from nonirradiated materials of certain woody plants, *Am. J. Bot.*, 37, 520, 1950.

633. **Kelly, C. D., Pady, S. M., and Polunin, N.**, Aerobiological sampling methods from aircraft, *Can. J. Bot.*, 29, 206, 1951.

634. **Kendrick, W. B.**, The time factor in the decomposition of coniferous leaf litter *Can. J. Bot.*, 37, 907, 1959.

635. **Kendrick, W. B.**, Soil fungi of a copper swamp, *Can. J. Microbiol.*, 8, 639, 1962.

636. **Kendrick, W. B.**, Fungi associated with breakdown of pine leaf litter in the organic horizon of a podzol, *Mycopathol. Mycol. Appl.*, 19, 241, 1963.

637. **Kendrick, W. B., Ed.**, *Taxonomy of Fungi Imperfecti*, University of Toronto Press, Toronto, 1971.

638. **Kendrick, W. B. and Bärlocher, F.**, *Mycology* (trans. of Müller, E. and Löeffler, W.),Georg Thieme, Stuttgart, 1976.

639. **Kendrick, W. B. and Burges, A.**, Biological aspects of the decay of *Pinus sylvestris* leaf litter, *Nova Hedwigia Z. Kryptogamenkd.*, 4, 313, 1962.

640. **Kendrick, W. B. and Weresub, L.**, Attempting neo-adansonian computer taxonomy at the ordinal level in the Basidiomycetes, *Sept. Zool.*, 15, 307, 1966.

641. **Kern, F. D.**, The importance of taxonomic studies of the fungi, *Torreya*, 43, 65, 1943.

642. **Kessler, A.**, Survey of air-borne pollen and fungus spores in Israel, 1951—1952, *Ann. Allergy*, 11, 322, 1953.

643. **Keys, T. F., Halderson, A. M., Rhodes, H. H., Roberts, G. D., and Fifer, E. Z.**, Nosocomial outbreak of *Rhizopus* infections associated with Elastoplast wound dressings, *Morbidity Mortality Weekly Rep.*, 27, 33, 1978.

644. **Kiffer, E. and Mangenot, F.**, Activities cellulolytiques de quelques sols forestieres, *Ann. Inst. Pasteur (Paris)*, 115, 582, 1968.

645. **Kirk, T. K.**, Effects of microorganisms on lignin, *Annu. Rev. Phytopathol.*, 9, 185, 1971.

646. **Kishimoto, R. A. and Baker, G. E.**, Pathogenic and potentially pathogenic fungi isolated from beach sands and selected soils of Oahu, Hawaii, *Mycologia*, 61, 539, 1969.

647. **Kjøller, A. and Ødum, S.**, Evidence for longevity of seeds and microorganisms in permafrost, *Arctic*, 24, 230, 1974.

648. **Klausmeier, R. E., Osmon, J. L., and Wells, D. R.**, The effect of trinitrotoluene on microorganisms, *Dev. Ind. Microbiol.*, 15, 309, 1974.

649. **Klein, R. M. and Cronquist, A.**, A consideration of the evolutionary and taxonomic significance of some biochemical, micromorphological, and physiological characters in the Thallophytes, *Q. Rev. Biol.*, 42, 105, 1967.

650. **Kleinschmidt, W. J. and Ellis, L. F.**, Fungal viruses and interferon induction, *Dev. Ind. Microbiol.*, 16, 128, 1975.

651. **Koch, W. J.**, *Fungi in the Laboratory, A Manual and Text*, University of North Carolina Student Stores, Chapel Hill, 1972.

652. **Kohlmeyer, J.**, Deterioration of wood by marine fungi in the deep sea, *Mater. Perform. Deep Sea, Special Tech. Publ.*, 445, 20, 1969

653. **Kohlmeyer, J.**, The role of marine fungi in the penetration of calcareous substances, *Am. Zool.*, 9, 741, 1969.

654. **Kohlmeyer, J.**, New genera and species of higher fungi from the deep sea (1615—5313 m), *Rev. Mycol.*, 41, 198, 1977.

655. **Kole, A. P.**, Flagella, in *The Fungi, An Advanced Treatise*, Vol. 1, Ainsworth, G. C. and Sussman, A. S., Eds., Academic Press, New York, 1965, 77.

656. **Kolkwitz, R.**, Oekologie der Saprobien, *Schriftenr. Ver. Wasser, Boden, Lufthyg. Berlin Dahlem*, p. 4, 1950.
657. **Kolkwitz, R. and Marsson, M.**, Biologie des Trinkwassers, Abwassers und der Vorfluter, in *Hubner, Gruber und Ficker's Handbuch der Hygiene*, Vol. 2, Hirzel, Leipzig, 1911.
658. **Kotlaba, F.**, Ekologicko-sociologická studie o mykofloře "Sokěslavských blat", *Preslia*, 25, 305, 1953.
659. **Kramer, C. L. and Pady, S. M.**, Inhibition of growth of fungi on rose bengal media by light, *Trans. Kansas Acad. Sci.*, 64, 110, 161.
660. **Kramer, C. L. and Pady, S. M.**, A new 24-hour spore sampler, *Phytopathology*, 56, 517, 1966.
661. **Kramer, C. L. and Pady, S. M.**, Viability of airborne spores, *Mycologia*, 60, 448, 1968.
662. **Kubiena, W. L.**, *Micropedology*, Collegiate Press, Ames, Iowa, 1938.
663. **Kuehn, H. H. and Gunderson, M. F.**, Psychrophilic and mesophilic fungi in fruit-filled pastries, *Appl. Microbiol.*, 10, 354, 1962.
664. **Kuehn, H. H. and Gunderson, M. F.**, Psychrophilic and mesophilic fungi in frozen food products, *Appl. Microbiol.*, 11, 352, 1963.
665. **Kühnelt, W.**, Über ein Einfluss der Mycels von *Clitocybe infundibuliformis* auf die Streufauna, in *Soil Organisms*, Doeksen, J. and vander Drift, J., Eds., North-Holland, Amsterdam, 1963, 281.
666. **Kurtzman, C. P., Wickerham, L. J., and Hesseltine, C. W.**, Yeasts from wheat and flour, *Mycologia*, 62, 542, 1970.
667. **Kwon-Chung, K. J.**, *Emmonsiella capsulata:* perfect state of *Histoplasma capsulatum*, *Science*, 177, 368, 1972.
668. **Kwon-Chung, K. J., Weeks, R. J., and Larsh, H. W.**, Studies on *Emmonsiella capsulata (Histoplasma capsulatum)*, *Am. J. Epidemiol.*, 99, 44, 1974.
669. **Ladisch, M. R., Gong, C. S., and Tsao, G. T.**, Corn crop residues as a potential source of single cell protein: kinetics of *Trichoderma viride* cellobiase action, *Dev. Ind. Microbiol.*, 18, 157, 1977.
670. **Lalor, J. H., Chinnici, J. P., and Llewellyn, G. C.**, Effects of a fungal metabolite, aflatoxin B_1, on larval viability and gross morphology in *Drosophila melanogaster*, *Dev. Ind. Microbiol.*, 17, 443, 1976.
671. **Lange, L.**, The distribution of Macromycetes in Europe, *Dan. Bot. Ark.*, 30, 1, 1974.
672. **Lange, M.**, Macromycetes. III. Ecological and plant geographical studies. II. De Botaniscke Expedition Til Vastgrønland 1946, *Medd. Groenl.*, 148(2), 67, 1957.
673. **Lange, M.**, *Kuehneromyces mutabilis* with morchelloid hymenophore, *Bull. Soc. Linn. Lyon*, 43 (special number), 223, 1974.
674. **Langeron, M. and Vanbreuseghem, R.**, *Précisede Mycologie*, 2nd ed., Masson et Cie, Paris, 1952.
675. **Langeron, M. and Vanbreuseghem, R.**, *Outline of Mycology*, (trans. by J. Wilkinson), Sir Isaac Pitman & Sons, London, 1965.
676. **Lanjouw, J., Ed.**, *International Code of Botanical Nomenclature*, Edinburgh Edition, Regnum Vegetabile, 1966, 46.
677. **Large, E. C.**, *The Advance of the Fungi*, Henry Holt, New York, 1940.
678. **LaRocca, P. T. and La Rocca, R.**, Environmental air in the aseptic processing of food, *Dev. Ind. Microbiol.*, 16, 339, 1975.
679. **Larsh, H. W., Cozad, G. C., Hinton, A., and Furcolow, M. L.**, The mouse as an aid in the isolation of *Histoplasma capsulatum* and the effect of adjuvants, *U.S. Public Health Serv. Public Health Monogr.*, 39, 86, 1956.
680. **Larsh, H. W. and Hinton, A.**, Soil studies, *U.S. Public Health Serv. Public Health Monogr.*, 39, 251, 1956.
681. **Larsh, H. W., Hinton, A., and Furcolow, M. L.**, Laboratory studies of *Histoplasma capsulatum*. III. Efficiency of the flotation method in isolation of *Histoplasma capsulatum* from the soil, *U.S. Public Health Serv. Public Health Monogr.*, 39, 74, 1956.
682. **Laurence, J. F. and Powell, J. A.**, Host relationships in North American fungus feeding moths (Decophoridae, Oinophilidae, Tineidae), *Bull. Mus. Comp. Zool. Harv. Univ.*, 138(2), 29, 1969.
683. **Leach, C. M.**, The sporulation of *Helminthosporium oryzae* as affected by exposure to near-ultraviolet radiation and dark periods, *Can. J. Bot.*, 39, 705, 1961.
684. **Leach, C. M.**, Sporulation of diverse species of fungi under near-ultraviolet radiation, *Can. J. Bot*, 40, 151, 1962.
685. **Leach, C. M.**, Practical guide to the effects of visible and ultraviolet light on fungi, in *Methods in Microbiology*, Vol. 4, Booth, C., Ed., Academic Press, New York, 1971, 609.
686. **Lee, B. K. H. and Baker, G. E.**, Fungi associated with the roots of red mangrove, *Rhizohora mangle*, *Mycologia*, 65, 894, 1973.
687. **Leonian, L. H.**, A study of factors promoting pycnidia-formation in some Sphaeropsidales, *Am. J. Bot.*, 11, 19, 1924.

688. **Leonian, L. H. and Lilly, V. G.,** Studies on the nutrition of fungi. IV. Factors influencing the growth of some thiamin-requiring fungi, *Am. J. Bot.*, 27, 18, 1940.
689. **Leukel, R. W.,** The present status of seed treatment, with special reference to cereals, *Bot. Rev.*, 2, 498, 1936.
690. **Levine, M. and Schoenlein, H. W.,** *A Compilation of Culture Media for the Cultivation of Microorganisms*, Williams & Wilkins, Baltimore, 1930.
691. **Li, C. Y., Lu, K. C., Trappe, J. M., and Bollen, W. B.,** *Poria weirii*-inhibiting and other phenolic compounds in roots of red alder and Douglas fir, *Microbios*, 5, 65, 1972.
692. **Lichtwardt, R. W.,** The Trichomycetes: what are their relationships?, *Mycologia*, 65, 1, 1973.
693. **Lichtwardt, R. W.,** Trichomycetes, in *Recent Advances in Aquatic Mycology*, Jones, E. B. G., Ed., Elek Science, London, 1976, 651.
694. **Licitis, R.,** Air-borne pollen and spores sampled at five New Zealand stations, 1951—52, *N.Z. J. Sci. Technol.*, 34, 289, 1953.
695. **Liles, M. P.,** A study of the life history of the forked fungus beetle, *Bolitotherus cornutus* (Panzer) (Coleoptera: Tenebrionidae), *Ohio J. Sci.*, 56, 329, 1956.
696. **Lilly, V. G.,** Fungi for thiamin (vitamin B_1) assay, *Proc. W. Va. Acad. Sci.*, 13, 72, 1940.
697. **Lilly, V. G. and Barnett, H. L.,** Growth rate and vitamin deficiencies of various fungi, *Proc. W. Va. Acad. Sci.*, 20, 27, 1947.
698. **Lilly, V. G. and Barnett, H. L,** *Physiology of the Fungi*, McGraw-Hill, New York, 1951.
699. **Lilly, V. G. and Barnett, H. L.,** The utilization of sugars by fungi, *W. Va. Agric. Exp. Stn. Bull.*, 362T, 1953.
700. **Lind, J.,** Studies on the geographical distribution of arctic circumpolar micromycetes, *K. Dan. Vidensk. Selsk. Biol. Skr.*, 2, 2, 1934.
701. **Lindeberg, G.,** Über die Physiologische Ligninannuenden Bodenhymenomyceten, *Symb. Bot. Ups.*, 8(2), 1, 1944.
702. **Lindeberg, G.,** On the decomposition of lignin and cellulose in litter caused by soil-inhabiting soil Hymenomycetes, *Ark. Bot.*, 33A, 1, 1946.
703. **Lindegren, C. C.,** Breeding yeats for their new role in nutrition, *Mo. Bot. Gard. Bull.*, 34, 37, 1944.
704. **Lindegren, C. C.,** The induction of dormancy in vegetative yeast cells by fat and carbohydrate storage and the conditions for reactivation, *Arch. Biochem.*, 8, 119, 1945.
705. **Lindegren, C. C.,** An analysis of the mechanism of budding in yeasts and some observations on the structure of the yeast cell, *Mycologia*, 37, 767, 1945.
706. **Lindegren, C. C.,** *The Yeast Cell, Its Genetics and Cytology*, Educational Publishers St. Louis, 1949.
707. **Lindegren, C. C.,** Food yeast, *Dev. Ind. Microbiol.*, 7, 226, 1965.
708. **Lingappa, Y., Sussman, A. S., and Bernstein, I. A.,** Effect of light and media upon growth and melanin formation in *Aureobasidium pullulans* (deBy.) Arn. (= *Pullularia pullulans*), *Mycopathol. Mycol. Appl.*, 20, 111, 1963.
709. **Lindsey, B. I.,** A survey of methods used in the study of microfungal succession on leaf surfaces, in *Microbiology of Aerial Plant Surfaces*, Dickinson, C. H. and Preece, T. F., Eds., Academic Press, New York, 1976, 217.
710. **Link, G. C. C.,** Reproduction in the Thallophytes, with special reference to fungi, *Bot. Gaz.*, 88, 1, 1929.
711. **Litchfield, J. H. and Overbeck, R. C.,** Submerged culture growth of *Morchella* species in food processing wastes substrates, *Proc. 1st Int. Congr. Food Sci. Technol.*, 2, 511, 1965.
712. **Littman, M. L.,** Growth of pathogenic fungi on a new culture medium, *Am. J. Clin. Path.*, 18, 409, 1948.
713. **Llano, G. A. P.,** Lichens — their biological and economic significance, *Bot. Rev.*, 10, 1, 1944.
714. **Llewellyn, G. C.,** Sterigmatocystis toxicity in three rodents, *Mesocricetus auratus, Meriones unguiculatus*, and *Mus musculus*, *Dev. Ind. Microbiol.*, 16, 343, 1975.
715. **Llewellyn, G. C., Carlton, W. W., Robbers, J. E., and Hansen, W. G.,** The response of the Syrian hamster (*Mesocricetus auraus*) to chronic administration of aflatoxin B_1 and 17 beta-estradiol., *Dev. Ind. Microbiol.*, 15, 358, 1974.
716. **Lochhead, A. G.,** The nutritional classification of soil bacteria, *Proc. Soc. Appl. Bacteriol.*, 15, 15, 1952.
717. **Lockwood, L. B. and Moyer, A. J.,** The production of chemicals by filamentous fungi, *Bot. Rev.*, 4, 140, 1938.
718. **Lockwood, L. B. and Reeves, M. D.,** Some factors affecting the production of itaconic acid by *Aspergillus terreus*, *Arch. Biochem.*, 6, 455, 1945.
719. **Lodder, J., Ed.,** *The Yeasts, A Taxonomic Study*, 2nd ed., North-Holland, Amsterdam, 1971.
720. **Löliger, H.-C.,** *Pelztierkrankheiten*, Gustav Fischer Verlag, Stuttgart, 1970.
721. **Lohwag, H.,** *Anatomie der Asco- und Basidiomyzeten*, Verlag von Gebruder Borntraeger, Berlin, 1941.

722. **Lowe, J. L.**, The Polyporaceae of the world, *Mycologia*, 55, 1, 1963.
723. **Ludzack, F. J. and Ettinger, M. B.**, Biological oxidation of hexadecanol under laboratory conditions, *J. Am. Water Works Assoc.*, 49, 849, 1957.
724. **Ludzack, F. J. and Ettinger, M. B.**, Chemical structures resistant to aerobic biochemical stabilization, *J. Water Pollut. Control Fed.*, 32, 1173, 1960.
725. **Ludzack, F. J. and Kinkead, D.**, Persistence of oily wastes in polluted water under aerobic conditions, *Ind. Eng. Chem.*, 48, 263, 1956.
726. **Lukens, R. J.**, Controlled release of foliar protectants, *Dev. Ind. Microbol*, 16, 333, 1975.
727. **Lund, A.**, *Studies on the Ecology of Yeasts*, Ph.D. thesis, Munksgaard Publishers, Copenhagen, 1954.
728. **Maas, J. L. and Stuntz, D. E.**, Mycoecology on serpentine soil, *Mycologia*, 61, 1106, 1969.
729. **Macauley, B. J. and Thrower, L. B.**, Succession of fungi in leaf litter of *Eucalyptus regnans*, *Trans. Br. Mycol. Soc.*, 49, 509, 1966.
730. **Machlis, L.**, Growth and nutrition of water molds in the subgenus *Euallomyces*. I. Growth factor requirements, *Am. J. Bot.*, 40, 189, 1953.
731. **Machlis, L.**, Growth and nutrition of water molds in the subgenus *Euallomyces*. II. Optimal composition of the minimal medium, *Am. J. Bot.*, 40, 450, 1953.
732. **Machlis, L.**, Growth and nutrition of water molds in the subgenus *Euallomyces*. III. Carbon sources, *Am. J. Bot.*, 40, 460, 1953.
733. **Mackenthun, K. M., Ed.**, Temperature and Aquatic Life, Laboratory Investigations, Series 6, Technical Advisory and Investigational Branch, Federal Water Pollution Control Administration, Cincinnati, 1967.
734. **Mackenthun, K. M. and Keup, L. E.**, Biological problems encountered in water supplies, *J. Am. Water Works Assoc.*, 62, 520, 1970.
735. **Macrae, R.**, Interfertility studies and inheritance of luminosity in *Panus stypticus*, *Can. J. Res. Sect. C*, 20, 411, 1942.
736. **Madelin, M. F.**, Fungal parasites of invertebrates. I. Entomognous fungi, in *The Fungi, An Advanced Treatise*, Vol. 3, Ainsworth, G. C. and Sussman, A. S., Eds., Academic Press, New York, 1968, 227.
737. **Madelin, M. F.**, Fungi parasitic on other fungi and lichens, in *The Fungi, An Advanced Treatise*, Vol. 3, Ainsworth, G. C. and Sussman, A. S., Eds., Academic Press, New York, 1968, 253.
738. **Madri, P. P., Claus, G., and Moss, E. E.** Infectivity of pathogenic fungi in a simulated marine environment, *Rev. Biol. (Lisbon)*, 5, 371, 1966.
739. **Magnus, H. A.**, Mercury-tolerance in *Pyrenophora avenae* in Norway, *Sci. Rep. Agric. Coll. Norway*, 49(24), 1, 1970.
740. **Mahloch, J. L.**, Fungi present during the aerobic decomposition of selected solid-wastes substrates, *Dev. Ind. Microbiol.*, 13, 308, 1972.
741. **Mains, E. B.**, The genus *Gibellula* on spiders in North America, *Mycologia*, 49, 306, 1950.
742. **Major, J.**, Functional, factorial approach to plant ecology, *Ecology*, 32, 392, 1951.
743. **Maldagne, M. E. and Hilger, F.**, Observations faunistiques et microbiologiques dans quelques biotopes forestieres equatoriaux, in *Soil Organisms*, Doeksen, J. and van der Drift, J., Eds., North-Holland, Amsterdam, 1963, 368.
744. **Malone, J. P. and Muskett, A. E.**, Seed-borne fungi, *Proc. Int. Seed Test. Assoc.*, 29, 179, 1964.
745. **Mandels, G. R.**, Kinetics of fungal growth, in *The Fungi, An Advanced Treatise*, Vol. 1, Ainsworth, G. C. and Sussman, A. S., Eds., Academic Press, New York, 1965, 599.
746. **Mandels, M. and Reese, E. T.**, Inhibition of cellulase, *Annu. Rev. Phytopathol.*, 3, 85, 1965.
747. **Mangenot, F.**, Recherches Méthodiques sur les Champignons de Certains Bois en Décomposition, Librarie Générale de l'Enseignement, Paris, 1952.
748. **Mangenot, M. F.**, Étude microbiologiques des litieres, (Commentaires sur les donnees experimentales recueilles à l'E.N.S.A.N., depuis 1963.), *Bull. Ec. Nat. Super. Agron. Nancy*, 8, 113, 1968.
749. **Mankau, R. and Mankau, S. K.**, The role of mycophagous nematodes in the soil. I. The relationship of *Aphelenchus avenae* to phytopathogenic soil fungi, in *Soil Organisms*, North-Holland, Amsterdam, 1963, 271.
750. **Marks, O. C. and Koslowski, T. T.**, *Ectomycorrhizae, their Ecology and Physiology*, Academic Press, New York, 1973.
751. **Marsh, P. B., Taylor, E. E., and Bassler, L. M.**, A guide to the literature on certain effects of light on fungi: reproduction, morphology, pigmentation, and phototropic phenomena, *Plant Dis. Rep. Suppl.*, 261, 251, 1959.
752. **Marshall, K. C. and Alexander, M.**, Growth characteristics of fungi and Actinomycetes, *J. Bacteriol.*, 60, 412, 1960.
753. **Martin, D. F. and Committee**, Control of Turfgrass Pests, Bull. L-187 and Suppl., Cooperative Extension Service, Ohio State University, Columbus, 1977.

754. **Martin, G. W.**, The numbers of fungi, *Proc. Iowa Acad. Sci.*, 58, 175, 1951.
755. **Martin, G. W.**, Are fungi plants?, *Mycologia*, 47, 779, 1955.
756. **Martin, G. W.**, The origin and status of the fungi (with a note on the fossil record), in *The Fungi, An Advanced Treatise*, Vol. 3, Ainsworth, G. C. and Sussman, A. S., Eds., Academic Press, New York, 1968, 635.
757. **Martin, G. W. and Alexopoulos, C. J.**, *The Myxomycetes*, University of Iowa Press, Iowa City, 1969.
758. **Martin, J. P.**, Use of acid, rose bengal, and streptomycin in the plate method for estimating soil fungi, *Soil Sci.*, 69, 215, 1950.
759. **Martin, S. M., Ed.**, *Culture Collections: Perspectives and Problems*, University of Toronto Press, Toronto, 1961
760. **Marx, D. H. and Daniel, W. J.**, Maintaining cultures of ectomycorrhizal and plant pathogenic fungi in sterile water cold storage, *Can. J. Microbiol.*, 22, 338, 1976.
761. **Mason, H. L.**, Pleistocene flora of the Tomales formation, *Carnegie Inst. Washington Publ.*, 415, 83, 1934.
762. **Mason, H. L. and Langenheim, J. H.**, Language analysis and the concept *environment*, *Ecology*, 38, 325, 1957.
763. **Matta, A.**, Microbial penetration and immunization of uncongenial host plants, *Annu. Rev. Phytopathol.*, 9, 387, 1971.
764. **Matseeb, V. A.**, Prediction of fruiting of edible mushrooms, *Mikol. Fitopatol.*, 6, 358, 1972.
765. **Maunsell, K.**, Quantitative Aspects of Allergy to House-Dust, in *1st Int. Cong. Allergy*, S. Karger, Basel, 1951, 306.
766. **Maunsell, K.**, Sensitization risk from inhalation of fungal spores, *J. Laryngol. Otol.*, 68, 765, 1954.
767. **May, K. R.**, Physical aspects of sampling airborne microbes, in *Airborne Microbes*, Gregory, P. H. and Monteith, J. L., Eds., Cambridge University Press, Cambridge, 1967, 60.
768. **Mayaudon, J. and Simonart, P.**, Étude de la décomposition de la matiere organique dans le sol au moyen de carbone radioactif. III. Décomposition des substances solubles dialysables des proteines et des hemicelluloses, *Plant Soil*, 11, 170, 1959.
769. **Mayaudon, J. and Simonart, P.**, Étude de la décomposition de la matiere organique dans le sol au moyen de carbone radioactif. V. Décomposition de cellulose et de lignine, *Plant Soil*, 11, 181, 1959.
770. **Mazur, P.**, Survival of fungi after freezing and desiccation, in *The Fungi, An Advanced Treatise*, Vol. 3, Ainsworth, G. C. and Sussman, A. S., Eds., Academic Press, New York, 1968, 325.
771. **McCallan, S. E. A.**, The nature of fungicidal action of copper and sulphur, *Bot. Rev.*, 15, 629, 1949.
772. **McCallan, S. E. A.**, Determination of individual fungus spore volumes and their size distribution, *Contrib. Boyce Thompson Inst.*, 19, 303, 1958.
773. **McCubbin, W. A.**, Preventing plant disease introduction, *Bot. Rev.*, 12, 101, 1946.
774. **McGinnis, M. R., Padhye, A. A., and Ajello, L.**, Storage of stock cultures of filamentous fungi, yeasts, and some aerobic Actinomycetes in sterile distilled water, *Appl. Microbiol.*, 28, 218, 1974.
775. **McIlveen, W. D. and Cole, H., Jr.**, Spore dispersal of Endogonaceae by worms, ants, wasps, and birds, *Can. J. Bot.*, 54, 1486, 1976.
776. **McKnight, K. H.**, Cultural studies on *Psilocybe*. I. Variation in a new species of the *P. coprophila* group, *Bull. Torrey Bot. Club*, 98, 4, 1971.
777. **Melin, E. and Nannfeldt, J. A.**, Researches into the blueing of ground woodpulp, *Sven. Skogsvardsfoeren. Tidskr.*, 397, 1934.
778. **di Menna, M. E.**, A search for pathogenic species of yeasts in New Zealand soils, *J. Gen. Microbiol.*, 12, 54, 1955.
779. **Menzies, G. D.**, Survival of microbial plant pathogens in soil, *Bot. Rev.*, 29, 79, 1963.
780. **Merchant, I. A. and Parker, R. A.**, *Veterinary Bacteriology and Virology*, 5th ed., Iowa State College Press, Ames, 1956.
781. **Metzler, D., Ritter, C., and Culp, R. L.**, Effect of standard water purification processes on the removal of *Histoplasma capsulatum* from water, *Am. J. Public Health*, 44, 1305, 1954.
782. **Metzler, D. F., Ritter, C., and Culp, R. L.**, Combined effect of water purification processes on the removal of *Histoplasma capsulatum* from water, *Am. J. Public Health*, 46, 1571, 1956.
783. **Meyer, E.**, The preservation of dermatophytes at subfreezing temperatures, *Mycologia*, 47, 664, 1955.
784. **Meyer, J. A.**, Moisissures du sol et des litières de la region de Yamgambi (Congo Belge), *Publ. Inst. Natl. Étude Agron. Congo Ser. Sci.*, 75, 1, 1959.
785. **Meyer, J. A.**, Ecologie et sociologie des microchampignons du sol de la Cuvette centrale Congolaise, *Publ. Inst. Natl. Étude Agron. Congo Ser. Sci.*, 101, 1, 1963.
786. **Meyers, S. P., Ahearn, D. G., Alexander, S. K., and Cook, W. L.**, *Pichia spartinae*, a dominant yeast of the *Spartina* salt marsh, *Dev. Ind. Microbiol.*, 16, 262, 1975.
787. **Meyl, A. H.**, Die Nematodenfauna höherer Pilze in Laub- und Nädelwaldern zwischen Braunschweig und dem Harz, *Mycopathol. Mycol. Appl.*, 7, 1, 1954.

788. **Middleton, J. T.**, The taxonomy, host range and geographic distribution of the genus *Pythium*, *Mem. Torrey Bot. Club*, 20, 1, 1943.

789. **Mikola, P.**, Experiments on the ability of forest soil Basidiomycetes to decompose litter material, *Commun. Inst. Forest. Fenn.*, 42(7), 1, 1954.

790. **Mikola, P.**, Experiments on the rate of decomposition of forest litter, *Commun. Inst. Forest. Fenn.*, 43(1), 1, 1954.

791. **Mikola, P.**, Studies on the decomposition of forest litter by Basidiomycetes, *Commun. Inst. Forest. Fenn.*, 48(2), 1, 1956.

792. **Milbury, W. F., Riedel, G. W., and Jones, P. H.**, Yeast production using conventional activated sludge methods for treatment of corn processing industrial wastes, in 3rd Can. Symp. Water Pollution Res., Toronto, Canada, 1968.

793. **Millbank, J. W.**, Nitrogen fixation in moulds and yeasts — a reappraisal, *Arch. Mikrobiol.*, 68, 32, 1969.

794. **Millar, C. E. and Turk, L. M.**, *Fundamentals of Soil Science*, John Wiley & Sons, New York, 1943.

795. **Miller, D. D. and Golding, N. S.**, The gas requirements of molds. V. The minimum oxygen requirements for normal growth and for germination of six mold cultures, *J. Dairy Sci.*, 32, 101, 1949.

796. **Miller, H. A. and Halls, L. K.**, Fleshy fungi commonly eaten by southern wildlife, *U.S. For. Serv. Res. Pap. SO*, No. 49, 1969.

797. **Miller, J. J. and Webb, N. S.**, Isolation of yeasts from soil with the aid of acid, rose bengal, and oxgall, *Soil Sci.*, 77, 197, 1954.

798. **Miller, O. K. and Laursen, G. A.**, Below ground fungal biomass on U.S. tundra biome sites at Barrow, Alaska, in *Soil Organisms and Decomposition in Tundra*, Tundra Biome Steering Committee, Stockholm, 1974, 151.

799. **Miller, P. R. and O'Brien, M.**, Plant disease forecasting, *Bot. Rev.*, 18, 547, 1952.

800. **Miller, R. L. and Committee**, Home Vegetable Garden Insect Control, Bull. No. 498, Cooperative Extension Service, Ohio State University, Columbus, 1977.

801. **Miller, R. L. and Ornamental Insect Control Recommendations Committee**, Insect and Mite Control on Ornamentals, Bull. No. 504, Cooperative Extension Service, Ohio State University, Columbus, 1977.

802. **Miller, R. L. and Committee**, Vegetable Insect and Disease Control, Commercial Growers, Bull. No. 459, Cooperative Extension Service, Ohio State University, Columbus, 1977.

803. **Millner, F. D., Marsh, P. B., Snowden, R. B., and Parr, J. F.**, Occurrence of *Aspergillus fumigatus* during composting of sewage sludge, *Appl. Environ. Microbiol.*, 34, 765, 1977.

804. **Milne, L. and Milne, M.**, *The Arena of Life, The Dynamics of Ecology*, Doubleday/Natural History Press, Garden City, N.Y., 1971.

805. **Mirams, R. V.**, A comparison of the gravity and impact methods of collecting spores and pollen grains from the atmosphere at Wellington, N. Z., *N.Z. J. Sci. Technol. Sect. B.*, 34, 378, 1953.

806. **Molchenov, A. A.**, Rate of decomposition of dead trees of pine and spruce, *Dokl. Akad. Nauk. SSSR*, 56, 869, 1947; *For. Abstr.*, 9, 2534, 1947.

807. **Moore, R. T.**, The ultrastructure of fungal cells, in *The Fungi, An Advanced Treatise*, Vol. 1, Ainsworth, G. C. and Sussman, A. S., Eds., Academic Press, New York, 1965, 95.

808. **Moreau, C.**, Les mycotoxines dans les produits lactiers, *Le Lait, Rev. Gen. Questions Lactieres*, 56, 286, 1976.

809. **Moreau, C.**, Variations de la pollution fongique des Arachides et des leurs tourteaux de la recolte à la consommation, *Rev. Mycol.*, 40, 97, 1976.

810. **Moreau, C., Debaste, H., and Bernard, F.**, Le *Geotrichum candidum* Link, cause possible de graves enterites chez les pintadeaux, *Mycopathol. Mycol. Appl.*, 53, 125, 1974.

811. **Moreau, M. and Moreau, F.**, Première contribution à l'étude de la microflora des dunes, *Rev. Mycol.*, 6, 49, 1941.

812. **Morita, R. Y.**, The physical environment for fungal growth. 2. Hydrostatic pressure, in *The Fungi, An Advanced Treatise*, Vol. 1, Ainsworth, G. C. and Sussman, A. S., Eds., Academic Press, New York, 1965, 551.

813. **Morrison, S. M., Elmund, G. K., Grant, D. W., and Smith, V. J.**, Protein production from feedlot wastes, *Dev. Ind. Microbiol.*, 18, 145, 1977.

814. **Morrow, M. B. and Lowe, E. P.**, Molds in relation to asthma and vasomotor rhinitis, *Mycologia*, 35, 638, 1943.

815. **Moser, M.**, Untersuchungen über den Einfluss der Waldbränden auf die Pilz-vegetation. I, *Sydowia, Ann. Myc.*, (Ser. 2), 3, 336, 1949.

816. **Mounce, I. and Macrae, R.**, The behavior of paired monosporous mycelia on *Lenzites saepiaria* (Wulf.) Fr., *L. trabea* (Pers.) Fr., *L. thermophila* Falck, and *Trametes americana* Overh., *Can. J. Res. Sect. C*, 14, 215, 1936.

817. **Müeller, E. and Löeffler, W.**, *Mykologie, Grundriss für Naturwissenschaften und Mediziner*, Georg Thieme Verlag, Stuttgart, 1968.

818. **Müller, G.**, Occurrence of dermatophytes in the soils of European beaches, *Sci. Total Environ.*, 2, 116, 1973.

819. **Mulinge, S. K.**, Development of coffee berry diseases in relation to the stage of berry growth, *Ann. Appl. Biol.*, 65, 269, 1969.

820. **Mulinge, S. K. and Apinis, A. E.**, Occurrence of thermophilous fungi in stored moist barley grains, *Trans. Br. Mycol. Soc.*, 53, 361, 1969.

821. **Mulinge, S. K. and Chesters, C. G. C.**, Methods of isolating the microflora of moulding high moisture barley in partially sealed silos, *Ann. Appl. Biol.*, 65, 285, 1970.

822. **Murdoch, D. I., Folnazzo, J. F., and Troy, V. S.**, Evaluation of plating media for citrus concentrates, *Food Technol. (Chicago)*, 6, 181, 1952.

823. **Murray, S., Woodbine, M., and Walker, T. K.**, Microbiological synthesis of fat: the formation of fat from sucrose, *J. Exp. Bot.*, 4, 251, 1953.

824. **Muskett, A. E.**, Seed borne fungi and their significance, *Trans. Br. Mycol. Soc.*, 33, 1, 1950.

825. **Neal, J. L., Wright, E., and Bollen, W. B.**, Burning Douglas fir slash: physical, chemical, and microbial effects in the soil, *Oreg. Stat Univ. For. Res. Lab. Res. Pap.*, 1, 1, 1965.

826. **Nelson, E. E.**, Survival of *Poria weirii* on wood buried in urea-amended forest soil, *Phytopathology*, 65, 501, 1975.

827. **Nelson, E. E.**, Effect of urea on *Poria weirii* and soil microbes in an artificial system, *Soil Biol. Biochem.*, 8, 51, 1976.

828. **Nelson, E. E. and Hartman, T.**, Estimating spread of *Poria weirii* in a high-elevation mixed conifer stand, *J. For.*, p. 141, March 1975.

829. **Nespiak, A.**, The investigations on the character of the correlations between the higher fungi and wood associations in the National Park of Bialowieza, *Monogr. Bot.*, 8, 1, 1959.

830. **Nickerson, W. J. and Edwards, G. A.**, Studies on the physiological bases of morphogenesis in fungi. I. The respiratory metabolism of dimorphic pathogenic fungi, *J. Gen. Physiol.*, 33, 41, 1949.

831. **Nilsson, S.**, Freshwater Hyphomycetes, *Symb. Bot. Ups.*, 18, 1, 1, 1964.

832. **Nobles, M. K.**, Studies in forest pathology. VI. Identification of cultures of wood rotting fungi, *Can. J. Res. Sect. C*, 26, 281, 1948.

833. **Nobles, M. K.**, Cultural characters as a guide to the taxonomy and phylogeny of the Polyporaceae, *Can. J. Bot.*, 36, 883, 1958.

834. **Nobles, M. K.**, Identification of cultures of wood rotting Hymenomycetes, *Can. J. Bot.*, 43, 1097, 1965.

835. **Nobles, M. K.**, Cultural characters as a guide to the taxonomy of the Polyporaceae, in *Evolution of the Higher Basidiomycetes*, Petersen, R. H., Ed., University of Tennessee Press, Knoxville, 1971, 169.

836. **Nolard-Tintigner, N.**, Étude experimentale sur l'epidemiologie et la pathologenie de la saprolegniose chez *Lebistes reticulatus* Peters et *Xiphophorus helleri* Heckel, *Acta Zool. Patho. Antverp*, 57, 1, 1973.

837. **Norris, L. A.**, Degradation of herbicides in the forest floor, in *Tree Growth and Forest Soils*, Youngberg, C. T. and Davey, C. B., Eds., Oregon State University Press, Corvallis, 1970, 397.

838. **Nuss, I.**, *Zur Okologie der Porlinge*, Bibliotheca Mycologica, J. Cramer, Vaduz, Liechtenstein, 1975, 45.

839. **Nyms, E. J., Auguière, J. P., and Wiaux, A. L.**, Taxonomic value of the property of fungi to assimilate hydrocarbons, *Antonie van Leeuwenhoek J. Microbiol. Serol.*, 34, 441, 1968.

840. **Ofosu-Asiedu, A. and Smith, R. S.**, Degradation of three softwoods by thermophilic and thermotolerant fungi, *Mycologia*, 65, 420, 1973.

841. **Ofosu-Asiedu, A. and Smith, R. S.**, Some factors affecting wood degradation by thermophilic and thermotolerant fungi, *Mycologia*, 65, 87, 1973.

842. **Ogawa, J. M., Hall, D. H., and Koepsell, P. A.**, Spread of pathogens within crops as affected by life cycle and environment, in *Airborne Microbes*, Gregory, P. H. and Monteith, J. L., Eds., Cambridge University Press, Cambridge, 1967, 247.

843. **Ogden, E. C. and Lewis, D. M.**, Airborne pollen and fungus spores of New York State, *N.Y. State Mus. Sci. Serv. Bull.*, p. 378, 1960.

844. **Okudaira, M., Kurata, M., and Sakabe, F.**, Studies on the fungal flora of the lung of human necropsy cases: a critical survey in connection with the pathogenesis of opportunistic fungus infections, *Mycopathologia*, 61, 3, 1977.

845. **Olive, L. S.**, The structure and behavior of fungus nuclei, *Bot. Rev.*, 10, 439, 1953.

846. **Olive, L. S.**, *The Mycetozoans*, Academic Press, New York, 1975.

847. **Olson, J. S. and Crossley, D. A., Jr.**, Tracer studies of the breakdown of forest litter, in *Radioecology*, Klechkovskii, V. M., et al., Eds., Halsted Press, New York, New York, 1963, 411.

848. **Oudemans, C. A. J. A.**, *Enumeratio systematica fungorum*, 5 vol., Martinus Nijhoff, The Hague, 1919—1924.
849. **Outschoorn, A. S**, USP standards for sterilization and sterility testing, *Dev. Ind. Microbiol.*, 18, 387, 1977.
850. **Overholts, L. O.**, Geographic distribution of some American Polyporaceae, *Mycologia*, 31, 629, 1939.
851. **Pady, S. M.**, An improved slit sampler for aerobiological investigations, *Trans. Kans. Acad. Sci.*, 57, 157, 1954.
852. **Pady, S. M.**, Quantitative studies of fungus spores in the air, *Mycologia*, 49, 339, 1957.
853. **Pady, S. M.**, A continuous spore sampler, *Phytopathology*, 49, 757, 1959.
854. **Pady, S. M. and Johnston, C. O.**, The concentration of airborne rust spores in relation to epidemiology of wheat rust in Kansas in 1954, *Plant Dis. Rep.*, 39, 463, 1955.
855. **Pady, S. M. and Kapica, L.**, Air-borne fungi in the arctic and other parts of Canada, *Can. J. Bot.*, 31, 309, 1953.
856. **Pady, S. M. and Kapica, L.**, Fungi in air over the Atlantic Ocean, *Mycologia*, 47, 34, 1955.
857. **Pady, S. M., Kramer, C. L., Pathak, V. K., Morgan, F. L., and Bhatti, M. A.**, Periodicity in airborne rust urediospores, *Phytopathology*, 55, 132, 1965.
858. **Pady, S. M., Kramer, C. L., and Wiley, B. J.**, Kansas aeromycology. XII. Materials, methods, and general results of diurnal studies, 1959—1960, *Mycologia*, 54, 168, 1962.
859. **Page, R. M.**, The physical environment for fungal growth. 3. Light, in *The Fungi, An Advanced Treatise*, Vol. 1, Ainsworth, G. C. and Sussman, A. S., Eds., Academic Press, New York, 1965, 559.
860. **Painter, H. A.**, Factors affecting the growth of some fungi associated with sewage purification, *J. Gen. Microbiol.*, 10, 177, 1954.
861. **Palmer, C. M.**, A composite rating of algae tolerating organic pollution, *J. Phycol.*, 5, 78, 1969.
862. **Panasenko, Y. T.**, Ecology of microfungi, *Bot. Rev.*, 33, 189, 1967.
863. **Park, D.**, The importance of antibiotics and inhibiting substances, in *Soil Biology*, Burges, A. and Raw, F., Eds., Academic Press, New York, 1967, 435.
864. **Park, D.**, The ecology of terrestrial fungi, in *The Fungi, An Advanced Treatise*, Vol. 3, Ainsworth, G. C. and Sussman, A. S., Eds., Academic Press, New York and London, 1968, 5.
865. **Park, D.**, Methods of detecting fungi in organic detritus in water, *Trans. Br. Mycol. Soc.*, 58, 261, 1972.
866. **Park, D.**, On the ecology of heterotrophic microorganisms in fresh water, *Trans. Br. Mycol. Soc.*, 58, 291, 1972.
867. **Parker-Rhodes, A. F.**, The Basidiomycetes of Skokholm Island. II. Genetical implications of spore measurements in two species, *New Phytol.*, 48, 382, 1949.
868. **Parker-Rhodes, A. F.**, The Basidiomycetes of Skokholm Island. III. Genetic isolation in *Panaeolus papilonaceus*, *New Phytol.*, 49, 328, 1950.
869. **Parker-Rhodes, A. F.**, The Basidiomycetes of Skokholm Island. IV. A case of hybridization in *Psilocybe (Deconia)*, *New Phytol.*, 49, 335, 1950.
870. **Parker-Rhodes, A. F.**, The Basidiomycetes of Skokholm Island. V. An elementary theory of anemophilous dissemination, *New Phytol.*, 50, 84, 1951.
871. **Parker-Rhodes, A. F.**, The Basidiomycetes of Skokholm Island. VII. Some floristic and ecological calculations, *New Phytol.*, 50, 227, 1951.
872. **Parker-Rhodes, A. F.**, The Basidiomycetes of Skokholm Island. X. Population densities, *New Phytol.*, 52, 273, 1953.
873. **Parker-Rhodes, A. F.**, The Basidiomycetes of Skokholm Island. XI. Intramycelial variation in *Hygrocybe turmunda* var. *lepida*, *New Phytol.*, 53, 145, 1954.
874. **Parker-Rhodes, A. F.**, Deme structure in higher fungi: *Mycena galopus*, *Trans. Br. Mycol. Soc.*, 37, 314, 1954.
875. **Parker-Rhodes, A. F.**, The Basidiomycetes of Skokholm Island. XII. Correlation with the chief plant associations, *New Phytol.*, 54, 259, 1955.
876. **Parker-Rhodes, A. F.**, Fairy ring kinetics, *Trans. Br. Mycol. Soc.*, 38, 59, 1955.
877. **Parker-Rhodes, A. F.**, Statistical aspects of fungus forays, *Trans. Br. Mycol. Soc.*, 38, 283, 1955.
878. **Parkinson, D.**, Soil microorganisms and plant roots, in *Soil Biology*, Burges, A. and Raw, F., Eds., Academic Press, New York, 1967, 449.
879. **Parkinson, D.**, Techniques for the study of soil fungi, *Bull. Ecol. Res. Comm., Stockholm*, 17, 29, 1973.
880. **Parkinson, D. and Coups, E.**, Microbial activity in a podzol, in *Soil Organisms*, Doeksen, J. and van der Drift, J., Eds., North-Holland, Amsterdam, 1963, 167.

881. **Parkinson, D., Gray, T. R. G., and Williams, S. T., Eds.,** *Methods for Studying the Ecology of Soil Microorganisms,* International Biological Programme Handbook 19, Blackwell Scientific, Oxford, 1971.

882. **Parkinson, D., Taylor, G. S., and Pearson, R.,** Studies on fungi in the root region. I. The development of fungi on young roots, *Plant Soil,* 19, 332, 1963.

883. **Parkinson, D. and Thomas, A.,** A comparison of methods for the isolation of fungi from rhizospheres, *Can. J. Microbiol.,* 11, 1001, 1965.

884. **Parkinson, D. and Thomas, A.,** Studies on fungi in the root region. VIII. Qualitative studies on fungi in the rhizosphere of dwarf bean plants, *Plant Soil,* 31, 299, 1969.

885. **Parmeter, J. R., Jr., Ed.,** *Rhizoctonia solani: Biology and Pathology,* University of California Press, Berkeley, 1970.

886. **Pass, T. and Griffin, G. J.,** Exogenous carbon and nitrogen requirements for conidial germination by *Aspergillus flavus, Can. J. Microbiol.,* 18, 1453, 1972.

887. **Pass, T. and Griffin, G. J.,** Interaction of pH and temperatures with exogenous carbon and nitrogen nutrition in conidial germination by *Aspergillus flavus, Phytopathology,* 64, 1151, 1974.

888. **Paterson, R. A.,** Lacustrine fungal communities, in *The Structure and Function of Fresh-Water Communities,* Virginia Polytechnic Institute and State University, Blacksburg, 1970, 209.

889. **Pelhate, J.,** Mycoflora des maise fourrages ensilés: determinism de son évolution, *Rev. Mycol.,* 39, 65, 1974.

890. **Pennycook, S. R. and Newhook, F. J.,** Dial periodicity and circadian rhythm of ballistospore discharge in the Sporobolomycetaceae, *Trans. Br. Mycol. Soc.,* 63, 237, 1974.

891. **Perlman, D.,** On the nutrition of *Memnoniella echinata* and *Stachybotrys atra, Am. J. Bot.,* 35, 36, 1948.

892. **Perlman, D.,** Effects of minor elements on the physiology of the fungi, *Bot. Rev.,* 15, 195, 1949.

893. **Perlman, D.,** Some mycological aspects of penicillin production, *Bot. Rev.,* 16, 449, 1950.

894. **Person, G.,** Genetical adjustment of fungi to their environment, in *The Fungi, An Advanced Treatise,* Vol. 3, Ainsworth, G. C. and Sussman, A. S., Eds., Academic Press, New York, 1968, 395.

895. **Persoon, C. H.,** *Synopsis methodica fungorum,* apud Henricum Dietrich, Gottingae, 1801.

896. **Petersen, F. M.,** Studies on the ecology of some species of the Pezizales, *Bot. Tidsskr.,* 62, 312, 1967.

897. **Petersen, F. M.,** Danish fireplace fungi, an ecological investigation on fungi on burns, *Dan. Bot. Ark.,* 27(3), 1, 1970.

898. **Petersen, F. M.,** The macromycetes in a burnt forest area in Denmark, *Bot. Tidsskr.,* 66, 228, 1971.

899. **Petersen, R. H.,** The rust fungus life cycle, *Bot. Rev.,* 40, 453, 1974.

900. **Peterson, A. F.,** Mycoflora of the normal eye and effects of contact lenses, *Dev. Ind. Microbiol.,* 17, 17, 1976.

901. **Peterson, E. A.,** Observations of fungi associated with plant roots, *Can. J. Microbiol.,* 4, 257, 1958.

902. **Peterson, G. W. and Smith, R. S., Jr., Eds.,** Forest Nursery Diseases in the United States, Agricultural Handbook No. 470, U.S. Forest Service, Washington, D.C., 1975, 1.

903. **Phaff, H. J. and Ahearn, D. G.,** *Rhodotorula,* in *The Yeasts,* Lodder, J., North-Holland, Amsterdam, 1971, 1187.

904. **Phaff, H. J., Miller, M. W., and Mrak, E. M.,** *The Life of Yeasts, Their Nature, Activity, Ecology, and Relation to Man,* Harvard University Press, Cambridge, 1966.

905. **Phaff, H. J., Miller, M. W., Yoneyama, M., and Soneda, M.,** A comparative study of the yeast florae associated with trees on the Japanese Islands and on the West Coast of North America, *Ferment. Technol. Today Proc.,* 4, IFS, 759, 1972.

906. **Phaff, H. J., Mrak, E. M., and Williams, O. B.,** Yeasts isolated from shrimp, *Mycologia,* 44, 431, 1952.

907. **Phaff, H. J., Yoneyama, M., and do Carmo-Sousa, L.,** A one-year quantitative study of the yeast flora in a single slime flux on *Ulmus carpinifolia* Gled., *Riv. Patol. Veg.,* Series 3, 4(4), 485, 1964.

908. **Phaff, H. J. et al.,** Studies on the ecology of *Drosophila* in the Yosemite region of California. II. Yeasts found in the alimentary canal of *Drosophila, Ecology,* 37, 533, 1956.

909. **Pharmaceutical Society of Great Britain,** Antibiotics: A Survey of Their Properties and Uses, *The Pharmaceutical Press,* London, 1952.

910. **Pidacks, C. et al.,** Coprogen, the isolation of a new growth factor required by *Pilobolus* species, *J. Am. Chem. Soc.,* 75, 6064, 1953.

911. **Pieczynska, E.,** Periphyton as food of aquatic animals, *Wiad. Ekol.,* 16, 133, 1970.

912. **Pieczynska, E.,** Periphyton in the trophic structure of fresh water ecosystems, *Pol. Arch. Hydrobiol.,* 17, 141, 1970.

913. **Pinto-Lopes, J.,** Variation of the antibiotic power of the Hymenomycetes with the different stages of their life cycle, *Port. Acta Biol.,* 1, 406, 1946.

914. **Pipes, W. O.,** Carnivorous plants in activated sludge, *Eng. Bull. Purdue Univ. Eng. Ext. Ser.,* 49(4), 647, 1965.

915. **Pipes, W. O. and Jenkins, D.**, *Zoophagus* in activated sludge — second observation, *Int. J. Air Water Pollut.*, 9, 495, 1965.
916. **Pirozynski, K. A.**, Geographical distribution of fungi, in *The Fungi, An Advanced Treatise*, Vol. 3, Ainsworth, G. C. and Sussman, A. S., Eds., Academic Press, New York, 1968, 487.
917. **Pirozynski, K. A. and Malloch, D. W.**, The origin of land plants: a matter of mycotrophism, *Bio-Systems*, 6, 153, 1975.
918. **Ponnamperuma, C.**, *The Origins of Life*, World of Science Library, Thames and Hudson, London, 1972.
919. **Pontecorvo, G.**, The parasexual cycle in fungi, *Annu. Rev. Microbiol.*, 10, 393, 1956.
920. **Porter, C. L. and Carter, J. C.**, Competition among fungi, *Bot. Rev.*, 4, 165, 1938.
921. **Potter, L. F. and Baker, G. E.**, The role of fish as conveyors of microorganisms in aquatic environments, *Can. J. Microbiol.*, 7, 595, 1961.
922. **Powell, C. C.**, Disease Control on Woody Ornamentals in the Nursery, Bull. No. 571, Cooperative Extension Service, Ohio State University, Columbus, 1976.
923. **Powell, C. C. and Lindquist, R. K.**, Insect, Mite, and Disease Control, Bull. No. 538, Cooperative Extension Service, Ohio State University, Columbus, 1976.
924. **Powell, N. T.**, Interrelations between nematodes and fungi in disease complexes, *Annu. Rev. Phytopathol.*, 9, 253, 1971.
925. **Prát, S. and Rypáček, V.**, *Studies about Humus*, Publishing House of the Czechoslovak Academy of Sciences, Prague, 1962.
926. **Preece, T. F. and Dickinson, C. H., Eds.**, *Ecology of Leaf Surface Microorganisms*, Academic Press, New York, 1971.
927. **Prescott, S. C. and Dunn, C. G.**, *Industrial Microbiology*, 3rd ed., McGraw-Hill, New York, 1959.
928. **Preston, E. A. and Wick, W. E.**, Laboratory evaluation of cephalosporin antibiotics, *Dev. Ind. Microbiol.*, 16, 190, 1975.
929. **Pugh, G. J. F.**, Ecology of fungi in developing coastal soils, in *Soil Organisms*, Doeksen, J. and van der Drift, J., Eds., North-Holland, Amsterdam, 1963, 439.
930. **Pugh, G. J. F.**, Terrestrial fungi, in *Biology of Plant Litter Decomposition*, Dickinson, C. H. and Pugh, G. J., Eds., Academic Press, New York, 1974, 303.
931. **Quispel, A.**, The mutual relations between algae and fungi in lichens, *Recl. Trav. Bot. Neerl.*, 40, 413, 1945.
932. **Rainbow, C. and Rose, A. H.**, *Biochemistry of Industrial Microorganisms*, Academic Press, New York, 1963.
933. **Ramsbottom, J.**, *Mushrooms and Toadstools*, Collins, London, 1953.
934. **Rand, M. C., Greenberg, A. E., and Taras, M. J., Eds.**, *Standard Methods for the Examination of Water and Wastewater*, 14th ed., American Public Health Association, American Water Works Association, and Water Pollution Control Federation, Washington, D.C., 1976, Section 915, 982.
935. **Ranzoni, F. V.**, Nutrient requirements for two species of aquatic Hyphomycetes, *Mycologia*, 43, 130, 1951.
936. **Ranzoni, F. V.**, The aquatic Hyphomycetes of California, *Farlowia*, 4, 353, 1953.
937. **Rao, A. S.**, Fungal populations in the rhizosphere of peanut (*Arachis hypogaea* L.), *Plant Soil*, 17, 260, 1962.
938. **Rao, B. V., Dias, F. F., and Bhat, J. V.**, A method for the preferential isolation of yeasts, *J. Indian Inst. Sci.*, 48, 20, 1966.
939. **Raper, J. R.**, A method of freeing fungi from bacterial contamination, *Science*, 85, 342, 1937.
940. **Raper, J. R.**, Role of hormones in the sexual reaction of heterothallic Achlyas, *Science*, 89, 321, 1939.
941. **Raper, J. R.**, Sexual hormones in *Achlya*, *Am. Sci.*, 39, 110, 1951.
942. **Raper, J. R.**, Chemical regulators of sexual processes in the Thallophytes, *Bot. Rev.*, 18, 447, 1952.
943. **Raper, J. R.**, Life cycles, sexuality, and sexual mechanisms in the fungi, in *Sex in Microorganisms*, Wenrich, D. H., Lewis, I. F., and Raper, J. R., Eds., American Association for the Advancement of Science, Washington, D.C., 1954, 42.
944. **Raper, J. R.**, *Genetics of Sexuality in Higher Fungi*, Ronald Press New York, 1966.
945. **Raper, J. R.**, On the evolution of fungi, in *The Fungi, An Advanced Treatise*, Vol. 3, Ainsworth, G. C. and Sussman, A. S., Eds., Academic Press, New York, *1968*, 677.
946. **Raper, J. R. and Flexer, A. S.**, The road to diploidy with emphasis on a detour, in *Organization and Control in Prokaryotic and Eukaryotic Cells*, Charles, H. P. and Knight, B. C., Ed., Cambridge University Press, Cambridge, 1970, 401.
947. **Raper, K. B.**, A decade of antibiotics in America, *Mycologia*, 44, 1, 1952.
948. **Raper, K. B. and Fennell, D. I.**, *The Genus Aspergillus*, Williams & Wilkins, Baltimore, 1965.
949. **Raper, K. B. and Thom, C.**, *A Manual of the Penicillia*, Williams & Wilkins, Baltimore, 1949.

950. **Reed, G. M.**, Physiological specialization of the parasitic fungi, *Bot. Rev.*, 1, 119, 1935.
951. **Reed, G. M.**, Physiological specialization of the parasitic fungi. II, *Bot. Rev.*, 12, 141, 1946.
952. **Reese, E. T., Cravetz, H., and Mandels, G. R.**, Activity of fungi on oils, *Farlowia*, 4, 409, 1955.
953. **Reese, E. T. and Downing, M. H.**, Activity of the Aspergilli on cellulose, cellulose derivatives and wool, *Mycologia*, 43, 16, 1951.
954. **Reese, E. T., Levinson, H. S., Downing, M. H., and White, W. L.**, Quartermaster culture collection, *Farlowia*, 4, 45, 1950.
955. **Reichenbach-Klinke, H.-H.**, Uber einige bisher unbekannte Hyphomyceten bei verschiedenen Susswasser- und Meersfischen, *Mycopathol. Mycol. Appl.*, 7, 333, 1956.
956. **Reichert, L.**, Fungi and plant diseases in relation to plant geography, *Trans. N.Y. Acad. Sci.*, 20 (Ser. 2), 333, 1958.
957. **Remsberg, E. E.**, Studies in the genus, *Typhula, Mycololgia*, 32, 52, 1940.
958. **Rettow, G. R. and Thompson, F. G.**, *Manual of Mushroom Culture*, Mushroom Supply Co., Toughkenamon, Penn., 1948.
959. **Reynolds, E. S.**, *Pullularia* as a cause of deterioration of paint and plastic surfaces in south Florida, *Mycologia*, 42, 433, 1950.
960. **Riccioti, L. P. and Kenny, G. M.**, The effect of light on the growth of beta-lactam antibiotic-producing fungi, *Dev. Ind. Microbiol.*, 17, 233,1976.
961. **Rice, M. A.**, The cytology of host-parasite relations, *Bot. Rev.*, 1, 327, 1935.
962. **Rice, M. A.**, The cytology of host-parasite relations. II, *Bot. Rev.*, 11, 288, 1945.
963. **Richards, M. and Haskins, R. H.**, Extracellular lysine production by various fungi, *Can. J. Microbiol.*, 3, 543, 1957.
964. **Richardson, M. J.**, Studies on *Russula emetica* and other agarics in a Scots pine plantation, *Trans. Br. Mycol. Soc.*, 55, 217, 1970.
965. **Ringrose, R. G.**, Nutritive properties of *Torula* yeast for poultry, Engr. Exp. Sta., *Univ. N. Hamp., Engineering Publ.*, 8, 75, 1949.
966. **Rippon, J. W. and Anderson, D. N.**, Metabolic rate of fungi as a function of temperature and oxidation-reduction potential (EH), *Mycopathol. Mycol. Appl.*, 40, 349, 1970.
967. **Ritter, C.**, Studies on the viability of *Histoplasma capsulatum* in tap water, *Am. J. Public Health*, 44, 1299, 1954.
968. **Ritzman, E. G.**, Wood yeast as a protein supplement, *Guernsey Breeder's J.*, November 15, 1948.
969. **Ritzman, E. G.**, Wood yeast protein as a feed for livestock, Eng. Exp. Sta., *Univ. N. Hamp., Engineering Publ.*, 7, 1948.
970. **Robbins, W. J.**, A survey of the growth requirements of some Basidiomycetes, *Mycologia*, 42, 470, 1950.
971. **Robbins, W. J.**, Growth requirements of dermatophytes, *Ann. N.Y. Acad. Sci.*, 50, 1357, 1950.
972. **Robbins, W. J. and Kavanaugh, V.**, Vitamin deficiencies of the filamentous fungi, *Bot. Rev.*, 8, 411, 1942.
973. **Robbins, W. J., et al.**, A survey of some wood destroying and other fungi for antibacterial activity, *Bull. Torrey Bot. Club*, 72, 165, 1945.
974. **Robinson, R. A.**, *Plant Pathosystems*, Springer-Verlag, New York, 1976.
975. **Rogers, D. P.**, A comparison of evolutionary tendancies in plants, fungi and animals, *Bull. Torrey Bot. Club*, 75 (Abstr.), 442, 1948.
976. **Romano, A. H.**, Fumaric acid fermentation of spent sulphite liquor, *Tappi*, 41, 687, 1958.
977. **Romano, A. H.**, Conversion of spent sulphite liquor sugars to fumaric acid by *Rhizopus* species, *Eng. Bull. Purdue Univ. Eng. Ext. Ser.*, 44, 711, 1960.
978. **Romano, A. H.**, Dimorphism, in *The Fungi, An Advanced Treatise*, Vol. 2, Ainsworth, G. C. and Sussman, A. S., Eds., Academic Press, New York, 1966, 181.
979. **Rooney, H. M. and McKnight, K. H.**, Aquatic Phycomycetes of Lily Lake, Utah, *Great Basin Nat.*, 32, 181, 1972.
980. **Rose, A. H.**, *Chemical Microbiology*, Butterworths, London, 1965.
981. **Rose, R. E. and Miller, J. G.**, Some sampling variations in soil fungal numbers, *J. Gen. Microbiol.*, 10, 1, 1954.
982. **Rossmoore, R. W. and Holtzman, G. H.**, Growth of fungi in cutting fluids, *Dev. Ind. Microbiol.*, 15, 273, 1974.
983. **Rossmoore, R. W. and Treusch, P. L.**, Comparison of several media for enumeration of cutting fluid fungi, *Dev. Ind. Microbiol.*, 16, 475, 1975.
984. **Rosswall, T., Ed.**, *Modern Methods in the Study of Microbial Ecology*, Swedish National Science Research Council, Ecological Research Communication 17, 1, 1973.
985. **Rothwell, D. F. and Hortenstine, G. G.**, Composted municipal refuse: its effects on carbon dioxide, nitrate, fungi and bacteria in Arredondo fine sand, *Agron. J.*, 61, 837, 1969.

986. **Routien, J. B.**, Observations on *Gastrella luteophila, Mycologia,* 31, 416, 1939.
987. **Routien, J. B. and Findlay, A. C.**, Problems in the search for microorganisms producing antibiotics, *Bacteriol. Rev.*, 16, 51, 1952.
988. **Roxburgh, J. M., Spencer, J. F. T., and Sallans, H. R.**, Submerged culture fermentation: factors affectng the production of ustilagic acid by *Ustilago zeae, Agric. Food Chem.*, 9, 1121, 1954.
989. **Ruinen, J.**, The phyllosphere. 1. An ecologically neglected milieu, *Plant Soil,* 15, 81, 1961.
990. **Rypáček, V.**, *Biologie Dřevokazných Hub,* Ceskoslovenske Akademie Ved., Prague, 1957.
991. **Saccardo, P. A.**, *Sylloge Fungorum Omnium Hucusque Cognitorum,* 25 vol, Pavia, Italy, 1882—1931.
992. **Sagara, N.**, Ecology of Ureophilous Fungi. An introduction to experimental ecology of fungi by chemical treatment of soil, in Proc. 12th Ann. Meeting Mycol. Soc. Japan (Abstr.), Tokyo, 1968.
993. **Sagara, N.**, Proteophilous fungi and fireplace fungi, (a preliminary report), *Trans. Mycol. Soc. Jpn.*, 14, 41, 1973.
994. **Sagara, N.**, Studies on the Ammonogenous Fungi, Ph.D. thesis, Kyoto University, Kyoto, Japan,
995. **Sagara, N.**, Ammonia fungi — a chemoecological grouping of terrestrial fungi, *Contrib. Biol. Lab. Kyoto Univ.*, 24, 205, 1975.
996. **Sagara, N.**, Presence of a mammalian carcass indicated by fungal fruiting bodies, *Nature (London),* 262, 316, 1976.
997. **Sagara, N. and Hamada, H.**, Responses of higher fungi to some chemical treatments of forest ground, *Trans. Mycol. Soc. Jpn.*, 6, 71, 1975.
998. **Saito, T.**, Microbiological decomposition of beech litter, *Ecol. Rev.*, 14, 141, 1956.
999. **Saito, T.**, The characteristic features of fungi taking part in the decomposition of beech litter, *Sci. Rep. Tohoku Univ. Ser. 4,* 24, 73, 1958.
1000. **Saito, T.**, An approach to the mechanism of the microbial decomposition of beech litter, *Sci. Rep. Tohoku Univ. Ser. 4,* 26, 125, 1960.
1001. **Saito, T.**, Coactions between litter decomposing Hymenomycetes and their associated microorganisms during decomposition of beech litter, *Sci. Rep. Tohoku Univ. Ser. 4,* 31, 255, 1965.
1002. **Saito, T.**, Sequential pattern of decomposition of beech litter with special reference to microbial succession, *Ecol. Rev.*, 16, 245, 1966.
1003. **Saito, T.**, Chemical changes in beech litter under microbiological decomposition, *Ecol. Rev.*, 14, 209, 1967.
1004. **Saksena, K. A.**, Isolation and large scale purification of mushroom viruses, *Dev. Ind. Microbiol.*, 16, 134, 1975.
1005. **Sanders, F. E., Mosse, B., and Tinker, P. S.**, *Endomycorrhizas,* Academic Press, New York, 1975.
1006. **Sargeant, K., Sheridan, A., O'Kelly, J., and Carnaghan, R. A.**, Toxicity associated with certain samples of groundnuts, *Nature (London),* 192, 1096, 1961.
1007. **Savile, D. B. O.**, A phylogeny of the Basidiomycetes, *Can. J. Bot.*, 33, 60, 1955.
1008. **Savile, D. B. O.**, Possible interrelationships between fungal groups, in *The Fungi, An Advanced Treatise,* Vol. 3, Ainsworth, G. C. and Sussman, A. S., Eds., Academic Press, New York, 1968, 649.
1009. **Schade, A. L.**, The nutrition of *Leptomitus, Am. J. Bot.*, 27, 376, 1940.
1010. **Schade, A. L. and Thiman, K. V.**, The metabolism of the water mold, *Leptomitus lacteus, Am. J. Bot.*, 27, 659, 1940.
1011. **Schafer, J. F.**, Tolerance to plant disease, *Annu. Rev. Phytopathol.*, 9, 235, 1971.
1012. **Scheerpeltz, O. and Höfler, K.**, *Käfer und Pilze,* Verlag für Jugend und Volk, Wien, 1948.
1013. **Scherr, G. H. and Weaver, R. H.**, The dimorphism phenomenon in yeasts, *Bacteriol. Rev.*, 17, 51, 1953.
1014. **Schmidt, E.**, The production and utilization of food and feed yeast in Germany, in Proc. 6th Meet. Tech. Panel on Wood Chemistry, FAO, Stockholm, 1953.
1015. **Schmidt, E. L. and Bankole, R. O.**, The use of fluorescent antibody with the buried slide technique, in *Soil Organisms,* Doeksen, J. and van der Drift, J., Eds., North-Holland, Amsterdam, 1963, 197.
1016. **Schopf, J. M.**, Variable coalification: the processes involved in coal formation, *Econ. Geol.*, 32, 207, 1948.
1017. **Schütte, K. H.**, Translocation in the fungi, *New Phytol.*, 55, 164, 1956.
1018. **Schwartz, R. M. and Dayhoff, M. O.**, Origins of prokaryotes, eukaryotes, mitochondria, and chloroplasts, *Science,* 199, 395, 1978.
1019. **Scott, R. A.**, *Cryptocalyx* — a new genus of fungi (Aspergillaceae) from the Eocene of Oregon, *Am. J. Bot.*, 43, 589, 1956.
1020. **Scott, W. W. and O'Bier, A. H.**, Aquatic fungi associated with diseased fish and fish eggs, *Prog. Fish Cult.*, 1962, 3, 1962.

1021. **Seaver, F. J.**, *The North American Cup-fungi (Operculatae)*, published by the author, New York, 1928.
1022. **Seaver, F. J.**, An unwelcome guest, *Mycologia*, 27, 83, 1935.
1023. **Seeliger, H. P. R.**, Serology as an aid to taxonomy, in *The Fungi, An Advanced Treatise*, Vol. 3, Ainsworth, G. C. and Sussman, A. S., Eds., Academic Press, New York, 1968, 597.
1024. **Seymour, R. L.**, The genus *Saprolegnia*, *Nova Hedwigia Z. Kryptogamenkd.*, 19, 1, 1970.
1025. **Shanks, R. E. and Olson, J. S.**, First-year breakdown of leaf litter in southern Appalachian forests, *Science*, 134, 194, 1961.
1026. **Shanor, L.**, Some observations and comments on the Laboulbeniales, *Mycologia*, 47, 1, 1955.
1027. **Shantz, H. L. and Piemeisel, R. L.**, Fungus fairy rings in eastern Colorado and their effect on vegetation, *J. Agric. Res. Washington, D.C.*, 11, 191, 1917.
1028. **Shaw, C. G.**, Host fungus index for the Pacific Northwest. I. Hosts, *Wash. Agric. Exp. Stn. Bull.*, 765, 1, 1973.
1029. **Shaw, C. G.**, Host fungus index for the Pacific Northwest. II. Fungi, *Wash. Agric. Exp. Stn. Bull.*, 766, 1, 1973.
1030. **Shell, G. L. and Boyd, J. L.**, Composting Dewatered Sewage Sludge, Publ. No. 1937, Bureau of Solid Waste Management, Environmental Control Administration, Division of Health Service, Public Health Service, U.S. Department of Health, Education, and Welfare, Public Health Service, Washington, D.C., 1969, 1.
1031. **Shepherd, C. J.**, The genome as a component of the ecosystem, in *Microbial Ecology*, Williams, R. E. O. and Spicer, E. C., Eds., Cambridge University Press, Cambridge, 1957, 1.
1032. **Sherbakoff, C. D.**, Breeding for resistance to *Fusarium* and *Verticillium* wilts, *Bot. Rev.*, 15, 177, 1949.
1033. **Sherwood, M. and Carroll, G.**, Fungal succession on needles and young twigs of old-growth Douglas fir, *Mycologia*, 66, 499, 1974.
1034. **Siang, W.**, Are aquatic Phycomycetes present in the air?, *Nature (London)*, 164, 1010, 1949.
1035. **Silverman, F. N., Schwarz, J., Lahey, M. E., and Carson, R. P.**, Histoplasmosis, *Am. J. Med.*, 1955, 410, 1955.
1036. **Simmonds, P. M.**, Rootrots of cereals, *Bot. Rev.*, 7, 308, 1941.
1037. **Simmonds, P. M.**, Rootrots of cereals. II, *Bot. Rev.*, 19, 131, 1953.
1038. **Simmons, E. G.**, Culture studies in the genera *Pleospora, Clathrospora*, and *Leptosphaeria*, *Mycologia*, 44, 230, 1952; Culture studies in the genera *Pleospora, Clathrospora* and Leptosphaeria. II, *Mycologia*, 46, 184, 1954.
1039. **Simmons, E. G.**, Typification of *Alternaria, Ulocladium*, and *Stemphylium*, *Mycologia*, 59, 67, 1967.
1040. **Simons, R. D. G. Ph.**, *Medical Mycology*, Elsevier, Amsterdam, 1954.
1041. **Simonson, L. G. and Liberta, A. E.**, New sources of fungal dextranase, *Mycologia*, 67, 845, 1975.
1042. **Simonson, L. G., Liberta, A. E., and Richardson, A.**, Characterization of extracellular dextranase from *Fusarium moniliforme*, *Appl. Microbiol.*, 30, 855, 1975.
1043. **Sinden, J. W.**, Ecological control of pathogens and weed-molds in mushroom culture, *Annu. Rev. Phytopathol.*, 9, 411, 1971.
1044. **Sinden, J. W.**, Pathogenicity of mushroom viruses, *Dev. Ind. Microbiol.*, 16, 123, 1975.
1045. **Sinden, J. W., Mix, A. J., and Siu, R. G. H.**, Effect of environment and mineral nutrition on cellulolytic activity of fungi, *Res. Rep. QM Gen. Lab. Microbiol. Ser.*, 8, 1948.
1046. **Singer, R.**, Cryptogamic flora of the Arctic: Fungi, *Bot. Rev.*, 20, 451, 1954.
1047. **Singer, R.**, *The Agaricales in Modern Taxonomy*, 3rd ed., J. Cramer Verlag, Vaduz, 1975.
1048. **Singer, R. and Moser, M.**, Forest mycology and forest communities in South America. I. The early fall aspect of the mycoflora of the Cordillera Pelada (Chile), with a mycogeographic analysis and conclusions regarding the heterogeneity of the Valdivian District, *Mycopathol. Mycol. Appl.*, 26, 129, 1965.
1049. **Singh, N. and Webster, J.**, Effect of dung extracts on the fruiting of *Pilobolus* species, *Trans. Br. Mycol. Soc.*, 67, 377, 1976.
1050. **Siu, R. G. H.**, *Microbial Decomposition of Cellulose*, Reinhold, New York, 1951.
1051. **Siu, R. G. H. and Reese, E. T.**, Decomposition of cellulose by microorganisms, *Bot. Rev.*, 19, 377, 1953.
1052. **Siu, R. G. H. and Reese, E. T.**, Proposal for a system of biological nomenclature with special reference to microorganisms, *Farlowia*, 4, 399, 1955.
1053. **Skinner, C. E., Emmons, C. W., and Tsuchiya, H. M.**, *Henrici's Yeasts, Molds, and Actinomycetes*, 2nd ed., John Wiley & Sons, New York, 1947.
1054. **Skinner, C. E. and Huxley, M. J.**, *Rhodotorula glutinis*, *Mycologia*, 48, 371, 1956.
1055. **Skinner, F. A., Jones, P. C. T., and Mollison, J. E.**, A comparison of a direct- and a plate-counting technique for the quantitative estimation of soil microorganisms, *J. Gen. Microbiol.*, 6, 261, 1952.

1056. **Skou, J. P.**, Microbiological studies in connection with irradiation of carrots, *Riso*, 16, 79, 1960.
1057. **Skou, J. P.**, *Aureobasidium pullulans* (deBy.) Arnaud — a common and very radio-resistant fungus on fresh fruits and vegetables, *Radiat. Preserv. Foods Proc.*, 2, 63, 1963.
1058. **Sladecek, V.**, Water quality system, *Verh. Internat. Verein. Limnol.*, 16, 809, 1966.
1059. **Sladecek, V.**, The measures of saprobity, *Verh. Internat. Verein. Limnol.*, 17, 546, 1969.
1060. **Sladecek, V. and Sladecekova, A.**, Relationship between wet weight and dry weight of the periphyton, *Limnol. Oceanogr.*, 8, 309, 1963.
1061. **Sladecekova, A.**, Limnological investigation methods for the periphyton ("Aufwuchs") community, *Bot. Rev.*, 28, 286, 1962.
1062. **Sladecekova, A.**, Aquatic Deuteromycetes as indicators of starch campaign pollution, *Int. Rev. Gesamten Hydrobiol.*, 48, 35, 1963.
1063. **Sladecekova, A.**, The significance of the periphyton in reservoirs for theoretical and applied limnology, *Verh. Internat. Verein. Limnol.*, 16, 753, 1966.
1064. **Smit, J. and Wieringa, K. T.**, Microbiological decomposition of litter, *Nature (London)*, 171, 794, 1953.
1065. **Smith, A. H.**, *Mushrooms in their Natural Habitats*, 2 vol., Sawyer's, Inc., Portland, Ore., 1949.
1066. **Smith, A. L.**, *Lichens*, Cambridge University Press, Cambridge, 1921.
1067. **Smith, A. M.**, Drying and wetting sclerotia promotes biological contol of *Sclerotium rolfsii* Sacc., *Soil Biol. Biochem.*, 4, 119, 1972.
1068. **Smit, A. M.**, Nutrient leakage promotes biological control of dried sclerotia of *Sclerotium rolfsii* Sacc., *Soil Biol. Biochm.*, 4, 125, 1972.
1069. **Smith, A. M.**, Biological control of fungal sclerotia in soil, *Soil Biol. Biochem.*, 4, 131, 1972.
1070. **Smith, A. M. and Griffin, D. M.**, Oxygen and the ecology of *Armillariella elegans*, Heim, *Aust. J. Biol. Sci.*, 24, 231, 1971.
1071. **Smith, G.**, Mycology and the war, *Trans. Br. Mycol. Soc.*, 29, 1, 1946.
1072. **Smith, G.**, *An Introduction to Industrial Mycology*, 5th ed., Edward Arnold Ltd., London, 1960.
1073. **Smith, G.**, Industrial mycology, *Research*, 15, 336, 1962.
1074. **Smith, G.**, Fungi under domestication, in *The Fungi, An Advanced Treatise*, Vol. 3, Ainsworth, G. C. and Sussman, A. S., Eds., Academic Press, London, 1968, 273.
1075. **Smith, N. G., Bourguin, A. W., Crow, S. A., and Ahearn, D. G.**, Effect of heptachlor on hexadecane utilization by selected fungi, *Dev. Ind. Microbiol.*, 17, 331, 1976.
1076. **Smith, R. A.**, Fungi surviving in a latex paint film after surface disinfection, *Dev. Ind. Microbiol.*, 18, 565, 1977.
1077. **Smith, R. S.**, Colonization and Degradation of Outside Stored Softwood Chips by Fungi, Res. Note No. R-83, Department of Forest Products, Royal College of Forestry, Stockholm, 1973, 1.
1078. **Smith, R. S.**, Automatic respiration analyses of the fungistatic potential of wood preservatives, including an oxathiin, *For. Prod. J.*, 25, 48, 1975.
1079. **Smith, R. S.**, Deterioration of pulpwood by fungi and its control, *Trans. Tech. Sect.*, 1(2), 33, 1975.
1080. **Smith, R. S. and Purdy, W. C.**, Studies of sewage purification. IV. The use of chlorine for the correction of sludge bulking in the activated sludge process, *U.S. Public Health Serv. Public Health Rep.*, 1746, 617, 1936.
1081. **Smith, R. S. and Swann, G. W.**, Colonization and degradation of western red cedar shingles and shakes by fungi, *Mater. Org.*, 3, 253, 1974.
1082. **Snyder, W. C. and Hansen, H. N.**, Advantages of natural media and environments in the culture of fungi, *Phytopathology*, 37, 420, 1937.
1083. **Snyder, W. C. and Hansen, H. N.**, The species concept in *Fusarium*, *Am. J. Bot.*, 27, 64, 1940.
1084. **Snyder, W. C. and Hansen, H. N.**, Variation and speciation in the genus, *Fusarium*, *Ann. N.Y. Acad. Sci.*, 60, 16, 1954.
1085. **Snyder, W. C., Patrick, Z. A., and Weinhold, R. A.**, Biological Consequences of Plant Residue Decomposition in Soil, No. EPA-R2-72-107, Environmental Protection Agency, Washington, D.C., 1972.
1086. **Soprunov, F. L.**, *Predacious Hyphomycetes and their Application in the Control of Pathogenic Nematodes*, (trans.), Israel Program for Scientific Translation, Jerusalem, 1966.
1087. **Sowden, F. J. and Ivarson, K. C.**, Decomposition of forest litter. II. Changes in the nitrogenous constituents, *Plant Soil*, 11, 249, 1959.
1088. **Sparrow, F. K.**, Phycomycetes from the Douglas Lake region of northern Michigan, *Mycologia*, 44, 759, 1952.
1089. **Sparrow, F. K.**, *Aquatic Phycomycetes*, 2nd ed., University of Michigan Press, Ann Arbor, 1960.
1090. **Sparrow, F. K., Jr.**, Ecology of freshwater fungi, in *The Fungi, An Advanced Treatise*, Vol. 3, Ainsworth, G. C. and Sussman, A. S., Eds., Academic Press, New York, 1968, 41.
1091. **Spencer, J. F. T., Gorin, P. A. J., and Gardner, N. R.**, Yeasts isolated from the South Saskatchewan, a polluted river, *Can. J. Microbiol.*, 16, 1051, 1970.

1092. **Spendlove, J. C.,** Industrial, agricultural, and municipal microbial aerosol problems, *Dev. Ind. Microbiol.*, 15, 20, 1974.

1093. **Sporne, K. R.,** On the phylogenetic classification of plants, *Am. J. Bot.*, 46, 385, 1959.

1094. **Sprague, R.,** *Septoria* disease of Gramineae in the western United States, *Oreg. State Monogr. Stud. Bot.*, 1944.

1095. **Sreeramulu, T. and Vital, B. P. R.,** Periodicity in the airborne spores of the false smut fungus, *Ustilaginoidea virens, Trans. Br. Mycol. Soc.*, 49, 443, 1966.

1096. **Stakman, E. C. and Christensen, C. M.,** Aerobiology in relation to plant disease, *Bot. Rev.*, 12, 205, 1946.

1097. **Standard, P. G. and Ahearn D. G.,** Effect of alkylbenzene sulfonates on yeasts, *Appl. Microbiol.*, 20, 646, 1970.

1098. **Stangl, J.,** Das Pilzwachstum in alluvialen Schotterebenen und seine Abhängigkeit von Vegetationsgesellschaften, *Z. Pilzkd.*, 36, 209, 1970.

1099. **States, J.,** Basidiocarp development in *Gloeophyllum (Lenzites) saepiarium* in the natural environment, *Can. J. Bot.*, 50, 1017, 1972.

1100. **States, J.,** Normal basidiocarp development in *Gloeophyllum (Lenzites) sepiarium* in culture, *Mycologia*, 67, 1166, 1975.

1101. **Stauffer, J. F. and Backus, M. P.,** Spontaneous mutation in selected stocks of *Penicillium chrysogenum, Ann. N.Y. Acad. Sci.*, 60, 35, 1954.

1102. **Stebbins, M. E. and Robbins, W. J.,** Mineral oil and preservation of fungus cultures, *Mycologia*, 41, 632, 1949.

1103. **Steinberg, R. A.,** Growth of fungi in synthetic nutrient solutions, *Bot. Rev.*, 5, 327, 1939.

1104. **Steinberg, R. A.,** Growth of fungi in synthetic nutrient solutions. II, *Bot. Rev.*, 16, 208, 1950.

1105. **Steinhaus, E. A.,** *Insect Microbiology,* Comstock Press, Ithaca, 1946.

1106. **Sternberg, T. H. and Newcomer, V. D., Eds.,** *Therapy of Fungus Diseases, An International Symposium,* Little, Brown, Boston, 1955.

1107. **Stevens, J. W.,** Preparation of dehydrated agar media containing orange juice serum, *Food Technol.*, 8, 88, 1954.

1108. **Stevens, N. E.,** Departures from ordinary methods in controlling plant diseases, *Bot. Rev.*, 4, 429, 1938.

1109. **Stevens, N. E. and Nienow, I.,** Plant disease control by unusual methods, *Bot. Rev.*, 13, 116, 1947.

1110. **Stevens, N. E. and Wood, J. I.,** Recent fluctuations in plant diseases in the United States, *Bot. Rev.*, 3, 277, 1937.

1111. **Stevens, R. B., Ed.,** *Mycology Guidebook,* University of Washington Press, Seattle, 1974.

1112. **Stevenson, J. A.,** Fungi of Puerto Rico and the American Virgin Islands, *Contrib. Reed Herbarium,* Clyde F. Reed, Baltimore, 1975, 23.

1113. **Stoddard, E. M. and Dimond, A. E.,** The chemotherapy of plant diseases, *Bot. Rev.*, 15, 345, 194.

1114. **Storck, R. and Alexopoulos, C. J.,** Deoxyribonucleic acid of fungi, *Bact. Rev.*, 34, 126, 1970.

1115. **Stover, R. H.,** Flood fallowing for eradication of *Fusarium oxysporum* f. *cubense.* II, *Soil Sci.*, 77, 401, 1954.

1116. **Stover, R. H.,** Fusarial wilt (Panama disease) of bananas and other *Musa* species, *Commonw. Mycol. Inst. Phytopathol. Pap.*, 4, 1962.

1117. **Stover, R. H.,** Flood-fallowing for eradication of *Fusarium oxysporum* f. *cubense.* III. Effect of oxygen on fungus survival, *Soil Sci.*, 80, 397, 1955.

1118. **Stover, R. H.,** *Banana, Plantain and Abaca Diseases,* Commonwealth Mycological Institute, Kew, 1972.

1119. **Stover, R. H., Thornton, N. C., and Dunlap, V. C.,** Flood-fallowing for eradication of *Fusarium oxysporum* f. *cubense.* I, *Soil Sci.*, 76, 225, 1953.

1120. **Sugihara, T. F. and Humfeld, H.,** Submerged culture of the mycelium of various species of mushrooms, *Appl. Microbiol.*, 2, 170, 1954.

1121. **Sundström, K.-R.,** Studies of the physiology, morphology and serology of *Exobasidium, Symb. Bot. Ups.*, 18, 3, 1, 1964.

1122. **Sussman, A. S.,** Dormancy and spore germination, in *The Fungi, An Advanced Treatise,* Vol. 2, Ainsworth, G. C. and Sussman, A. S., Eds., Academic Press, New York, 1966, 733.

1123. **Sussman, A. S.,** Longevity and survivability of fungi, in *The Fungi, An Advanced Treatise,* Vol. 3, Ainsworth, G. C. and Sussman, A. S., Eds., Academic Press, New York, 1968, 447.

1124. **Sussman, A. S. and Halvorson, H. O.,** *Spores: Their Dormancy and Germination,* Harper & Row, New York, 1966.

1125. **Sussman, A. S., Lingappa, Y., and Bernstein, I. A.,** Effect of light and media upon growth and melanin formation in *Cladsporium mansonii, Mycopathol. Mycol. Appl.*, 20, 307, 1963.

1126. **Swaebly, N. A., Christensen, C. M., and Grahek, T. A.,** Tests of different media for the collection and identification of air-borne saprophytic fungi, *J. Allergy*, 21, 404, 1950.

1127. **Swift, M. J.**, Estimtion of mycelial growth during decomposition of plant litter, in *Modern Methods in the Study of Microbial Ecology*, Swedish National Science Research Council, Ecological Research Communication, 17, 323, 1973.
1128. **Szabo, I. M.**, The pattern of microbial communities, *Microbial Communities in a Forest-Rendzina Ecosystem*, Akademiei Kiado, Budapest, 1974.
1129. **Szabo, I., Marton, M., and Partai, G.**, Micro-milieu studies in the A-horizon of a mull-like Rendzina, in *Soil Micromorphology*, Jongerius, A., Ed., American Elsevier, New York, New York, 1964, 33.
1130. **Tabak, H. and Cooke, W. B.**, The effects of gaseous environments on the growth and metabolism of fungi, *Bot. Rev.*, 34, 126, *1968*.
1131. **Tabak, H. and Cooke, W. B.**, Growth and metabolism of fungi in an atmosphere of nitrogen, *Mycologia*, 60, 115, 1968.
1132. **Taligoola, H. K., Apinis, A. E., and Chesters, C. G. G.**, Microfungi colonizing collapsed aerial parts of *Phragmites communis* Trin. in water, *Nova Hedwigia Z. Kryptogamenkd.*, 23, 465, 1972.
1133. **Taligoola, H. K., Apinis, A. E., and Chesters, C. G. C.**, Microfungi colonizing the inflorescences of *Phragmites communis* Trin., *Nova Hedwigia Z. Kryptogamenkd.*, 26, 483, 1975.
1134. **Tapke, V. F.**, Environment and the cereal smuts, *Bot. Rev.*, 14, 359, 1948.
1135. **Tarrant, R. F., Moore, D. G., Bollen, W. B., and Loper, B. R.**, Pesticides in soil: DDT residues in forest floor and soil after aerial spraying, Oregon, 1965—1968, *Pestic. Monit. J.*, 6, 65, 1972.
1136. **Tarrant, R. F. and Trappe, J. M.**, The role of *Alnus* in improving the forest environment, *Plant Soil*, (Special Vol.), 335, 1971.
1137. **Tawes, J. W., III, Hoke, G. D., Jr., and Llewellyn, G. C.**, Contraception, fertilization, and teratology in rats intubated with the mycotoxin, aflatoxin B_1, *Dev. Ind. Microbiol.*, 18, 711, 1977.
1138. **Tayama, H. K.**, Tips on Growing Bedding Plants, Bull. No. MM-265, Cooperative Extension Service, Ohio State University, Columbus, 1976.
1139. **Thiers, H. D.**, The status of the genus *Suillus* in the United States, in *Studies in the Higher Fungi*, J. Cramer, Vaduz, 1975, 247.
1140. **Thom, C.**, Fungi in Cheese Ripening: Camembert and Roquefort, Bull. No. 82, United States Department of Agriculture, Bureau of Animal Industry, Washington, D.C., 1906.
1141. **Thom, C.**, The evolution of species concepts in *Aspergillus* and *Penicillium*, *Ann. N.Y. Acad. Sci.*, 60, 24, 1954.
1142. **Thom, C. and Raper, K. B.**, *A Manual of the Aspergilli*, Williams & Wilkins, Baltimore, 1945.
1143. **Thomas, E. A.**, Uber die Biologie von Flechtenbildern, *Beitr. Kryptogamenflora Schweiz*, 9, 1, 1939.
1144. **Tichy, V.**, An experimental study of fungus succession in wood, *Ceska Mykol.*, 29, 90, 1975.
1145. **Togashi, K.**, *Biological Characters of Plant Pathogens — Temperature Relations*, Meibundo, Tokyo, 1949.
1146. **Transeau, E. N.**, Fruiting patterns of *Coprinus variegatus* Pk., *Am. J. Bot.*, 36, 596, 1949.
1147. **Trappe, J. M.**, Fungus associates of ectotrophic mycorrhizae, *Bot. Rev.*, 28, 538, 1962.
1148. **Trappe, J. M.**, Mycorrhizal hosts and distribution of *Cenococcum graniforme*, *Lloydia*, 27, 100, 1964.
1149. **Trappe, J. M.**, Mycorrhiza-forming Ascomycetes, in Mycorrhizae, *U.S. Dep. Agric. Misc. Publ.*, 1189, 19, 1969.
1150. **Trappe, J. M.**, Studies on *Cenococcum graniforme*. I. An efficient method for isolation from sclerotia, *Can. J. Bot.*, 47, 1389, 1969.
1151. **Trappe, J. M.**, Regulation of soil organisms by red alder; a potential biological system for control of *Poria weirii*, *Managing young forests in the Douglas fir region*, 3, 35, 1972.
1152. **Trappe, J. M., Li, C. Y., Lu, K. C., and Bollen, W. B.**, Differential response of *Poria weirii* to phenolic acids from Douglas fir and red alder roots, *For. Sci.*, 19, 191, 1973.
1153. **Trappe, J. M. and Maser, C.**, Germination of spores of *Glomus macrocarpus* (Endogonaceae) after passage through a rodent digestive tract, *Mycologia*, 68, 433, 1976.
1154. **Trappe, J. M., Stahly, E. A., Benson, N. R., and Duff, D. M.**, Mycorrhizal deficiency of apple trees in high arsenic soils, *Hortic. Science*, 8, 52, 1973.
1155. **Trappe, J. M. and Strand, R. F.**, Mycorrhizal deficiency in a Douglas fir forest nursery, *For. Sci.*, 15, 381, 1969.
1156. **Tresner, H. D., Backus, M. P., and Curtis, J. T.**, Soil microfungi in relation to the hardwood forest continuum in southern Wisconsin, *Mycologia*, 46, 314, 1954.
1157. **Tribe, H. T.**, Ecology of microorganisms in soils as observed during their development on buried cellulose film, in *Microbial Ecology*, Williams, R. E. O. and Spicer, E. C., Eds., Cambridge University Press, Cambridge, 1957, 287.
1158. **Tribe, H. T.**, Interactions of soil fungi on cellulose film, *Trans. Br. Mycol. Soc.*, 49, 457, 1966.
1159. **Tsao, P. H.**, Selective media for isolation of pathogenic fungi, *Annu. Rev. Phytopathol.*, 6, 157, 1970.

1160. **Tubaki, K. and Ito, T.**, Fungi inhabiting in brackish water, *Rep. Tottori Mycol. Inst.*, 10, 523, 1973.

1161. **Tubaki, K., Ito, T., and Matsuda, Y.**, Aquatic sediment as a habitat of thermophilic fungi, *Ann. Microbiol. (Japan)*, 24, 199, 1974.

1162. **Tubaki, K. and Yokoyama, T.**, Successive fungal flora on sterilized leaves in the litter of forests. I, *Res. Commun. Inst. Ferment. Osaka*, 5, 24, 1971.

1163. **Tulasne, L. R. and Tulasne, C.**, *Selecta Fungorum Carpologia*, 3 vol., Imperial Press, Paris, 1861—1865.

1164. **Turner, J. N.**, Measurement of mould growth on solid substrates, *Trans. Br. Mycol. Soc.*, 36, 159, 1953.

1165. **Tyrell, D.**, Biochemical systematics and fungi, *Bot. Rev.*, 35, 305, 1969.

1166. **Ubrizsy, G.**, Recherches phytosociologiques sur la vegetation mycologiques de Nyirseg, *Acta Geobot.*, 5, 251, 1942.

1167. **Ubrizsy, G.**, Recherches sur les relations transpiratives du *Psalliota campestris* et d'autres champignons, *Kalonlenymot Agraricidomanyi Egyotem Kert-es Szologszdasagtudomanyi Karanek Közlemenyi*, 12, 1, 1948.

1168. **Ubrizsy, G.**, Zoenologische Untersuchungen an bodenbeiobenden Grosspilzen einiger Waldtypen in Ungarn, in Omagiu lui Tradeau Savulescu en Prilejul Implinirii a 70 de ani, *Editura Academiei, Republicii Populare Romine*, 1959, 783.

1169. **Ubrizsy, G.**, Mycological investigations in some Hungarian forest types and special sites. I, *Acta Phytopathol. Acad. Sci. Hung.*, 1, 75, 1966; Mycological investigations in some Hungarian forest types and special sites. II, *Acta Phytopathol. Acad. Sci. Hung.*, 1, 277, 1966.

1170. **Ubrizsy, G.**, Mycoflora of Vas County, Western Hungary, with special reference to the researches of Carolus Clusius (1526—1609), *Acta Phytopathol. Acad. Sci. Hung.*, 4, 261, 1969.

1171. **Updegraff, D. M., Espinosa, R., Griffin, L., King, J., Schneider, S., and Rolz, C.**, The production of animal feed by the growth of fungi on wastes from the coffee and rum-distilling industries, *Dev. Ind. Microbiol.*, 14, 317, 1973.

1172. **Vaartaja, O.**, Chemical treatment of seed beds to control nursery diseases, *Bot. Rev.*, 30, 1, 1964.

1173. **Vaartaja, O. and Salisbury, P. J.**, Mutual effects *in vitro* of microorganisms isolated from tree seedlings, nursery soil, and forests *For. Sci.*, 11, 160, 1965.

1174. **Vaicum, L., Gruia, E., and Godeau, S.**, The determination of some enzymatic activities as a research method of the activated sludge, *Rev. Roum. Biochim.*, 2, 87, 1965.

1175. **Vallentyne, J. R.**, Environmental biophysics and microbial ubiquity, *Ann. N.Y. Acad. Sci.*, 108, 342, 1963.

1176. **van Beverwijk, A. L.**, Observations on submerged water moulds in the Netherlands, *Antonie van Leeuwenhoek J. Microbiol. Serol.*, 14, 223, 1948.

1177. **Vanbreuseghem, R.**, Intérét théorique et pratique d'un nouveau dermatophyte isolé du sol: *Keratinomyces ajelloi* gen. nov., sp. nov., *Bull. Cl. Sci. Acad. R. Belg.*, 38 (Ser. 5), 158, 1952.

1178. **Vanbreuseghem, R., Abraham, J. E., Brickaye-van Geestruicken, M., Cisek, K., de Vroey, Ch., Kameswaren, S., Maertens, K., Muyunga, C., and Nolard-Tintiger, N.**, Eye and nose contamination by fungi in Africa, Asia, and Europe, *Arogya — J. Health Sci.*, 1, 29, 1975.

1179. **Van der Plank, J. E.**, Spread of plant pathogens in space and time, in *Airborne Microbes*, Gregory, P. H. and Monteith, J. L., Eds., Cambridge University Press, Cambridg, 1967, 227.

1180. **Van der Plank, J. E.**, *Principles of Plant Infection*, Academic Press, New York, 1975.

1181. **van Niel, C. B.**, Natural selection in the microbial world, *J. Gen. Microbiol.*, 13, 201, 1955.

1182. **Vanterpool, T. C. and Macrae, R.**, Notes on the Canadian tuckahoe, its occurrence in Canada and the interfertility of its perfect state, *Polyporus tuberaster* Jacq. ex Fr., *Can. J. Bot.*, 29, 147, 1951.

1183. **Vaughn, E. K., Newhook, J. F., and Robertson, G. I.**, The influence of germinating seeds and developing roots on re-establishment of microorganisms in fumigated soil, *Ann. Appl. Biol.*, 58, 389, 1966.

1184. **Verrall, A. F.**, The control of fungi in lumber during air seasoning, *Bot. Rev.*, 11, 398, 1945.

1185. **Viro, P. J.**, Investigations on forest litter, *Commun. Inst. For. Fenn.*, 45(6), 1, 1955.

1186. **Vogt, K. and Staffeldt, E. E.**, Fungal degradation of pure and mixed carbon substrates, *Dev. Ind. Microbiol.*, 18, 571, 1977.

1187.

1188. **de Vries, G. A. and Cormane, R. H.**, A study on the possible relationships between certain morphological and physiological properties of *Aspergillus fumigatus* Fres. and its presence in, or on, human and animal (pulmonary) tissue, *Mycopathol. Mycol. Appl.*, 39, 241, 1969.

1189. **Wagener, W. W. and Davidson, R. W.**, Heart rots of living trees, *Bot. Rev.*, 20, 61, 1954.

1190. **Wagoner, P. E.**, Microclimate and plant disease, *Annu. Rev. Phytopathol.*, 3, 103, 1965.

1191. **Waid, J. S.**, Distribution of fungi within the decomposing tissues of ryegrass roots, *Trans. Br. Mycol. Soc.*, 40, 351, 1957.

1192. **Waid, J. S.**, Physiological and biochemical adjustment of fungi to their environment, in *The Fungi, An Advanced Treatise*, Vol. 3, Ainsworth, G. C. and Sussman, A. S., Eds., Academic Press, New York, 1968, 289.
1193. **Waid, J. S., Preston, K. J., and Harris, P. J.**, A method to detect metabolically active microorganisms in leaf litter habitats, *Soil Biol. Biochem.*, 3, 235, 1971.
1194. **Waksman, S. A.**, Soil fungi and their activities, *Soil Sci.*, 2, 103, 1916.
1195. **Waksman, S. A.**, Is there any fungus flora of the soil?, *Soil Sci.*, 3, 565, 1917.
1196. **Waksman, S. A.**, *Principles of Soil Microbiology*, 2nd ed., Williams & Wilkins, Baltimore, 1932.
1197. **Waksman, S. A.**, The microbiology of cellulose decomposition and some economic problems involved, *Bot. Rev.*, 6, 637, 1940.
1198. **Waksman, S. A.**, Three decades with soil fungi, *Soil Sci.*, 58, 89, 1944.
1199. **Waksman, S. A. and Fred, E. B.**, A tentative outline of the plate method for determining the number of microorganisms in the soil, *Soil Sci.*, 14, 27, 1922.
1200. **Walker, J. C.**, Disease resistance in vegetable crops. II, *Bot. Rev.*, 19, 606, 1953.
1201. **Walker, N.**, Soil microorganisms and plant protection chemicals, in *Soil Biology*, Burges, A. and Raw, F., Eds., Academic Press, New York, 1967, 493.
1202. **Walton, L. B.**, Studies concerning organisms occurring in water supplies with particular reference to those found in Ohio. 1, *Ohio Biol. Surv. Bull.*, 24, 1, 1930.
1203. **Wang, C. J. K.**, Fungi of Pulp and Paper in New York, Tech. Publ., State University of New York, College of Forestry, 1965, 87.
1204. **Warcup, J. H.**, The ecology of soil fungi, *Trans. Br. Mycol. Soc.*, 34, 376, 1951.
1205. **Warcup, J. H.**, Studies on the growth of Basidiomycetes in the soil, *Ann. Bot. (London)*, 15, 305, 1951.
1206. **Warcup, J. H.**, On the origin of colonies of fungi developing on soil dilution plates, *Trans. Br. Mycol. Soc.*, 38, 298, 1955.
1207. **Warcup, J. H.**, Isolation of fungi from hyphae present in the soil, *Nature (London)*, 175, 953, 1955.
1208. **Warcup, J . H.**, Fungi in soil, in *Soil Biology*, Burges, A. and Raw, F., Eds., Academic Press, New York, 1967, 51.
1209. **Watson, A. G.**, The effect of soil fungicide treatments on the inoculum potential of spermosphere fungi and damping-off, *N.Z. J. Agric. Res.*, 9, 931, 1966.
1210. **Watson, A. G.**, Seasonal variation in the inoculum potentials of spermosphere fungi, *N.Z. J. Agric. Res.*, 9, 956, 1966.
1211. **Weaver, R. H.**, Ecological Study on the Effects of Strip Mining on the Microbiology of Streams, Res. Rep. 181-35, Water Resources Institute, University of Kentucky, Lexington, 1968.
1212. **Webley, D. M., Eastwood, D. J., and Gimingham, C. H.**, Development of a soil microflora in relation to plant succession on sand-dunes, including the "rhizosphere" flora associated with colonizing species, *J. Ecol.*, 40, 168, 1952.
1213. **Webster, J.**, Succession of fungi on decaying cocksfoot culms. I and II, *J. Ecol.*, 44, 517, 1956; 45, 1, 1957.
1214. **Webster, J.**, Experiments with spores of aquatic Hyphomycetes. I. Sedimentation and impaction on smooth surfaces, *Ann. Bot. (London)*, 23, 595, 1959.
1215. **Webster, J.**, *Introduction to Fungi*, Cambridge University Press, Cambridge, 1970.
1216. **Webster, J. and Dix, N. J.**, Succession of fungi on decaying cocksfoot culms. III. A comparison of the sporulation and growth of some primary saprophytes on stem, leaf blade, and leaf sheath, *Trans. Br. Mycol. Soc.*, 43, 85, 1960.
1217. **Wehmeyer, L. E.**, A biologic and phylogenetic study of the stromatic Sphaeriales, *Am. J. Bot.*, 13, 575, 1926.
1218. **Wehmeyer, L. E.**, The genera *Leptosphaeria*, *Pleospora*, and *Clathrospora* in Mount Rainier National Park, *Mycologia*, 44, 621, 1952.
1219. **Wehmeyer, L. E.**, *A World Monograph of the Genus Pleospora and Its Segregates*, University of Michigan Press, Ann Arbor, 1961.
1220. **Weindling, R.**, Association effects of fungi, *Bot. Rev.*, 4, 475, 1938.
1221. **Westerdijk, J.**, The concept "association" in mycology, *Antonie van Leeuwenhoek J. Microbiol. Serol.*, 15, 187, 1949.
1222. **Weston, W. H.**, Mycology during the last fifty years, in *Fifty Years of Botany*, Steere, W. C., Ed., McGraw-Hill, New York, 1958.
1223. **Wheeler, B. E. J.**, Fungal parasites of plants, in *The Fungi, An Advanced Treatise*, Vol. 3, Ainsworth, G. C. and Sussman, A. S., Eds., Academic Press, New York, 1968, 179.
1224. **Wheeler, H. E.**, Genetics and evolution of heterothallism in *Glomerella*, *Phytopathology*, 44, 352, 1954.
1225. **Wheeler, H.**, *Plant Pathogenesis*, Springer-Verlag, New York, 1975.

1226. **Whetzel, H. H.**, A synopsis of the genera and species of the Sclerotiniaceae, a family of stromatic inoperculate Discomycetes, *Mycologia*, 37, 648, 1945.

1227. **Whitaker, A.**, Amino acid transport in the fungi, an essay, *Trans. Br. Mycol. Soc.*, 61, 365, 1976.

1228. **White, W. L.**, History, distribution and economic significance of the cellulose destroying fungus *Memnoniella echinata*, *Farlowia*, 3, 399, 1949.

1229. **White, W. L.**, Fungi in relation to the degradation of woollen fabrics, *Mycologia*, 42, 199, 1950.

1230. **White, W. L. and Siu, R. G. H.**, Resistance of resin-impregnated cotton fabrics to microorganisms, *Ind. Eng. Chem.*, 39, 1628, 1947.

1231. **White, W. L. et al.**, Assay of cellulolytic activity of fungi isolated from fabrics and related materials exposed to the tropics, *Mycologia*, 40, 34, 1948.

1232. **White, W. L., Siu, R. G. H., and Reese, E. T.**, The black Aspergilli in relation to cellulosic substrates, *Bull. Torrey Bot. Club*, 75, 604, 1948.

1233. **Whitehouse, H. L. K.**, Heterothallism and sex in the fungi, *Biol. Rev.*, 24, 411, 1949.

1234. **Whitehouse, H. L. K.**, Multiple allelomorph heterothallism in the fungi, *New Phytol.*, 48, 212, 1949.

1235. **Whitehouse, H. L. K.**, The significance of some sexual phenomena in the fungi, *Indian Phytopathol.*, 4, 91, 1951.

1236. **Whitehouse, H. L. K.**, A survey of heterothallism in the Ustilaginales, *Trans. Br. Mycol. Soc.*, 34, 340, 1951.

1237. **Whittaker, R. M.**, On the broad classification of organisms, *Q. Rev. Biol.*, 34, 210, 1959.

1238. **Whittaker, R. M.**, New concepts of kingdoms of organisms, *Science*, 163, 150, 1969.

1239. **Wickerham, L. J.**, Taxonomy of yeasts, *U.S. Dep.. Agric. Tech. Bull.*, No. 1029, 1951.

1240. **Wicklow, M. C., Bollen, W. B., and Denison, W. C.**, Comparison of soil microfungi in 40-year-old stands of pure alder, pure conifer, and alder-conifer mixtures, *Soil Biol. Biochem.*, 6, 73, 1974.

1241. **Wieringa, K. T.**, A method for isolating and counting pectolytic microbes, in Proc. 4th Int. Congr. Microbiology, Copenhagen, 1949, 482.

1242. **Wieringa, K. T.**, On the dissimilation of pectic substances in nature, in *Atti del 6th Congr. Int. Microbiologia*, Roma, 6(16), 281, 1953.

1243. **Wieringa, K. T.**, The micro-organisms decomposing pectic substances in the dew retting process of flax, *Neth. J. Agric. Sci.*, 4, 204, 1956.

1244. **Wilcoxson, R. D. and Sadia, T. W.**, Translocation in fungi, *Bot. Rev.*, 34, 32, 1968.

1245. **Wiley, A. J., Dubey, G. A., Lueck, B. F., and Hughes, L. P.**, *Torula* yeast grown on sulphite liquor, *Ind. Eng. Chem.*, 42, 1830, 1950.

1246. **Wiley, A. J., Holderby, J. M., and Fries, K. W.**, Food and feed yeast in the U.S.A., in Proc. 6th Meet. Panel on Wood Chemistry, FAO, Stockholm, 1954.

1247. **Wilhelm, S.**, The dual phenomenon in the dermatophytes, *Mycologia*, 39, 716, 1947.

1248. **Wilkins, W. H.**, Investigation into the production of bacteriostatic substances by fungi: a revision of the testing method, *Ann. Appl. Biol.*, 36, 257, 1949.

1249. **Wilkinson, V. and Lucas, R. L.**, Effects of herbicides on the growth of soil fungi, *New Phytol.*, 68, 709, 1969.

1250. **Williams, A. J., Wallace, R. H., and Clark, D. S.**, Changes in the yeast population on Quebec apples during ripening, *Can. J. Microbiol.*, 2, 645, 1956.

1251. **Williams, L. G. and Mount, D. M.**, Influence of zinc on periphytic communities, *Am. J. Bot.*, 52, 26, 1965.

1252. **Williams, R. E. O. and Spicer, C. C., Eds.**, *Microbial Ecology*, Cambridge University Press, Cambridge, 1957.

1253. **Williams, R. H., Stang, E. J. and Janson, B. F.**, Commercial Fruit Spray Recommendations for Ohio, Bull. No. 506, Cooperative Extension Service, Ohio State University, Columbus, 1976.

1254. **Williams, S. E. and Aldon, E. F.**, Endomycorrhizal (Vescicular-arbuscular) associations of some arid zone shrubs, *Southwest. Nat.*, 20, 437, 1976.

1255. **Williams, S. T.**, The distribution of fungi in the horizons of a podzolized soil, in *Soil Organisms*, Doeksen, J. and van der Drift, J., Eds., North-Holland, Amsterdam, 1963, 158.

1256. **Willis, M. J., Ed.**, Proc. Conf. Histoplasmosis, Public Health Monogr. No. 39, 1956.

1257. **Willoughby, L. G.**, The occurrence and distribution of reproductive spores of Saprolegniales in fresh water, *J. Ecol.*, 50, 733, 1968.

1258. **Willoughby, L. G.**, Some observations on the locations of sites of fungal activity at Blelham Tarn, *Hydrobiologia*, 25, 352, 1965.

1259. **Willoughby, L. G.**, Decomposition of litter in fresh water, in *Biology of Plant Litter Decomposition*, Dickinson, C. H. and Pugh, G. J., Eds., Academic Press, New York, 1974, 659.

1260. **Wilson, H. A.**, A method of inoculating soil media with soil for the isolation of fungi, *Soil Sci. Soc. Am. Proc.*, 21, 289, 1957.

1261. **Windisch, S.**, Uber die Grundlagen der Oekologie der Hefen, *Oecol. Plant.*, 3, 69, 1968.

1262. **Winner, H. I. and Hurley, R.**, *Candida albicans*, Little, Brown, Boston, 1964.

1263. **Winters, H. and Guidetti, G.**, Extracellular enzymes from fungal isolates involved in the biodeterioration of paint film, *Dev. Ind. Microbiol.*, 17, 173, 1976.

1264. **Winters, H., Isquith, J. R., and Goll, M.**, A study of the ecological succession in biodeterioration of a vinyl acrylic paint film, *Dev. Ind. Microbiol.*, 17, 167, 1976.

1265. **Witkamp, M.**, Seasonal Fluctuations of the Fungus Flora in Mull and Mor of an Oak Forest, Ph.D. thesis, Landbouwhogenschool, Arnhem, Holland, 1960.

1266. **Witkamp, M.**, Factors influencing microbial immobilization of certain radionuclides in leaf litter, *J. Tenn. Acad. Sci.*, 38 (Abstr.), 78, 1963.

1267. **Witkamp, M.**, Microbial populations of leaf litter in relation to environmental conditions and decomposition, *Ecology*, 44, 370, 1963.

1268. **Witkamp, M.**, Decomposition of leaf litter in relation to environment, microflora and microbial respiration, *Ecology*, 47, 194, 1966.

1269. **Witkamp, M.**, Macroflora, mycoflora, and soil relationships in a pine plantation, *Ecology*, 47, 238, 1966.

1270. **Witkamp, M. and van der Drift, J.**, Break down of forest litter in relation to environmental factors, *Plant Soil*, 15, 295, 1961.

1271. **Wogan, G. N., Ed.**, *Mycotoxins in Foodstuffs*, MIT Press, Cambridge, 1965.

1272. **Wohlrab, C. and Tuveson, R. M.**, Distribution of fungi in early stages of succession in Indiana dune sand, *Am. J. Bot.*, 52, 1050, 1965.

1273. **Wohlrab, G., Tuveson, R. W., and Olmsted, C. E.**, Fungal populations from early stages of succession in Indiana dune sand, *Ecology*, 44, 734, 1963.

1274. **Wolf, F. T.**, A contribution to the life history and geographic distribution of the genus *Allomyces*, *Mycologia*, 33, 158, 1941.

1275. **Wolf, F. T.**, The microbiology of the upper air, *Bull. Torrey Bot. Club*, 70, 1, 1943.

1276. **Woodbine, M., Gregory, M. E., and Walker, T. E.**, Microbiological synthesis of fat: a preliminary survey of the fat-producing moulds, *J. Exp. Bot.*, 2, 204, 1952.

1277. **Woods, D. D.**, The integration of research on the nutrition and metabolism of microorganisms, *J. Gen. Microbiol.*, 9, 151, 1953.

1278. **Woollett, L. and Hedrick, L. R.**, Ecology of yeasts in polluted water, *Antonie van Leeuwenhoek J. Microbiol. Serol.*, 36, 427, 1970.

1279. **Woollett, L. L., Hedrick, L. R., and Tarver, M. G.**, A statistical evaluation of the ecology of yeasts in polluted water, *Antonie van Leeuwenhoek J. Microbiol. Serol.*, 36, 437, 1970.

1280. **Wright, J. M.**, The production of antibiotics in soil. I. Production of gliotoxin by *Trichoderma viride*, *Ann. Appl. Biol.*, 41, 250, 1954.

1281. **Yamasato, K., Goto, S., Ohwada, K., Okuno, D., Araki, N., and Iizuka, H.**, Yeasts from the Pacific Ocean, *J. Gen. Appl. Microbiol.*, 20, 289, 1974.

1282. **Yarwood, C. E.**, Water content of fungus spores, *Am. J. Bot.*, 37, 626, 1950.

1283. **Yoneyama, M.**, Studies on natural habitats of yeasts, Bark-inhabiting yeasts, *J. Sci. Hiroshima Univ. Ser. B. Div. 2*, 8, 19, 1957.

1284. **Yusef, H. M.**, The requirements of some Hymenomycetes for essential metabolites, *Bull. Torrey Bot. Club*, 80, 43, 1953.

1285. **Zak, B.**, Pure culture synthesis of Pacific madrone ectomycorrhiza, *Mycologia*, 68, 362, 1976.

1286. **Zak, B.**, Pure culture synthesis of bear berry mycorrhiza, *Can. J. Bot.*, 54, 1297, 1976.

1287. **Zaleski, K., Domanski, S., and Wojciehowski, I. E.**, Fungi of state forests Zielonka (Province Poznan, Poland), collected in 1946—1947, *Acta Soc. Bot. Pol.*, 19, 101, 1948.

1288. **Zapater, R. C.**, *Atlas de Diagnostici Micologica*, 3rd ed., Editorial "El Ataneo", S. A., Buenos Aires, 1973.

1289. **Zeidberg, L. G.**, Soil studies, *U.S. Public Health Serv. Public Health Monogr.*, 39, 240, 1956.

1290. **Zeller, S. M.**, Studies in the physiology of the fungi. II. *Lenzites sepiaria* Fries, with a special reference to enzyme activity, *Ann. Mo. Bot. Gard.*, 3, 439, 1916.

1291. **Zjic, J. E. and Chiu, Y. S.**, Recovery of heavy metals by microbes, *Dev. Ind. Microbiol.*, 13, 91, 1972.

1292. **Zuck, R. K.**, Isolates intermediate between *Stachybotrys* and *Memnoniella*, *Mycologia*, 38, 69, 1946.

1293. **Zuck, R. K.**, Alternation of generations and the mode of nutrition, *Drew Univ. Stud.*, 6, 3, 1953.

INDEX

A

B

M

N

O

R

S